Effective Length and Notional Load Approaches for Assessing Frame Stability: Implications for American Steel Design

By the Task Committee on Effective Length
of the Technical Committee on Load and Resistance Factor Design
of the Technical Division
of The Structural Engineering Institute
of the American Society of Civil Engineers

Approved for publication by
The Structural Engineering Institute of the
American Society of Civil Engineers

Task Committee on Effective Length

Russell Q. Bridge
Murray J. Clarke
Roberto T. Leon
Eric M. Lui
Tauqir M. Sheikh
Donald W. White
Jerome F. Hajjar, Chairman

Published by
ASCE American Society of Civil Engineers
345 East 47th Street
New York, New York 10017-2398

Abstract:
This committee report elucidates several contemporary techniques for assessing column stability in the design of steel frame structures. The primary intent of the document is to provide an understanding of the strengths, limitations, and assumptions with respect to column and frame stability made in the American Institute of Steel Construction (AISC) *Load and Resistance Factor Design Specification for Structural Steel Buildings* (LRFD). The report focuses on three techniques for stability design. Two approaches are based on the use of effective length factors and are specifically outlined in sections of the Commentary of the Second Edition (1993) of the AISC LRFD Specification. The third method involves the use of a notional load approach for stability design (with the use of an effective length factor equal to one for all columns in the frame). While notional load approaches are not mentioned in the AISC specifications, this technique is commonly used in some form within several other design standards throughout much of the rest of the world. Examples are included to illustrate the procedures for both common and unusual conditions encountered in practice. This document is applicable to both unbraced and braced frames having either fully-restrained or partially-restrained connections. In addition, while most of the discussions pertain equally to either AISC *Allowable Stress Design* or LRFD practice, to provide focus to the report, all discussions are framed within the context of the more current LRFD specification. A practical introduction to this material is provided through the discussions of the assumptions, advantages and disadvantages of each of the methods, and the step-by-step examples. The more detailed discussions and derivations provide a reference regarding some of the more complex issues involved with design for stability.

Library of Congress Cataloging-in-Publication Data

Effective length and notional load approaches for assessing frame stability / by the Task Committee on Effective Length of the Technical Committee on Load and Resistance Factor Design of the Technical Division of the Structural Engineering Institute of the American Society of Civil Engineers.
p. cm.
Includes index.
ISBN 0-7844-0230-2
1. Building, Iron and steel. 2. Structural frames--Design. 3. Structural stability. 4. Load factor design. 5. Steel, Structural. I. Structural Engineering Institute. Technical Committee on Load and Resistance Factor Design. Task Committee on Effective Length.
TA684.E288 1997 97-7981
CIP

Library of Congress Catalog Card No: 97-7981
ISBN 0-7844-0230-2
Manufactured in the United States of America.

Cover design by William Raaum, Art Director, and Beth Desnick, Principal, Desnick and Nelson, Minneapolis, MN.

PREFACE

For decades, structural engineers have been exploring various approaches for assessing column and frame stability in the design of steel building structures. The methods have evolved over time, the procedures changing to account for common design practices and to incorporate the power of the computer. Yet engineers have always sought a balance between being practical in their computations, while still insuring that their stability assessment yields a safe design. Accounting for column stability during design is arguably one of the most complex computations undertaken by the structural engineer, who often must consider not only local member behavior, but the behavior of the complete frame system.

The approaches for considering column stability in the design of steel frames vary widely between different codes and specifications throughout the world. However, within the context of using elastic analysis, these procedures may be subdivided into two general categories: effective length approaches and notional load (or equivalent imperfection) procedures. To many designers, it is not clear what the strengths, limitations, and validity of these methods are in the context of the American Institute of Steel Construction (AISC) specifications.

In 1993, the ASCE Structural Engineering Institute Technical Committee on Load and Resistance Factor Design (LRFD) formed the Task Committee on Effective Length to write this report elucidating several contemporary techniques for assessing column stability in the design of steel frame structures. The primary intent of this report is to provide a thorough understanding of the assumptions with respect to column and frame stability made in the AISC specifications, and to present the derivation and use of one alternate approach that is in common use in some form within several other design standards. Examples are included to illustrate the procedures for both common and unusual conditions encountered in practice.

Many specific approaches for assessing member and frame stability have been proposed in the past that are not discussed in this report. These excellent contributions are recognized and acknowledged by this Task Committee. Other monographs, most notably those of the Structural Stability Research Council and the Council on Tall Buildings and Urban Habitat, provide such information. This document is more focused. At its heart, three techniques for stability design are discussed. Two approaches are based on the use of effective length factors and are specifically outlined in sections of the Commentary of the Second Edition (1993) of the AISC *Load and Resistance Factor Design Specification for Structural Steel Buildings.*

The third method involves the use of a notional load approach for stability design (with the use of an effective length factor equal to one for all columns in the frame). While notional load approaches are not mentioned in the AISC specifications, this technique is commonly used throughout much of the rest of the world. It is thus discussed in this report to provide balance and a more complete perspective on stability design within the context of the AISC specifications.

This report is meant to be applicable to both unbraced and braced frames having either fully-restrained or partially-restrained connections. In addition, while most of the discussions pertain equally to either AISC *Allowable Stress Design* or LRFD practice, to provide focus to the report, all discussions are framed within the context of the more current LRFD specification.

This Task Committee initially set out to write a concise report elucidating the derivations and practical uses of these three approaches to stability design. However, a much more in-depth and comprehensive document has resulted from the Task Committee effort. There are many subtleties associated with the practical application of any stability design procedure. In the past, these subtleties have often led to misconceptions and misinformation in design practice and in the stability design literature. The committee determined that many of these issues could not be clarified without providing a more detailed exposition of the approaches to stability design. Although the document as a whole is large, it is hoped that a practical introduction to this material is provided through the discussions of the assumptions, advantages, and disadvantages of each of the methods, and through the step-by-step examples, and that the more detailed discussions provide a valuable reference regarding some of the more complex issues involved with design for stability.

While all members of the Task Committee provided advice and comments on all chapters in the document, the primary authors of the individual chapters and their corresponding appendices were:

Chapter 1	D. W. White
Chapter 2	J. F. Hajjar and D. W. White
Chapter 3	E. M. Lui, R. T. Leon, and D. W. White
Chapter 4	M. J. Clarke and R. Q. Bridge
Chapter 5	D. W. White and J. F. Hajjar
Chapter 6	J. F. Hajjar, R. T. Leon, and T. Q. Sheikh
Chapter 7	J. F. Hajjar

This report received two comprehensive rounds of external review, the first in the summer of 1994, and the second in the fall of 1995. An Advisory

Group performed the primary review of this report and provided advice throughout the tenure of the Task Committee. The Advisory Group included:

Mr. William F. Baker	Skidmore, Owings & Merrill, Chicago, Illinois
Prof. Gregory G. Deierlein	Cornell University, Ithaca, New York
Mr. Nestor R. Iwankiw	American Institute of Steel Construction, Chicago, Illinois
Prof. D. J. Laurie Kennedy	University of Alberta, Edmonton, Alberta, Canada
Mr. William J. LeMessurier	LeMessurier Associates, Cambridge, Massachusetts
Prof. Joseph A. Yura	University of Texas, Austin, Texas

Comments on this report were also received from members of the AISC Specification Subcommittee on Stability, the ASCE Structural Engineering Institute Committee on Load and Resistance Factor Design, the Structural Stability Research Council Task Group 4 on Frame Stability and Columns as Frame Members, and additional individuals, including Mr. Stanley D. Lindsey and Dr. Arvind V. Goverdhan, Stanley D. Lindsey and Associates, Atlanta, Georgia, and Prof. Nicholas S. Trahair, University of Sydney, Australia. In addition, Prof. Duane S. Ellifritt, University of Florida, provided guidance in his role as chairman of the ASCE Structural Engineering Institute Committee on Metals. The Task Committee expresses their sincere gratitude for the extensive contributions of the Advisory Group and the other reviewers. It should be noted that these reviewers and the committees that they represent provided guidance and comments on this report, not endorsement of the contents of the report.

The members of the committee would also like to thank Mr. Robert D. Zofkie and Mr. Michael J. Squarzini, Thornton-Tomasetti Engineers, for their assistance with the research and computations required to prepare the report. The financial support of the American Institute of Steel Construction is also gratefully acknowledged.

By the Committee:

Russell Q. Bridge	University of Western Sydney, Kingswood, Australia
Murray J. Clarke	University of Sydney, Sydney, Australia

Jerome F. Hajjar, chairman	University of Minnesota, Minneapolis, Minnesota
Roberto T. Leon	Georgia Institute of Technology, Atlanta, Georgia
Eric M. Lui	Syracuse University, Syracuse, New York
Tauqir M. Sheikh	CBM Engineers, Inc., Houston, Texas
Donald W. White	Georgia Institute of Technology, Atlanta, Georgia

Minneapolis, Minnesota
January 1997

TABLE OF CONTENTS

NOTATION

The following notation is used in this report.

A, A_g	= Gross area of cross section
B_2	= Story sidesway amplification factor at factored load levels
B_{2s}	= Story sidesway amplification factor at service load levels
C_L	= Clarification factor
$(C_L)_{avg}$	= Weighted average value of C_L over the story
E	= Modulus of elasticity
F_a	= Allowable stress for column design
$F_{a(L)}$	= Allowable stress for column design based on actual column length
F_{rc}	= Compressive residual stress
F_{rt}	= Tensile residual stress
F_y	= Yield stress
G_A, G_B	= Ratio of column-to-beam stiffness at ends A and B of column
H	= Applied laterial load
H	= Shear in column
H_i	= Applied transverse load at level i
H_{not}	= Applied notional load
I	= Moment of inertia of cross section
I_c	= Moment of inertia of column
I_g	= Moment of inertia of girder
I'	= Modified moment of inertia of cross section
K	= Effective length factor
$[K]$	= Global tangent stiffness matrix of structure
K_A	= Approximate inelastic effective length factor
K_A	= Spring stiffness at end A of column
K_B	= Spring stiffness at end B of column
K_{C_L}	= Story-based effective length factor based on use of C_L
K_e	= Generic elastic effective length factor
$[K_e]$	= Global elastic stiffness matrix of structure
$[K_g]$	= Global geometric stiffness matrix of structure
K_i	= Generic inelastic effective length factor
K_{K_n}	= Effective length factor based upon story buckling load equaling sum of individual column buckling loads calculated using K_n
K_n	= Sidesway uninhibited nomograph effective length factor

K_{R_L} = Effective length factor based upon using R_L and a relatively conservative value of C_L

K_{story} = Generic story-based effective length factor

K_{system} = Effective length factor from eigenvalue system buckling analysis

K_x, K_y = Effective length factors for buckling about x and y axes

L = Member length

L_b = Beam or girder length

L_c = Length of column

L_g = Length of girder

L'_g = Modified length of girder

L_x, L_y = Unsupported column lengths about the x and y axes

M = Bending moment

M = Joint bending moment

M_A, M_B = Bending moment at ends A and B of the member

M_L = Larger column end moment

M_S = Smaller column end moment

M_b = Beam end bending moment

M_{bx} = Nominal member out-of-plane flexural strength for in-plane bending about the major principal x-axis.

M_c = Column end bending moment

M_{cn} = Nominal moment capacity of connection

M_{des} = Design connection moment

M_f = First-order bending moment

M^{fixed} = Fixed-end bending moment

M_g = Factored gravity load bending moment

M_n = Nominal flexural strength

M_{nx}, M_{ny} = Nominal flexural strength about the x and y axes of the cross section

M_p = Plastic bending moment

M_{pb} = Plastic bending moment of beam

M_{ser} = Service load connection moment

M_{sx} = Nominal section strength for flexure about the x-axis.

M_u = Required bending moment

M_{ux}, M_{uy} = Member maximum second-order elastic bending moment about the x and y axes of the cross section

M_1 = Member maximum first-order bending moment

M_2 = Member maximum second-order bending moment

N = Notional lateral load

N	= Ratio of the axial force in all columns in a story to the axial force in all non-leaning columns in a story
P	= Axial force
P	= Gravity load
P_e	= Euler buckling load based on $K = 1$
P_{cr_i}	= Critical buckling load for column i as approximated by a particular approach to computing effective length
$P_{cr_i}^{new}$	= Critical buckling load for column i after one or more members in a story have breached their allowable capacity limit
$P_{e\hat{\tau}}$	= Inelastic buckling load of column
$P_{e\hat{\tau}}(C_L)$	= Buckling load computed using K_{C_L}
$P_{e\hat{\tau}}(K_n)$	= Buckling load computed using K_{K_n}
$P_{e\hat{\tau}(nomo)}$	= Buckling load computed using K_n
$P_{e\hat{\tau}(RL)}$	= Buckling load computed using K_{R_L}
$P_{e\hat{\tau}(story)}$	= Euler buckling load computed using K_{story}
$P_{e\hat{\tau}(system)}$	= Euler buckling load computed using K_{system}
P_L	= First-order resistance to sidesway of column (i.e., force which produces unit rotation in column)
P_n	= Nominal axial strength
$P_{n(L)}$	= Nominal axial strength based on actual length
$P_{n(L)x}$	= Nominal axial strength based on actual length, relative to the x-axis
P_{nx}, P_{ny}	= Nominal axial strength for failure about the x and y axes
P_s	= Axial force at service loads
P_u	= Required axial force
P_y	= Squash load
Q	= Factored axial force in girder
Q_{cr}	= Estimated critical buckling capacity of girder
R_k	= Connection stiffness
$\overline{R_k}$	= Nondimensional connection stiffness
R_{kdes}	= Design connection stiffness
R_{kf}	= Transitional secant connection stiffness
R_{ki}	= Initial connection stiffness
R_{kt}	= Tangent connection stiffness
R_{kser}	= Service load connection stiffness
$R_{k,system}$	= System rotational stiffness
R_{ku}	= Ultimate connection stiffness
R_L	= Ratio of the axial force in all leaner columns in a story to the axial force in all columns in a story

R_{stiff} = Linear elastic stiffness of a cantilever column
S = Elastic section modulus
S = First-order member or story lateral stiffness
S = Standardization constants
S_L = Approximate ratio of the story yield strength to the elastic story buckling capacity
S_{ij} = Stability function
$S_{k,system}$ = System translational stiffness
V = Shear
Z = Plastic section modulus
d = Depth of a wide-flange member
e = Error in the column axial strength associated with the use of $K = 1$
e_{el} = Error in the column axial strength associated with the use of $K = 1$, compared to the column strength based on an elastic buckling analysis
e_{inel} = Error in the column axial strength associated with the use of $K = 1$, compared to the column strength based on an inelastic buckling analysis
e_o = Global initial frame non-verticality
f_a = Axial stress at service load conditions
h = Story height
k = First-order sidesway stiffness
k_c = Refinement factor for initial angular sway imperfection Ψ and notional load parameter ξ based on number of columns per plane
$[k_e]$ = Element elastic stiffness matrix of structure
$[k_g]$ = Element geometric stiffness matrix of structure
k_S = Refinement factor for notional load parameter ξ based on first-order member or story lateral stiffness
k_s = Refinement factor for initial angular sway imperfection based on number of stories
k_y = Refinement factor for notional load parameter ξ based on column yield stress
k_λ = Refinement factor for notional load parameter ξ based on "average" story slenderness
r = Radius of gyration of cross section
r_x, r_y = Radius of gyration about the x and y axes of the cross section

Δ = Relative transverse displacement between the ends of a member
Δ = Elastic second-order sway displacement
Δ_f = Elastic first-order sway displacement
Δ_o = Local story initial out-of-plumbness
Δ_{oh} = First-order interstory deflection due to $\sum_{non-leaner} H$ on story
Δ_{ph} = Second-order interstory deflection due to $\sum_{non-leaner} H$ on story
Ψ = Initial angular out-of-plumbness (sway) imperfection
Ψ_s = Initial angular story out-of-plumbness (sway) imperfection
α = Ratio of smaller and larger column end moments
α_A, α_B = Elastic rotational end-restraint stiffnesses at ends A and B of the member
$\bar{\alpha}_A$, $\bar{\alpha}_B$ = Normalized elastic rotational end-restraint stiffness at ends A and B of the member
β = First-order stiffness factor
β = End-moment ratio equal to the ratio of the smaller to the larger end-moment, taken as positive when the member is bent in double-curvature
γ = Strength parameter
γ = Stiffness factor of column
δ = Deflection of member axis from the rotated member chord
δ_o = Column initial out-of-straightness at mid-height
ε = Percent error in calculation of K_{story}
ε = Error in the LRFD beam-column interaction equation at its maximum design limit, associated with the use of $K = 1$
ε_{el} = Error in the LRFD beam-column interaction equation at its maximum design limit, associated with the use of $K = 1$, compared to the column strength based on an elastic buckling analysis
ε_{inel} = Error in the LRFD beam-column interaction equation at its maximum design limit, associated with the use of $K = 1$, compared to the column strength based on an inelastic buckling analysis
η = Shape factor of cross section
θ = Connection rotational deformation
θ_A, θ_B = Column rotation at ends A and B of beam
θ_b = Beam end rotation
θ_{des} = Connection rotational deformation at M_{des}
θ_f = Transitional connection rotational deformation

θ_r = Relative rotation of beam to column
θ_r = Flexural deformation of connection
θ_{ser} = Connection rotational deformation at M_{ser}
θ_u = Connection rotational deformation at M_u
κ = Parameter used to compute stability functions
λ_c = Column slenderness parameter based on effective length
$\lambda_{c(L)}$ = Modified column slenderness parameter based on actual length
$\lambda_{c(L)x}$ = Modified column slenderness parameter based on actual length, relative to x-axis
λ_{K_n} = Buckling parameter for story using approach of K_{K_n}
λ_{R_L} = Buckling parameter for story using approach of K_{R_L}
λ_{story} = Buckling parameter for story
λ_{system} = Buckling parameter (first buckling mode eigenvalue) of total frame system
ξ = Parameter defining magnitude of notional load
ξ_o = Simple notional load parameter (equal to 0.005)
ξ_M = Modified notional load parameter
ξ_R = Refined notional load parameter
σ_{rc} = Compressive residual stress at flange tips
τ = Inelastic stiffness reduction factor based on column effective length
$\hat{\tau}$ = Ratio of tangent modulus to elastic modulus
τ_L = Inelastic stiffness reduction factor based on column actual length
ϕ_b, ϕ_c = Resistance factors for bending and axial compression

METRIC CONVERSION

Distance	
1 inch	= 25.4 millimetres
1 foot	= 0.305 metres
Area	
1 square inch	= 645 square millimetres
1 square foot	= 0.093 square metres
Volume	
1 cubic inch	= 1.64×10^4 cubic millimetres
1 cubic foot	= 0.0283 cubic metres
Mass	
1 pound	= 0.454 kilograms
Force	
1 pound-force	= 4.448 newtons
1 kip	= 4.448 kilonewtons
Bending Moment	
1 pound-force-inch	= 0.113 newton-metres
1 kip-inch	= 0.113 kilonewton-metres
Stress	
1 pound-force per square inch	= 6.895 kilopascals
1 kip per square inch	= 6.895 megapascals

CHAPTER 1

INTRODUCTION

Although the subject of frame stability is a mature one, the approaches for consideration of stability in the design of frames vary widely among the present specifications and standards throughout the world. Furthermore, a large number of design methods have been proposed in the vast literature on frame stability, many of which have subtle differences and subtle implications with regard to their application. This report addresses two categories of approaches to frame stability design that are prevalent in one form or another in many of the current specifications and standards for steel frame design: *effective length* approaches, and *equivalent imperfection* approaches—referred to most often as *notional horizontal load* methods.

Both of these types of approaches are capable of accounting for the various deleterious effects on the strength of column and beam-column members that in general must be considered in design. These effects include initial residual stresses, member out-of-straightness, frame imperfections (i.e., column out-of-plumbness in regular rectangular frames, or errors in joint position from the ideal geometry in general non-regular types of framing), and incidental load eccentricities in the transfer of design forces to the members [SSRC, 1981]. Also, both types of approaches account for the interdependencies between the strength of individual members and the strength of the overall structural system. However, effective length and notional load procedures differ in the way that they approximate these effects.

In effective length methods, the above effects are accounted for predominantly within expressions for column strength, i.e., either single or multiple column strength curves. Each of the column and beam-column members within a frame is designed based on a column curve for an equivalent pin-ended member of length KL, where K is referred to as the effective length factor and L is the actual member length. Alternatively, in equivalent imperfection procedures, the actual column length is utilized in the design calculations, and artificially large frame imperfections are included in the analysis of the structural system. Often, the equivalent initial imperfection effects are modeled by "notional" horizontal loads. This type of approach accounts for the effects of residual stresses, out-of-straightness, errors in joint geometry, incidental load eccentricities, and interactions between member and system strength largely by affecting the applied moment term of the beam-column interaction equations.

1.1 Objectives and Scope

There has been much debate about the appropriateness of various methods for frame stability design, many of which fit within the effective length or notional horizontal load categories. Therefore, it is no surprise that present specifications and standards reflect varying philosophies associated with each of these types of approaches. For example, the AISC LRFD [AISC, 1993] and Japanese [AIJ, 1989] specifications require an effective length approach, whereas the Canadian limit states standard CSA-S16.1 [CSA, 1994], the Australian Specification AS4100 [SAA, 1990], and Eurocode 3 [CEN, 1993] either give preference to or require the use of notional horizontal load techniques. This report takes the position that both types of approaches are legitimate for frame stability design, and that both have distinct advantages as well as limitations.

The primary objectives of this report are: (1) to consolidate and present within a unified context the most up-to-date information on the basis, underlying assumptions, accuracy, and appropriate use of effective length and notional horizontal load methods, and (2) to compare and contrast the different types of procedures. The following two over-riding principles, as suggested by Kennedy, et al. [1993], are considered throughout:

> **"The formulations must embody the actual (perceived) behavior of the member [and ideally, the structural system] under the range of load and restraint conditions and at the same time be relatively simple to use. Simplicity of formulation need not lead to significant approximations, however. Design rules [and their associated analysis procedures] should neither be so simple nor so complex that the true behavior is obscured, but be clear and transparent, enabling the designer to understand the formulation readily."**

To paraphrase the above quotation, it can be stated that the consideration of stability in frame design should be as simple as possible, but no simpler than required to maintain reasonable accuracy and clarity. In this light, the report addresses, in addition to the proper application of effective lengths and notional horizontal loads, limits that express when these devices may be neglected entirely within a design. This is in keeping with the following disclaimer suggested by Higgins shortly after the effective length concept was first introduced within the AISC Specifications [Higgins, 1964]:

> **"It should not be construed from the following remarks that, from now on, all tier buildings, or even very many, must be designed on the basis of overall instability unless provided with an extensive bracing system."**

Many of the discussions and presentations within the report are general in nature. However, due to differences in matters such as selection of column curves, beam-column interaction formats, load and resistance factors, and specific implementations of effective length or notional load procedures, direct comparison and contrasting within the context of even a few of the most prominent specifications is difficult. Therefore, in this document, all the specific presentations of design procedures, design results, and their implications are based on AISC LRFD practice [AISC, 1994]. This allows for a thorough and consistent exposition of the effective length and notional load approaches, while avoiding the complex, intricate, and subtle differences between various standards. Also, this permits a direct focus on the appropriateness of both the effective length and notional horizontal load approaches for design, using AISC LRFD as the example. *In general, the concepts and procedures discussed are equally applicable to other limit states standards, as well as to Allowable Stress Design* [AISC, 1989].

Within the above context, the report is focused further as follows:

1. In all discussions of design equations, it is assumed that the error in determining second-order elastic forces is negligible. Issues associated with accurate calculation of these forces (either based on direct second-order analysis or by use of approximate amplification factors, and the appropriate use of effective length factors within amplifiers) are addressed elsewhere (e.g., [Gaiotti and Smith, 1989; Iffland, 1988; Liew et al., 1991; White and Hajjar, 1991]).

2. With the exception of idealized models for determining inelastic buckling loads and/or effective lengths, design usage of inelastic analysis is not considered. That is, the report addresses solely the issue of frame stability design based on either first-order linear elastic analysis with appropriate amplification factors, or direct second-order elastic analysis. "Exact" inelastic analysis results are presented in certain cases, but only for the purposes of calibrating the notional load approach and investigating the quality of the elastic analysis/design procedures.

   The primary implication of elastic analysis/design is that the design strength of the structural system is limited to the strength associated with the first achievement of a limit state within any of the members or other structural components of the frame. Any "reserve capacity" associated with inelastic redistribution of forces from the most critically loaded members is not considered. In this type of approach, the beam members within regular rectangular frames are commonly assumed to remain elastic for all loadings less than the factored design loads. Therefore, these members provide elastic restraint to the rigidly- (or semi-rigidly) connected column and beam-column members for all loadings less than or equal to the system design strength. However, the column and beam-column members, and furthermore any semi-rigid connections, in general can exhibit significant yielding prior to the frame reaching its elastic design capacity.

3. Direct consideration of the contribution of the slab to the strength and stiffness provided by the beams and beam-to-column connections is not addressed. The committee is not aware of any established values for these strengths and stiffnesses in fully-restrained frame construction. The reader is referred to [Leon et al., 1996], [Leon and Forcier, 1992] and [Zaremba, 1988] for discussion of these issues in the context of semi-rigid frames with composite beam-slab systems.

4. The primary emphasis is on assessment of in-plane stability of framing systems. However, issues regarding the proper use of effective length factors or notional loads when checking out-of-plane or spatial member strengths are addressed. This is necessary for the design examples and comparisons of methods to be useful within a reasonably broad practical context. The examples are limited to members of compact cross-section to avoid the additional complexity associated with accounting for local buckling in the determination of the member strengths, and to allow a more direct comparison of the qualities and limitations of the procedures for overall frame stability design. However, the procedures and examples are relevant to elastic design of members with non-compact or slender cross-section elements to the extent that local buckling effects may be included in determining the resistances.

5. Practical as well as extreme examples are provided that illustrate the key issues discussed. An extensive effort has been made to compare the effective length and notional horizontal load approaches directly for a large number of the examples cited within the engineering literature. The types of frames considered include fully-restrained (FR) and partially-restrained (PR) frames with and without leaning columns, and moderate and short height frames.

   Effective length procedures for braced frames are presented. However, sidesway-inhibited design examples in which K is less than one are not provided. This is in part due to the fact that the use of $K < 1$ for the design of column members has important ramifications on the second-order moments in the beams. The beam bending moments may be amplified significantly in frames designed with $K < 1$. In the opinion of the committee, inclusion of design examples with $K < 1$ would require that the proper calculation of these second-order moments be addressed.

6. Stability considerations under dynamic loading (seismic, blast, etc.) are not addressed. The committee recognizes that stability under seismic loading is an area where further research is needed.

7. With the exception of the above restrictions, when choices exist between simplicity of presentation and comprehensivity of the discussions, the committee has opted in most cases toward comprehensivity. This is particularly true with respect to discussion of the implications of different methods on design solutions. Detailed derivations of equations are not always presented, but an attempt is made in all cases to clearly

explain the basis for and underlying assumptions of the models. References are provided where detailed derivations are not included. Also, an attempt is made to address the accuracy and appropriate use of all the methods discussed. Step-by-step examples are provided to illustrate the application of each of the major techniques.

There are many subtleties associated with application of any stability design approach to specific situations. These subtleties often lead to misconceptions and misinformation in design practice, and in the stability design literature. In the opinion of the committee, many of these problems cannot be clarified and solved without providing a detailed exposition of the procedures. Although the document as a whole is large, it is hoped that the step-by-step examples will provide a useful basic introduction for the more casual reader, and that the detailed discussions will provide valuable information for resolving some of the complex issues involved with various design situations for the highly-technical reader.

The following section explains the development of the beam-column interaction equations of the AISC LRFD Specification [1993], and the history of effective length in the AISC Specifications. This is followed by a summary of key attributes and phenomena that may affect frame stability, and a simple example that highlights several of the issues addressed in this document. These sections lay the groundwork for understanding the proper application of the stability design procedures discussed in the subsequent chapters, within the context of the AISC LRFD Specification. Chapter 1 is closed by an overview of the contents of this publication, and an outline of state-of-the-art review references for specific topics not addressed.

1.2 The AISC LRFD Beam-Column Interaction Equations

The AISC LRFD Specification [1993] provides the following interaction equations for design of steel beam-columns:

$$\frac{P_u}{\phi_c P_n} + \frac{8}{9}\left[\frac{M_{ux}}{\phi_b M_{nx}} + \frac{M_{uy}}{\phi_b M_{ny}}\right] \leq 1.0 \quad \text{for} \quad \frac{P_u}{\phi_c P_n} \geq 0.2 \tag{1.1a}$$

$$\frac{P_u}{2\phi_c P_n} + \frac{M_{ux}}{\phi_b M_{nx}} + \frac{M_{uy}}{\phi_b M_{ny}} \leq 1.0 \quad \text{for} \quad \frac{P_u}{\phi_c P_n} < 0.2 \tag{1.1b}$$

where P_u is the required axial compressive strength or the axial load effect on the member being considered, P_n is the nominal compressive strength, M_{ux} and M_{uy} are respectively the maximum second-order elastic moments about the strong- and weak-axes of the cross-section within the unsupported length of the member being considered, M_{nx} and M_{ny} are the corresponding nominal flexural strengths, and ϕ_c and ϕ_b are the resistance factors for compression and flexure. The load effects P_u, M_{ux} and M_{uy} may be determined either directly from second-order elastic analysis or, for typical shear frames, by use of first-order elastic analysis with approximate amplifiers applied to the first-order moments [AISC, 1993].

Equations 1.1a and 1.1b describe a single interaction curve that characterizes the strengths of steel frame members associated with all the possible beam-column limit states. This includes the in-plane strength of members bent about their strong- or weak axes, the out-of-plane lateral-torsional strength of members loaded in-plane about their strong-axis of bending, and the spatial strength of members subjected to general biaxial bending and axial compression. The equations are capable of capturing this wide range of limit states by: (1) calculation of the nominal axial strength P_n as the smaller of the in-plane and out-of-plane strengths of the member if loaded solely by concentric axial force (i.e., zero applied moment), and (2) calculation of the nominal strong-axis flexural strength of the member M_{nx} (i.e., the strength for zero axial load) considering the effects of member inelasticity and lateral-torsional buckling as appropriate. Local buckling effects are accounted for in the calculation of the P_n, M_{nx}, and M_{ny} values for members with non-compact or slender cross-section elements [AISC, 1993]. It is important to note that, in many cases where the structural system is well proportioned and the columns are unsupported between the story-levels in the weak-axis direction, P_n will be controlled by out-of-plane flexural buckling. However, in-plane stability still must be checked.

1.2.1 Development of the AISC LRFD Interaction Equations—Guidelines

In the development of the AISC LRFD interaction equations, the following guidelines were established [LeMessurier, 1985; Yura, 1988; Liew et al., 1991]:

1. The equations should be general and applicable to a wide range of problems: members with various slenderness ratios having representative initial imperfections (i.e., bow and sweep) and residual stresses, subjected to strong- or weak-axis bending, sway or nonsway displacements, transverse lateral loads or end loads only, and with or without leaning column effects.

2. The equations should be based on the load effects obtained from second-order elastic analysis since, at the present time, second-order inelastic analysis is not readily accessible for design office use.

3. The equations should distinguish clearly between the second-order elastic load effects and the resistances such that the calculation of second-order forces (by direct analysis or approximate amplification of first-order load effects) can be clearly separated from the interpretation and design application of the equations.

4. The equations should predict identical ultimate strengths for problems in which the strengths are the same. For example, if matching geometric imperfections are assumed, the behavior in each of the problems shown in Fig. 1.1 is physically identical in both the elastic and inelastic ranges.

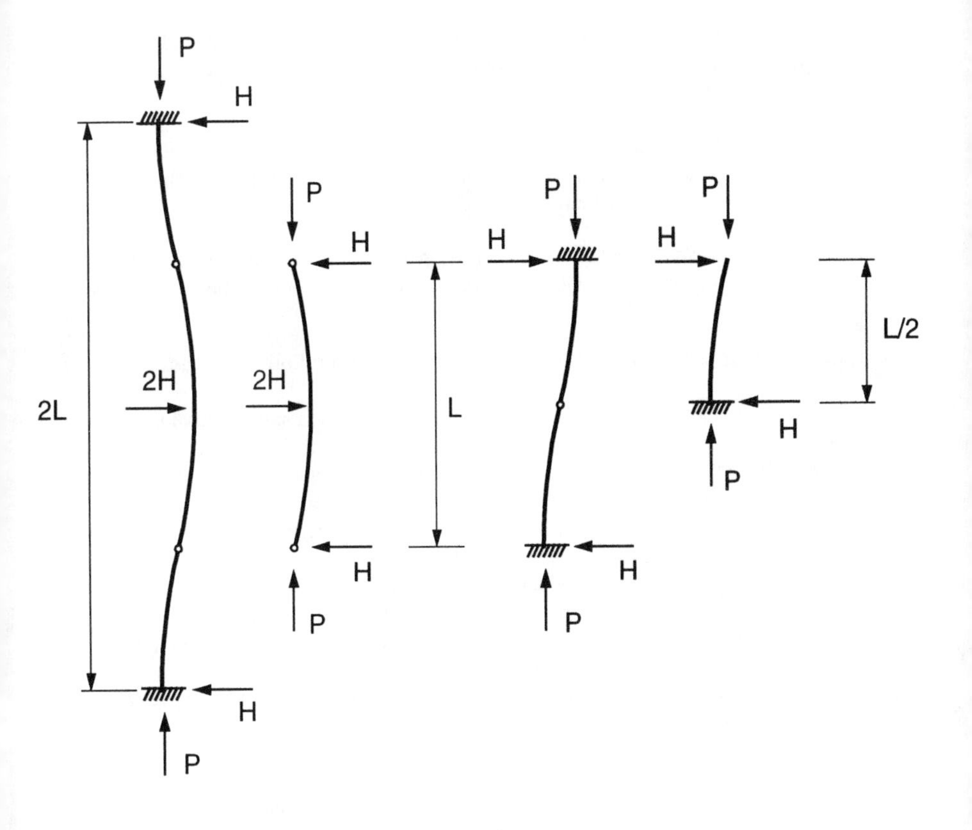

Figure 1.1. Beam-column problems that have identical solutions.

5. The equations should not necessarily be required to consider strength and stability separately, since in general, all columns of finite length fail by some combination of inelastic bending and stability effects.

6. The equations should be capable of capturing the limit state of pure stability under axial load, including the effects of restraint provided by *elastic* beam members to the *elastic* or *inelastic* columns.

7. The equations should not be more than five percent unconservative when compared to the strengths obtained from "exact" second-order inelastic solutions.

1.2.2 Development of the AISC LRFD Interaction Equations—Procedure

The AISC LRFD interaction equations were determined in part by curve fitting to strength curves obtained from "exact," two-dimensional, second-order inelastic analyses. In these analyses, the spread of plasticity, including the effects of initial residual stresses, is explicitly modeled. The primary source for these "plastic-zone" solutions was the research by Kanchanalai [1977]. In all of Kanchanalai's studies, the members were assumed to be perfectly straight and plumb. Also, a typical residual stress distribution for wide-flange sections (self-equilibrating, with a constant residual tension in the web of F_{rt}, and a linear variation of residual stress from F_{rt} at the web flange junction to a maximum compressive residual stress of $F_{rc} = 0.3F_y$ at the flange tips) was assumed. A W8x31 was utilized as the column section in all of Kanchanalai's analyses. The elastic modulus was taken as $E = 29{,}000$ *ksi (200 GPa)* and the yield stress was taken as $F_y = 36$ *ksi (250 MPa)*.

Given data from "exact" plastic-zone analyses such as Kanchanalai's, the AISC LRFD interaction equations can be arrived at based on the following procedure [LeMessurier, 1985; Yura, 1988; Liew et al., 1991]:

1. Ultimate strength curves from the second-order inelastic analysis are plotted in a non-dimensional form as the normalized axial load P/P_y versus the normalized *first-order* moment M_1/M_p. Typical results for beam-columns of the geometry shown in Fig. 1.1 are illustrated in Fig. 1.2 ($M_1 = HL/2$). For generation of the example strength plots, W8x31 sections bent about their weak axis, A36 steel, and $L/r = KL/r = 100$ ($2L/r = 200$ and $L/2r = 50$) are assumed.

2. The normalized *first-order* moment M_1/M_p, obtained from the plastic-zone analysis, is converted to a *second-order* elastic moment M_2/M_p based on the second-order elastic amplification for the case being considered. For the beam-columns shown in Fig. 1.1, the "exact" amplification factor is $\frac{\tan\alpha}{\alpha}$, where

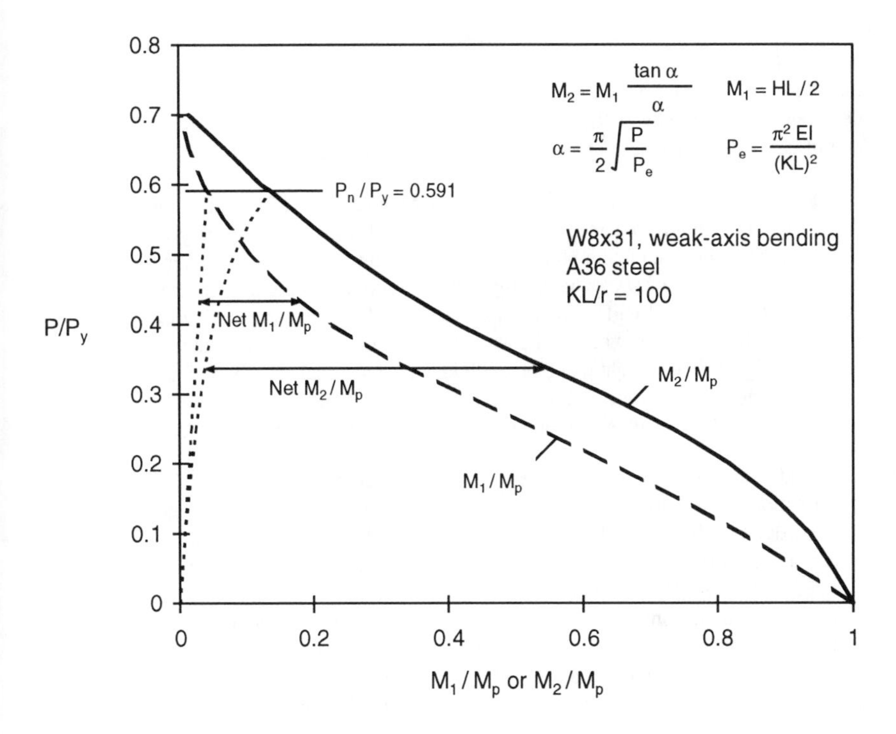

Figure 1.2. Illustration of procedure for determining the net M_1/M_p and M_2/M_p curves (W8x31, weak-axis bending, A36 steel, *KL/r = 100*) [LeMessurier, 1985].

$\alpha = \frac{\pi}{2}\sqrt{\frac{P}{P_e}}$ and $P_e = \frac{\pi^2 EI}{L^2}$. However, it should be noted that calculation of the exact second-order elastic moments is not required. In the actual development of the AISC LRFD interaction curves, a simple elastic $P\Delta$ amplification was assumed for sway frame cases [LeMessurier, 1994].

3. The column strength of an axially loaded member with $KL/r = 100$ is given by the AISC LRFD column formulas as $0.591P_y$. However, for axial loading only ($H = 0$), the column strength obtained from plastic-zone analysis of the members shown in Fig. 1.1 is $0.71P_y$. The discrepancy between these column strengths is due to the fact that the LRFD equations represent the "nominal strength" of the actual imperfect member, whereas the above plastic-zone analyses are exact solutions of perfectly straight members with idealized residual stresses and zero initial imperfections. In Fig. 1.2, the abscissa of the M_1/M_p curve at the ordinate of $P/P_y = 0.591$ is obviously "used up" by the nominal imperfection effects.

4. To include the nominal imperfection effects in the development of the specification beam-column interaction curves, the "used-up" first-order moment capacity due to column initial imperfections is assumed to vary linearly with the axial load, as shown in Fig. 1.2. The net usable first-order moment capacity (labeled as the "*Net* M_1/M_p" in the figure) is then obtained by subtracting the "used-up" moment from the first-order moment M_1/M_p. This value, when multiplied by the elastic moment amplification factor yields the net usable second-order moment capacity (shown as the "*Net* M_2/M_p" in Fig. 1.2). These results are plotted in Fig. 1.3. The AISC LRFD equations are obtained by curve fitting to the lower bound of the "*Net* M_2/M_p" curves for this and a large number of other potential design cases. The final AISC LRFD expressions are obtained by incorporating the resistance factors, ϕ_c and ϕ_b.

   It should be noted that, if exact plastic-zone analyses were conducted on the actual imperfect beam-column members, then the net M_1/M_p and M_2/M_p values could be obtained directly without the need for the above adjustments. However, in general the "exact" strength of an imperfect column, computed from a plastic-zone analysis, will not match exactly with the nominal strength specified by any selected column design formula. The AISC LRFD column curve is simply a reasonable, representative fit to the actual strengths for many types of columns. The LRFD equations are based on the strength of an "equivalent" simply-supported column of length KL, with a mean maximum out-of-straightness at its mid-length of approximately $KL/1500$ [1]. These equations form a lower bound to

---

[1] Since the out-of-straightness of $KL/1500$ is on the "equivalent" simply-supported column of length KL, it can be argued that this out-of-straightness accounts for both the effects of the *actual* column's out-of-straightness, as well as its out-of-plumbness. The out-of-plumb value can be taken as the out-of-straightness at the position L along the length of the equivalent column.

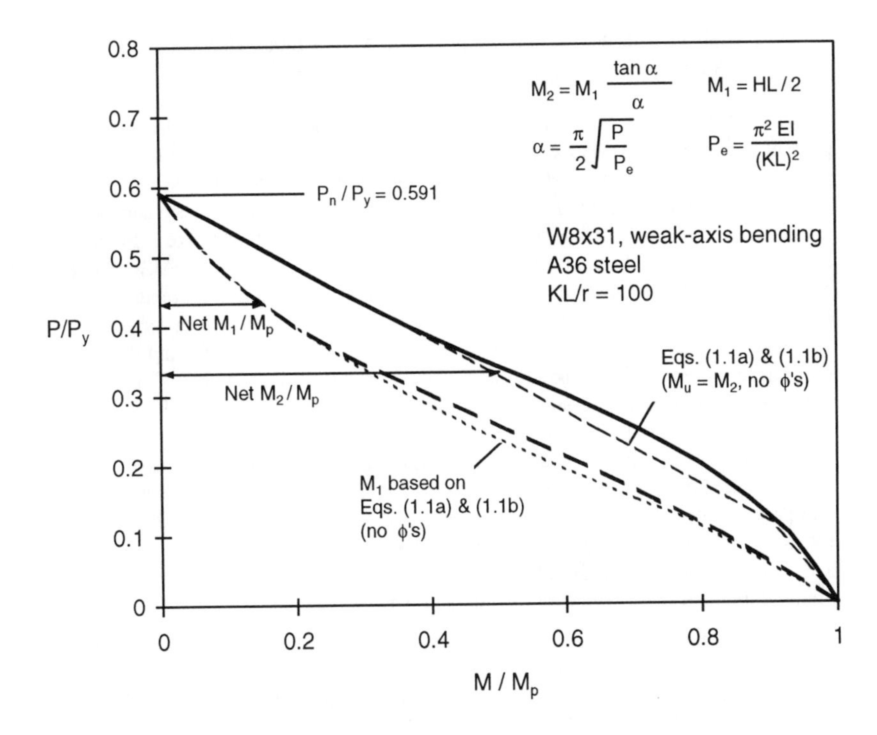

Figure 1.3. Example curve fitting of the AISC-LRFD beam-column equations to the "exact" strength curves (W8x31, weak-axis bending, A36 steel, *KL/r = 100*) [LeMessurier, 1985].

the available test data on simply-supported column members that satisfy the ASTM A6 maximum out-of-straightness limit of *L/1000* [Tide, 1985].

The LRFD interaction equations (without ϕ factors) are never more than five percent unconservative when compared to all the computed results -- adjusted as outlined above -- for in-plane strong- and weak-axis bending, including sway and non-sway cases, members with transverse loads or with end loads only, and systems with leaning columns. Furthermore, the weak-axis $L/r = 100$ case shown here is the most critical of all the Kanchanalai data considered in the development of the AISC LRFD beam-column expressions [LeMessurier, 1985]. The LRFD equations are an excellent fit to this curve. In general, the equations also give a superb fit for all strong-axis cases of the type shown in the figures for L/r from *0* to *100*, but they are very conservative for the weak-axis case when L/r ranges from *0* to *40*. This is due to the large shape factor and significant convexity of the cross-section full-plastification strength for weak-axis bending of wide-flange shapes. They are moderately conservative for both axes when L/r is greater than *120* [LeMessurier, 1985]. Also, for sidesway-inhibited cases, the LRFD equations tend to be somewhat more conservative for beam-columns subjected to reversed-curvature bending than for cases involving single-curvature moment diagrams [Clarke and Bridge, 1992]. This is due to the fact that the LRFD equations do not account for the effect of moment gradient on the shape of the interaction curve.

For limit states involving out-of-plane lateral-torsional failure, the AISC LRFD beam-column interaction equations appear to be quite accurate for beam-type wide-flange sections with larger r_x/r_y values. However, for typical column-type wide-flange sections with r_x/r_y values smaller than 2.0, they can be quite conservative compared to more complex expressions for lateral-torsional strength, such as contained in the current Australian Standard [SAA, 1990; White, 1993; White et al., 1993]. For spatially loaded beam-column members, the linear interaction relationships posed by Eqs. 1-1a and 1-1b can be very conservative compared to the nonlinear interaction relationships provided in Appendix H of the AISC LRFD Specifications [AISC, 1993]. However, the equations of the LRFD Appendix H are limited to braced frames only.

1.2.3 History of Effective Length in the AISC Specifications

The effective length concept was first introduced into the AISC Specifications in 1963 [AISC, 1963]. Prior to this date, ordinary column design was performed based on the actual length of the member. Subsequently, much research has ensued that has addressed the calculation of effective lengths for a wide range of member and structural system configurations encountered in practice. Major references to this work are outlined in Section 1.6.

The concept of inelastic effective length, which is required for accurate calculation of the maximum strength of stocky heavily-loaded columns that are restrained by elastic beams, first appeared in the eighth edition of the AISC Allowable Stress Design Manual [AISC, 1978]. Also, the eighth edition introduced the concept that overall sidesway stability effects could be neglected in frame design (i.e., second-order amplification could be neglected and $K = 1$ could be assumed) if a quantified amount of lateral frame stiffness is provided, subject to limits on several other design parameters.

The second edition of the AISC LRFD Specification [AISC, 1993] is the first to address explicitly the concept of leaning columns and their effect on lateral stability, although the knowledge of leaning column effects certainly is not new. Detailed equations, which are based on the concept of overall story stability in shear-type building frames, are outlined in the Specification commentary. Also, the second edition LRFD Specification no longer requires effective length factors to be greater than one in unbraced frames. This is in recognition of the fact that a beam-column of relatively small section, and/or one that is heavily-loaded, may in effect be "braced" by the sidesway flexural stiffness of other members within the framing system.

Up to the present time [1996], the AISC Specifications have not recognized the use of equivalent imperfection or notional load procedures for stability design.

1.3 Review of Key Effects on Frame Stability

As noted at the beginning of this chapter, both of the major approaches to frame stability design—effective length and notional horizontal load—account for the deleterious effects of residual stresses, out-of-straightness, errors in joint geometry, and incidental load eccentricity, and for the beneficial as well as deleterious interactions between member and system strength. To place these stability design procedures within a broad context, Tables 1.1 and 1.2 provide a somewhat comprehensive listing of behavioral phenomena and physical attributes that may affect the capacity of a frame. The primary effects addressed by the methods discussed in this report are listed in Table 1.1, whereas a categorization of other potential effects are summarized in Table 1.2. Other useful categories of member as well as system behavior and attributes are provided in [Birnstiel and Iffland, 1982], [SSRC, 1988], [McGuire, 1990], and [Ackroyd, 1991].

As indicated in the tables, the behavioral phenomena affecting the strength of a frame may be subdivided into two groups: geometric nonlinear and material nonlinear. For the purpose of the discussions within this document, frame stability design is defined as a design which appropriately accounts for the effects of significant geometric nonlinearities, and thus properly models the physical attributes that influence the geometric nonlinear behavior. Within the context of this definition, the calculation of effective length factors or the use of notional horizontal loads in the analysis can be neglected only when all the associated geometric nonlinear effects listed in Table 1.1 have a small influence on the strength of a frame and its members. As noted in [SSRC, 1976] and [SSRC, 1988], there are certain classes of

Table 1.1 In-plane behavioral phenomena and physical attributes generally addressed by the procedures considered in this report.

NONLINEAR BEHAVIORAL PHENOMENA

Geometric Nonlinear Phenomena:

- $P\Delta$ effect—effect of axial force acting through displacements associated with member chord rotation
- $P\delta$ effect—effect of axial force acting through displacements associated with member curvature
- Flexural buckling
- Interaction of member and frame stability

Material Nonlinear Phenomena:

- Distributed yielding along member lengths in members subjected to large axial loads (influenced by initial residual stresses)
- Plastic section strength under any combination of axial force and bending moment

PHYSICAL ATTRIBUTES

- Initial geometric imperfections
  - i. Erection of out-of-plumbness
  - ii. Member of out-of-straightness
  - iii. Incidental joint or load eccentricities
- Initial residual stresses from manufacturing and fabrication processes
- Member rotational and sidesway end restraint
- Leaning columns
- Connection and beam-column panel zone rigidity and strength

Table 1.2 Other behavioral phenomena and physical attributes which may influence frame stability

NONLINEAR BEHAVIORAL PHENOMENA

Geometric Nonlinear Phenomena:

- Bowing effect—effect of curvature on axial displacements at member ends
- Wagner effect—effect of bending moments and axial forces acting through displacements associated with member twisting
- Behavior of moments under finite three-dimensional rotations
- Out-of-plane lateral-torsional or torsional buckling of members
- Local buckling and local distortion
- Interaction of local and member stability

Material Nonlinear Phenomena:

- Inelastic redistribution of forces
- Plastic interaction of axial force, moment, and shear
- Effect of loading rate on yielding behavior
- Elastic unloading subsequent to plastic deformation
- Influence of loading sequence
- Strain-hardening and strain aging
- Cyclic plasticity effects (Bauschinger effect, cyclic hardening, deflection stability under repeated loading, and low-cycle fatigue)

PHYSICAL ATTRIBUTES

- Initial member out-of-plane sweep, twisting imperfections, and cross-section distortion
- Initial residual stresses from erection out-of-fit, construction sequencing, etc.
- Member torsional, warping, and out-of-plane end restraint
- Cross-section symmetry/non-symmetry
- Prismatic/non-prismatic member profile
- Stiffness and location through the cross-section depth of out-of-plane bracing
- Location of transverse member loads with respect to the cross-section depth and shear center (these attributes also influence out-of-plane behavior)
- Member shear and torsional flexibility
- Composite interconnection with floor slabs, and other interconnections with primary or secondary systems
- Finite joint size
- Variability of structural resistances
- Type of loading, static or dynamic

frames for which this is the case. For practical purposes, the in-plane strength of these types of frames is limited only by material nonlinear effects.

In general, both material and geometric nonlinear effects are important, and the overall frame stability behavior exhibits the following pattern [SSRC, 1976]:

> **As load is increased on a frame, members are stressed elastically until parts of the frame are strained into the inelastic region. The ultimate capacity is reached when the combination of progressive yielding, axial force, and joint displacements reduce the stiffness of the structure so that the frame becomes unstable. All tests show conclusively that unbraced frames are likely to fail through instability before the formation of a plastic mechanism and that any rational analysis and design procedure should attempt to include this effect....**

However, in elastic analysis and design, the inelastic reserve strength of the structure beyond the load level corresponding to a limit state in the most critically loaded structural component is not considered. The maximum design capacity for a given load combination is determined as the load associated with a limit state being reached in the "most critically loaded" component of the structural system. Any possible inelastic redistribution of forces is neglected in the design.

For routine elastic analysis and frame stability design, emphasis is often placed on the P-Δ and P-δ effects and their influence on the strength of a frame and its members (although in general, frame design cannot be restricted solely to consideration of these effects [SSRC, 1988]). The P-Δ and P-δ effects are illustrated for an arbitrary beam-column subjected to sidesway in Fig. 1.4. The result of these effects on the distribution and magnitude of the forces within the structure may be estimated either from a direct second-order elastic analysis or based on approximate amplification factors. It can be stated that in notional load procedures, the effect of the notional horizontal loads is to approximate the increased P-Δ and P-δ effects and increased in-plane forces due to residual stresses (i.e., distributed yielding) and geometric imperfections. In this type of design procedure, the second-order elastic analysis provides an estimate of the "actual" second-order inelastic forces within the system at the achievement of a limit state in the most critically loaded component of the structure.

With the exception of the calculation of the P-Δ and P-δ effects on the magnitude and distribution of forces within the structure as discussed above, and with minor exception, the consideration of bowing effects and/or three-dimensional elastic stability effects within the analysis (see Table 1.2), all the potential material and geometric nonlinear effects on the design strength are addressed solely within

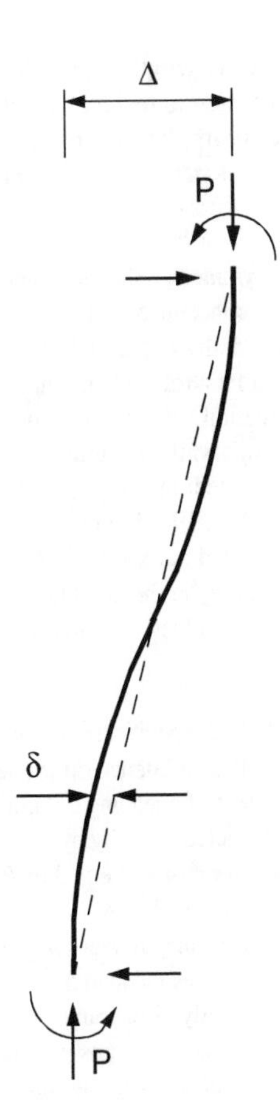

Figure 1.4. $P\Delta$ and $P\delta$ effects in a beam-column member.

specification design formulas[2]. For example, in the effective length approaches, the increased P-Δ and P-δ effects and increased in-plane forces due to residual stresses and geometric imperfections discussed above are accounted for implicitly through use of effective length factors-- the second-order elastic analysis is performed on the perfect structure geometry.

In the ideal case where the member cross-sections are symmetric, the initial geometry is perfectly straight and plumb, the axial loading is perfectly concentric, and three-dimensional deflections of the structural system are restrained, the P-Δ and P-δ effects eventually cause a bifurcation of the structural system into a flexural buckling mode as the loads on the structure are increased. Of course, pure bifurcation behavior does not exist in reality because of the existence of imperfections in the geometry, the loading, and the member properties. The limit states behavior of frames subjected only to gravity loads only approaches the bifurcation solution in certain cases [SSRC, 1988]. Frames often are subjected to substantial bending actions due to their primary loading, and in these situations, the P-Δ and P-δ effects tend to be evidenced primarily in the amplification of frame internal forces. These amplified forces then limit the overall strength that can be attained, as noted in the previous excerpt from [SSRC, 1976].

In the commentary of the second edition of the AISC LRFD Specification [1993], the contribution of leaning columns to the P-Δ effects (and thus their influence on frame stability) is given detailed attention. The primary stability effect of a leaning column is illustrated in Fig. 1.5. When the structure in this figure is deflected laterally by an amount Δ, the axial force in the leaning column (*c1*) creates a secondary moment equal to $P_1 \Delta$. For overall equilibrium of the leaning column, a shear force equal to $P_1 \Delta / L$ must be developed. Therefore, column *c2*, which provides the only resistance to sidesway for this example frame, must sustain a destabilizing shear force equal to this value in addition to the destabilizing effects of the gravity force that it supports directly. For frames in which a large percentage of the structural system involves the use of gravity columns with beam-to-column connections providing negligible moment restraint, the leaning column effects can increase the second-order forces within the structural system substantially. Also, if the structure is subjected predominantly to gravity loading, the buckling load of the structural system can be reduced significantly by the leaning column actions. The consideration of leaning column effects in stability design is addressed in numerous sections throughout the report.

---

[2] For frames with semi-rigid connections, material nonlinearities in the connections must in general be accounted for in some way within the analysis (e.g., through direct modeling of the connection nonlinearities, or by appropriate use of tangent and secant stiffnesses).

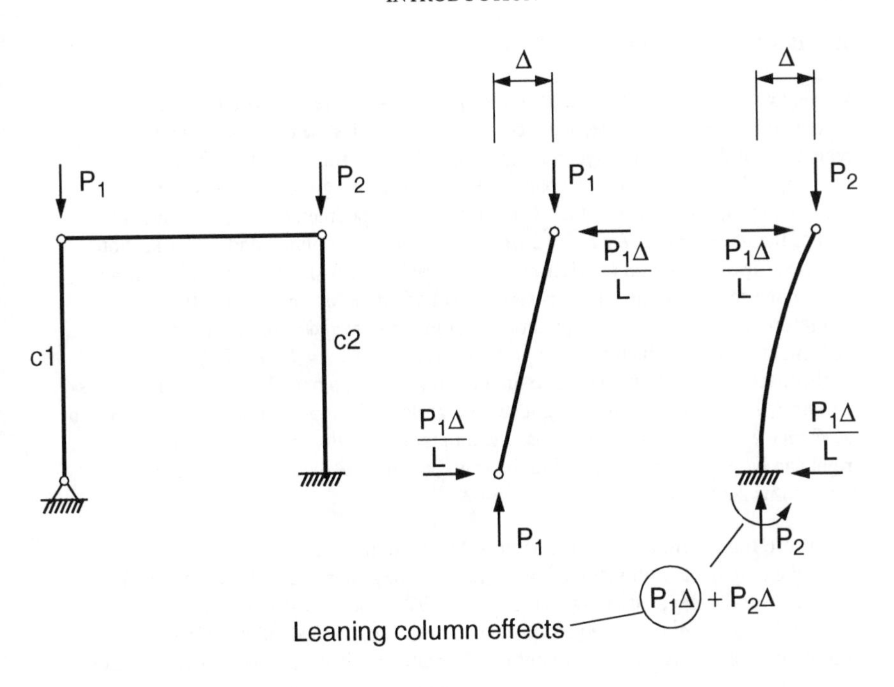

Figure 1.5. Leaning column effects on sidesway stability.

1.4 Illustrative Example

Figure 1.6 shows two versions of a simple portal frame composed of a leaning column on the left-hand side, and a column that provides the entire sidesway resistance of the structural system on the right. In the frame on the left, the lateral-resisting column is pinned at its base and restrained by a girder at its top. This column is proportioned such that its elastic stiffness parameter EI_c / L_{c1} is two times that of the girder (i.e., $EI_c / L_{c1} = 2\ EI_b / L_b$). The same column and girder sections are used for the frame on the right; however, the lateral-resisting column is twice as long in the second frame, and it is restrained both at its top and at its bottom. Therefore, the elastic stiffness parameters of the girders and of the lateral-resisting column are equal for each of these problems (i.e., $EI_c /L_{c2} = EI_c / (2L_{c1}) = EI_b / L_b$). In the discussions below, the frame on the left (with the pinned base) is referred to as *Frame 1*, and the frame on the right (with a girder at its top and bottom) is referred to as *Frame 2*. The two frames are identical in their behavior since the frame on the right can be generated by reflecting the frame on the left about a horizontal axis through its supports[3].

These frames are utilized in Chapters 2 and 4 to introduce the application of the effective length and notional load procedures. They are selected because they are two of several cases studied by Kanchanalai [1977] that are extremely sensitive in testing the ability of design equations to capture the combined effects of material and geometric nonlinearity on the strength. The stability effects are more severe in these frames than would be expected in most building designs. However, these frames are representative of feasible designs for cases such as industrial buildings with non-redundant lateral-resisting systems, small design lateral load, and large gravity loads [Springfield, 1991].

As noted in Section 1.2.2, the AISC LRFD beam-column equations were developed in part by curve fitting to strength plots developed from Kanchanalai's data. Initial geometric imperfections as well as strain-hardening of the structural steel were neglected in Kanchanalai's solutions. Also, his solutions assume a W8x31 section and a self-equilibrating residual stress distribution with a peak compressive residual stress of $0.3\ F_y$ at the flange tips, a linear variation through the flange width, and constant residual tension in the web. A yield stress of $F_y = 36\ ksi\ (250\ N/mm^2)$ is assumed.

Figure 1.7 compares two *incorrect* beam-column strength interaction curves to the "exact" Kanchanalai strength curve, adjusted to account for the nominal initial

---

[3] For this statement to be correct, the geometric imperfections in both the top and bottom halves of the frame on the right must match the geometric imperfections in the frame on the left. It can be argued that this situation is not likely for the two structures in their actual fabricated and erected configurations.

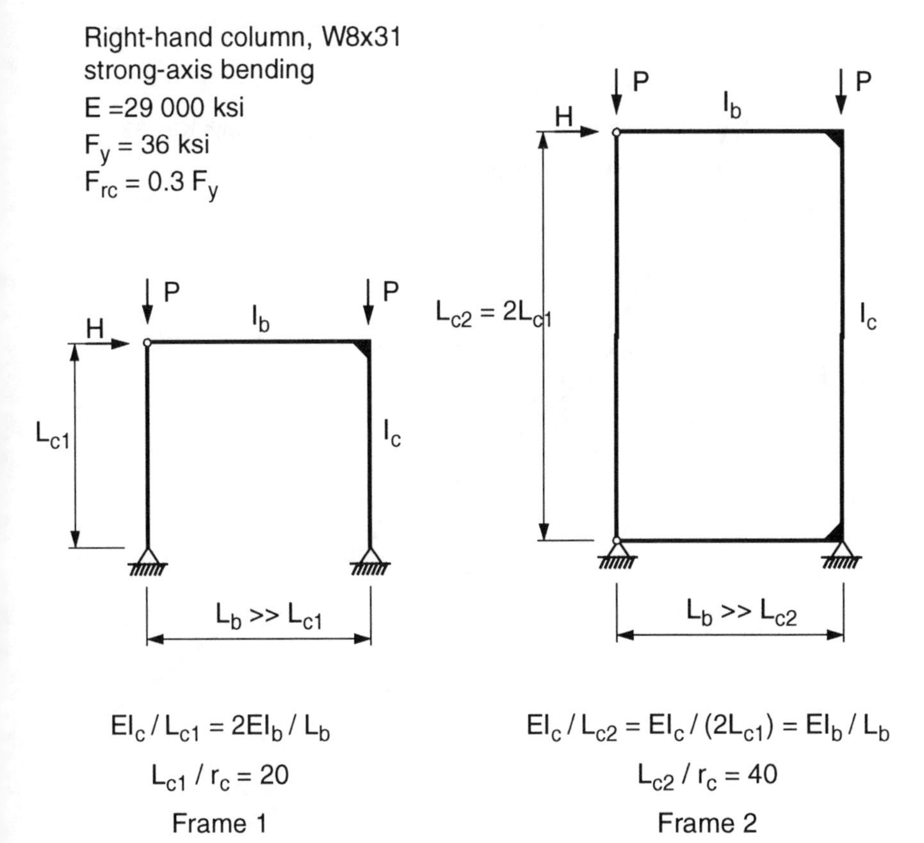

$EI_c / L_{c1} = 2EI_b / L_b$

$L_{c1} / r_c = 20$

Frame 1

$EI_c / L_{c2} = EI_c / (2L_{c1}) = EI_b / L_b$

$L_{c2} / r_c = 40$

Frame 2

Figure 1.6. Leaning column frames [Kanchanalai, 1977].

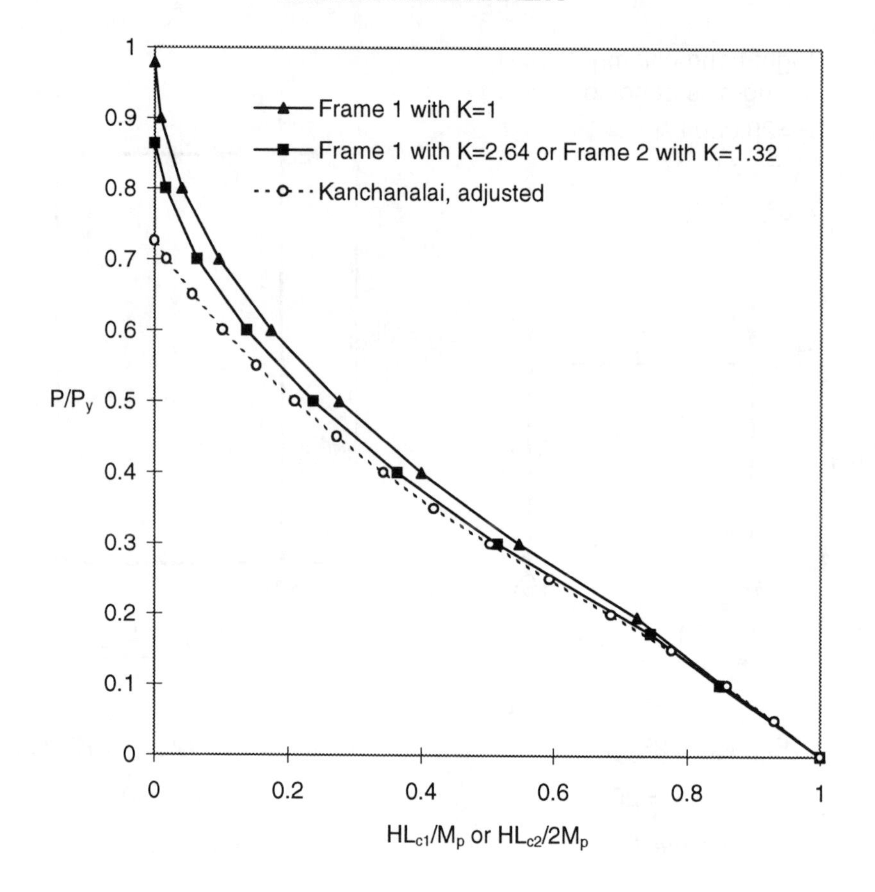

Figure 1.7. Comparison of AISC LRFD beam-column interaction curves with incorrect $K=1$ or $K=2.64$ for *Frame 1*, and $K = 1.32$ for *Frame 2*, to the adjusted Kanchanalai data

imperfection effects. The Kanchanalai curve is dashed in the figure. One of the incorrect curves is obtained for *Frame 1* using $K = 1$ without any notional load or equivalent geometric imperfection. The other is based on $K = 2.64$ for *Frame 1* (or $K = 1.32$ for *Frame 2*). These effective length factors are determined from the traditional sidesway-uninhibited alignment chart in the AISC Specifications [AISC, 1993], without any modifications to account for girder end conditions, leaning column effects, or column inelasticity. All the curves in Fig. 1.7, and in the subsequent figures, are plotted in terms of P/P_y versus the primary (i.e., *first-order*) bending moments HL_c / M_p. Therefore, all the curves in Fig. 1.7 and in the subsequent figures are analogous to the "*Net* M_1 / M_p" curve shown in Fig. 1.3.

Figure 1.7 first shows that, in general, the use of $K = 1$ without notional loads or equivalent geometric imperfections results in an unconservative design. For extreme cases, such as the example illustrated here, the unconservative error can be substantial[4]. This unconservatism exists both in Allowable Stress Design [Wood et al., 1976] and in Load and Resistance Factor Design [White and Hajjar, 1991; Liew et al., 1991]. *In general, calculation of second-order elastic forces does not remove the need for use of an effective length (or alternatively, the need for equivalent geometric imperfections or notional horizontal loads).* It is for this reason that the LRFD Specification requires an effective length in determining the nominal axial resistance, P_n, although the applied moment term in the AISC LRFD beam-column interaction equations is M_u, the maximum second-order elastic moment within the members. Nevertheless, as noted in Section 1.1, effective length factors and notional loads may be neglected in certain situations.

Secondly, the *(Frame 1 with K=2.64 or Frame 2 with K = 1.32)* curve in Fig. 1.7 illustrates that, for cases in which the underlying assumptions of the traditional effective length alignment charts are violated, the unconservative errors associated with these violations can be large[5]. Although second-order elastic moments are included in the calculations, the strength of the lateral-load resisting column with the nomograph effective length factors ($K = 2.64$ or 1.32) is still significantly overestimated. In fact, the beam-column interaction curve for *Frame* 2 with $K = 1.32$

---

[4] The maximum unconservative error for the $K=1$ curve shown in Fig. 1.7 is 35 percent. This corresponds to the case of pure gravity loading P, with zero lateral load H. The errors, measured along a radial line from the origin of the plots, are always smaller for combined axial compression and bending (for example, the error for the $K=1$ curve at $HL_c = 0.2\,M_p$ is 10 percent). For the most sensitive benchmark cases, it is common for the unconservative error based on $K=1$ to be in the order of 10 to 20 percent at a primary moment of $0.2M_p$.

[5] The AISC LRFD beam-column interaction curves for *Frame 1* and *Frame 2* are identical, when based on equivalent procedures for determining effective length factors.

is very similar to the interaction curve for this frame based on $K = 1$ (the peak P/P_y based on $K = 1$ is equal to 0.919 for this frame whereas the peak P/P_y based on $K = 1.32$ is 0.864).

Thirdly, the curves in Fig. 1.7 show that the unconservative errors in design of these frames can be limited by placing a maximum limit on the axial force P. This is accomplished implicitly within the Canadian Standard [CSA, 1994] by placing a maximum limit on the second-order amplification due to the P-Δ effects [White et al., 1991]. However, it has been shown that the current CSA limits on the second-order sidesway amplifer generally are not sufficient to hold the error in the design equations to acceptable values [Springfield, 1991; White et al., 1991] -- if notional horizontal loads are included with all load combinations, these limits are not needed at all.

The "correct" AISC-LRFD beam-column interaction equations for the example frames are shown with the "exact" adjusted Kanchanalai results in Fig. 1.8. These curves are based on *Frame* 2 and its exact elastic and inelastic effective length factors of 2.235 and 1.95 respectively. The "correct" design curves for *Frame* 1 are identical (the corresponding K values are 4.47 and 3.90). For the example frames, the LRFD equations match well with the adjusted Kanchanalai curves when the exact inelastic K factor is used. The LRFD design equations are slightly conservative compared to the Kanchanalai results for intermediate values of $HL_{c2} / 2M_p$. This conservatism for intermediate values of moment is not as large for certain sensitive benchmark problems (i.e., the exact solution is more concave), and greater conservatism is exhibited for other cases. It should be noted that the Kanchanalai curve is adjusted (to account for nominal geometric imperfection effects) such that it matches with the design strength based on $K = 1.95$ for the pure axial loading case (see Section 1.2.2).

In Fig. 1.8, the beam-column interaction curve based on the elastic effective length illustrates a situation that occurs often in building frames. The design based on elastic effective length can be significantly conservative compared to the "exact" adjusted strength, or compared to the design strength based on an inelastic effective length. In Fig. 1.8, the conservative error for the design curve based on elastic effective length is as large as eight percent.

Finally, Fig. 1.9 shows the results of the notional load procedure discussed later in the report for the example frames of Fig. 1.6. In addition to the actual lateral load H, a notional horizontal load of $0.004\Sigma P$ is applied to the frames where ΣP is the summation of the vertical load supported by the frame. The value 0.004 is developed in Chapter 4 as the "best" *constant* notional load parameter for A36 steel (0.005 is the appropriate value for $F_y = 50$ ksi). Whereas the design solutions are identical for both of the frames of Fig. 1.6 when an effective length procedure is employed, the notional load design results are slightly different for these frames. This is because the column strength is evaluated based on the actual column lengths (L_c and $2L_c$) in the notional load approach. Nevertheless, it can be argued that it is unlikely that the

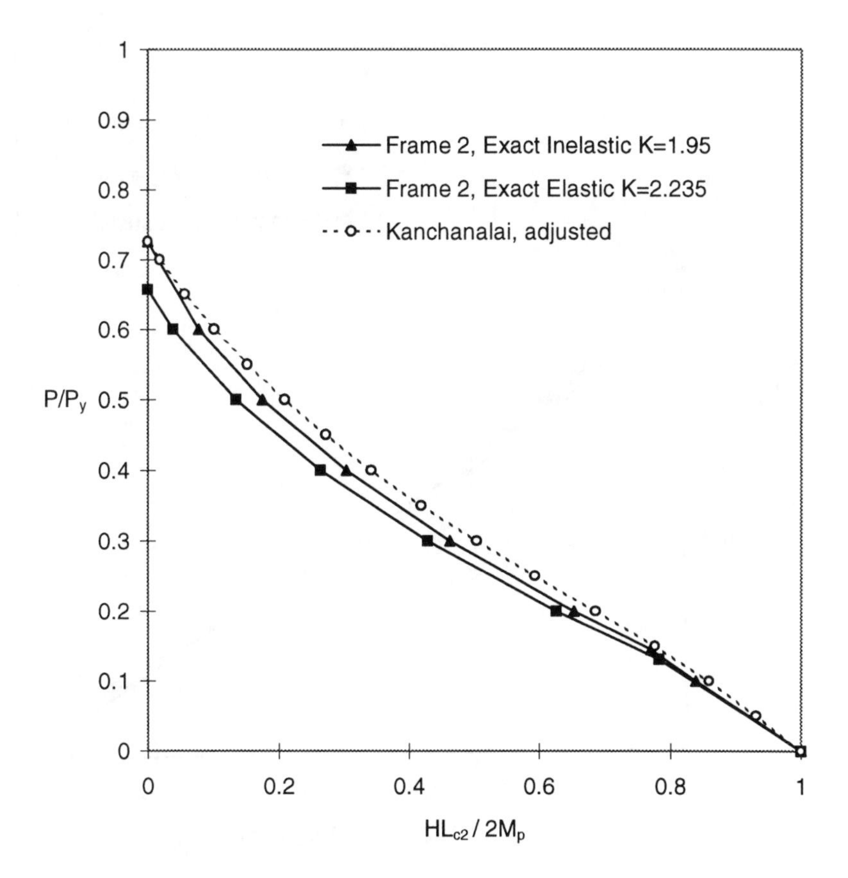

Figure 1.8. Comparison of AISC LRFD beam-column interaction curves with exact elastic and inelastic *K* factors to the adjusted Kanchanalai data.

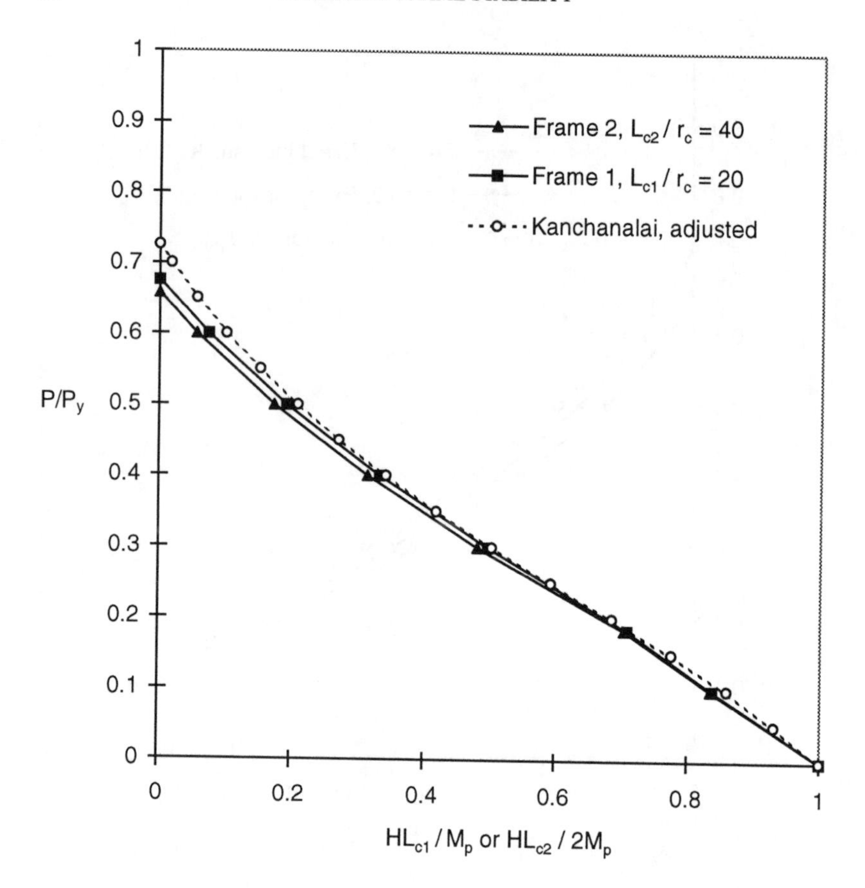

Figure 1.9. Comparison of AISC-LRFD beam-column interaction curves with notional horizontal load of $0.004P$ to the adjusted Kanchanalai data -- *Frame 1* ($L_{c1} / r_c = 20$) and *Frame 2* ($L_{c2} / r_c = 40$).

"real" geometric imperfections in an actual frame built to the specifications of *Frame 2* would match with the "real" geometric imperfections in an actual frame built to the specifications of *Frame 1*. Thus, the strengths for *Frame 1* and *Frame 2* should not be identical. Both of the notional load design solutions are shown in the figure.

The results based on the actual column length and a notional horizontal load are slightly conservative for the example frames compared to the adjusted Kanchanalai solution, particularly for large axial load and small applied bending moment. However, the largest conservative error for these frames is similar to that of the LRFD interaction equations based on the exact inelastic effective length (it should be recalled that the "exact" LRFD curve is least accurate at intermediate values of P/P_y and non-zero moment). Depending on the lateral-resisting column's cross-section type, errors of similar or greater magnitude can be associated with the use of a single column curve for design. The conservative error in the notional load approach is larger for some sensitive benchmark frames, particularly for sensitive frames in which the "exact" effective length factors are approximately equal to one. This conservatism can be alleviated at the expense of some additional complexity by use of non-constant equations for the notional load parameter.

1.5 Report Organization

The contents of this report are organized as follows. Chapter 2 provides a state-of-the-art overview of effective length calculation procedures. First, the formally sanctioned stability design approach in AISC LRFD [1993] is presented within a unified context by showing the relationship between the LRFD column strength and idealized buckling strengths. Concepts of inelastic effective length, by which the restraint offered to inelastic columns by elastic beams may be quantified, are reviewed, and the implications of inelastic effective lengths are explained. This is followed by a discussion of various models for calculation of the effective length (i.e., for calculating the buckling load), including buckling of an idealized subassemblage composed only of the column being investigated or designed and its immediately adjacent members, sidesway story buckling, and buckling of the full structural system. Finally, Chapter 2 presents a detailed exposition of two effective length procedures that are based on overall story sidesway buckling and are outlined in the LRFD Specification Commentary [AISC, 1993]. These specific effective length equations are developed, step-by-step practical procedures for use in design are presented, and the accuracy relative to the results from "exact" buckling analyses is discussed.

Chapter 3 follows with a presentation of concepts and procedures for calculation of effective length within partially-restrained (or semi-rigid) framing. One of the key issues addressed in this chapter is the appropriate use of the effective length concept for framing in which the restraining elements (i.e., the beams and their connections) are relatively flexible and exhibit significant nonlinearity, possibly even at service load levels. The proper modeling of the connections in determining effective length, and in performing the analysis to determine the load effects P_u and M_u, is discussed. Also, the issues involved with, the implications of, and the resolution of what has been referred in the literature to as the "*K* factor paradox" are discussed for partially-restrained frame construction.

Chapter 4 presents the motivation for and development of a notional horizontal load approach that may be utilized to account for frame stability within the context of AISC LRFD practice. This chapter considers directly the "exact" second-order inelastic behavior of geometrically imperfect members and frames, and addresses the question of how the influence of both member and story imperfections can best and most logically be taken into account in a design process based on second-order elastic analysis (or first-order elastic analysis with approximate amplification factors). Strong- and weak-axis in-plane and three-dimensional beam-column strength calculations are considered. A distinction is drawn between the model of an equivalent pin-ended imperfect beam-column (the effective length concept) and the physically imperfect beam-column, and comparisons are made between the strength predictions based on the effective length and notional horizontal load approaches.

Chapter 5 discusses conditions for which the use of effective lengths or notional horizontal loads may be neglected. First, limits that have been suggested in various design standards and specifications for definition of frames as "non-stability critical" are reviewed and placed within the context of LRFD practice. Next, the error in the LRFD *beam-column interaction equations* caused by use of $K=1$ without the application of notional loads, is outlined. This error is derived directly from the AISC LRFD equations based on one of the elastic story-based effective length approaches presented in Chapter 2. By inspection of these error relationships, general limits on the story amplification factor B_2 are deduced for which neither effective length nor notional load need to be considered in the design. Since these relations are based on elastic effective lengths, they are conservative for many design cases. In some cases substantial economy can be gained by considering inelastic effective lengths, and therefore, simple relationships based on the inelastic buckling behavior are desirable. Approximate equations for inelastic effective length factors, recently proposed by LeMessurier [1995], are outlined, and the implications of these equations on when $K=1$ can be used with small error are discussed. The chapter closes with recommended rules for when a design may be based on $K=1$ without consideration of notional horizontal loads.

Chapter 6 presents extensive examples, and compares and contrasts all of the techniques discussed in Chapters 2 through 5 based on these example problems. Practical as well as extreme examples are provided to illustrate the key issues discussed. The design examples demonstrate the appropriate use of each of the methods discussed in the previous chapters. An effort has been made to compare the effective length and notional horizontal load approaches directly for a large range of cases cited within the engineering literature. The types of frames considered include fully-restrained (FR) and partially-restrained (PR) frames with and without leaning columns, and moderate height and short frames.

Finally, Chapter 7 summarizes the strengths and limitations of both the effective length and the notional load types of procedures.

1.6 Other References

As noted in Section 1.1, the focus of this report is on the elucidation of general procedures for stability analysis and design within the effective length and equivalent geometric imperfection-notional horizontal load categories. The procedures discussed have grown from various incremental refinements on the extensive base of frame stability research over the past 40 years. Where possible, the developments are generalized to apply to a broad range of specific design procedures. Where specific procedures are presented, an attempt has been made to provide the most up-to-date versions of methods that are most amenable to general design use.

A number of important but specific issues and procedures in stability design are not addressed, or are given limited coverage. Some of these issues include, but are not limited to, approximate amplification of moments in tall "cantilever-type" building frames (as noted previously, appropriate calculation of second-order elastic forces is not addressed at all in this document), effects of column base restraint, effects of column splices, effects of variations in column axial forces due to wind pressure, development and use of approximate procedures applicable to special types of members or structural framing (e.g., slender structural systems, staggered truss systems, gabled frames, tapered columns, stepped columns, etc.), and presentations of analytical solutions (tables, etc.) providing effective lengths for various specific frame or member geometries. The reader is referred to [SSRC, 1976], [SSRC, 1988], [Iffland, 1988], [Iffland and Birnstiel, 1982], [Nair, 1975, 1983, 1987], [CTBUH, 1979], and [ECCS, 1976] for extensive state-of-the-art information on these and other related topics at the time of their publication.

References

Ackroyd, M. [1991], "Design Needs from Inelastic Analysis," *Proceedings, 1991 Annual Technical Session*, Structural Stability Research Council, Lehigh Univ., Bethlehem, PA, 41-51.

AISC [1963], *Specification for the Design, Fabrication, and Erection of Structural Steel for Buildings*, 6th Ed., American Institute of Steel Construction, New York, NY.

AISC [1978], *Specification for the Design, Fabrication, and Erection of Structural Steel for Buildings*, 8th Ed., American Institute of Steel Construction, New York, NY.

AISC [1989], *Specification for Structural Steel Buildings: Allowable Stress Design and Plastic Design*, 9th Ed., American Institute of Steel Construction, Chicago, IL.

AISC [1993], *Load and Resistance Factor Design Specification for Structural Steel Buildings*, 2nd Ed., American Institute of Steel Construction, Chicago, IL.

AISC [1994], *Manual of Steel Construction, Load and Resistance Factor Design*, 2nd Ed., American Institute of Steel Construction, Chicago, IL.

AIJ [1989], *Standard for Limit State Design of Steel Structures*, Architectural Institute of Japan, 1989.

Birnstiel, C. and Iffland, J.S.B. [1980], "Factors Influencing Frame Stability," *Journal of the Structural Division*, ASCE, 106(2), 491-504.

CEN [1993], *ENV 1993-1-1 Eurocode 3: Design of Steel Structures, Part 1.1 -- General Rules and Rules for Buildings*, European Committee for Standardization, Brussels, Belgium.

CTBUH [1979], *Structural Design of Tall Steel Buildings*, Council on Tall Buildings and Urban Habitat and American Society of Civil Engineers, Vol. SB, C.N. Gaylord and M. Watabe (ed.), Washington, D.C.

Clarke, M.J. and Bridge, R.Q. [1992], "The Inclusion of Imperfections in the Design of Beam-Columns," *Proceedings, 1992 Annual Technical Session*, Structural Stability Research Council, 327-346.

CSA [1994], *Limit States Design of Steel Structures*, CAN/CSA-S16.1-M94, Canadian Standards Association, Rexdale, Ontario.

ECCS [1976], *Manual on Stability of Steel Structures*, 2nd Ed., European Convention for Constructional Steelwork, ECCS-Committee 8-Stability 1976, Brussels, Belgium.

Gaiotti, R. and Smith, B.S. [1989], "P-Delta Analysis of Building Structures," *Journal of Structural Engineering*, ASCE, 115(4), 755-771.

Higgins, T.R. [1964], "Effective Column Length—Tier Buildings," *Engineering Journal,* American Institute of Steel Construction, 1(1), 12-15.

Iffland, J.S.B. [1988], "Stability Analysis and Design," *Proceedings*, Seminar on Structural Design of Tall Buildings, ASCE Metropolitan Section and Council on Tall Buildings, January 25 & 27, New York, NY, 38 pp.

Iffland, J.S.B. and Birnstiel, C. [1982], "Stability Design Procedures for Building Frameworks," *Research Report, AISC Project No. 21.62*, American Institute of Steel Construction, Chicago, IL.

Kanchanalai, T. [1977], "The Design and Behavior of Beam-Columns in Unbraced Steel Frames," *AISI Project No. 189, Report No. 2*, Civil Engineering/Structures Research Lab., Univ. of Texas, Austin, TX, 300 pp.

Kennedy, D.J.L., Picard, A. and Beaulieu, D. [1993], "Limit States Design of Beam-Columns: The Canadian Approach and Some Comparisons," *Journal of Constructional Steel Research,* 25(1 & 2), 141-164.

LeMessurier, W.J. [1985], Personal communication with W. McGuire.

LeMessurier, W.J. [1994], Personal communication with D.W. White.

LeMessurier, W.J. [1995], "Simplified K Factors For Stiffness Controlled Design," *Restructuring: America and Beyond*, Proceedings of Structures Congress XIII, ASCE, 1797-1812.

Leon, R.T. and Forcier, G.P. [1992], "Parametric Study of Composite Frames," *Connections in Steel Structures II*, R. Bjorhovde, et al. (ed.), AISC, 152-159.

Leon, R.T., Hoffman, J.J., and Staeger, T. [1996], *Partially Restrained Composite Connections*, Steel Design Guide Series No. 8, American Institute of Steel Construction, Chicago, IL.

Liew, J.Y.R., White, D.W. and Chen, W.F. [1991], "Beam-Column Design in Steel Frameworks - Insights on Current Methods and Trends," *Journal of Constructional Steel Research*, 18(4), 269-308.

McGuire, W. [1990], "Structural Analysis in Load and Resistance Factor Design," Introductory remarks at a seminar on *Frame Analysis in LRFD*, AISC National Steel Conference, 1990, American Institute of Steel Construction, Chicago, IL.

Nair, R.S. [1975], "Overall Elastic Stability of Multistory Buildings, *Journal of the Structural Division*, ASCE, 101(ST12), 2487-2503.

Nair, R.S. [1983], "A Simple Method of Overall Stability Analysis for Multi-story Buildings," *Developments in Tall Buildings*, Council on Tall Buildings and Urban Habitat, Hutchison and Ross, Stroudsburg, PA, 471-481.

Nair, R.S. [1987] "Tall Building Stability -- Practical Considerations", *Materials and Member Behavior*, Proceedings of Structures Congress '87, ASCE, 582-594.

SAA [1990], *AS4100-1990, Steel Structures*, Standards Association of Australia, Sydney, Australia.

Springfield, J. [1991], "Limits on Second Order Elastic Analysis," *Proceedings, 1991 Annual Technical Session*, Structural Stability Research Council, 89-99.

SSRC [1976], *Guide to Stability Design Criteria for Metal Structures*, 3rd Ed., B.G. Johnston, (ed.), Structural Stability Research Council, Wiley, 616 pp.

SSRC [1981], "Technical Memorandum No. 5: General Principles for the Stability Design of Metal Structures", *Civil Engineering*, ASCE, February.

SSRC [1988] *Guide to Stability Design Criteria for Metal Structures,* 4th Ed., T.V. Galambos (ed.), Structural Stability Research Council, Wiley, 786 pp.

Tide, R.H.R. [1985], "Reasonable Column Design Equations," *Proceedings, 1985 Annual Technical Session*, Structural Stability Research Council, 47-55.

White, D.W. and Hajjar, J.F. [1991], "Application of Second-Order Elastic Analysis in LRFD: Research to Practice," *Engineering Journal*, AISC, 28(4), 133-148.

White, D.W., Liew, J.Y.R., and Chen, W.F. [1991], "Consideration of Inelastic Stability in Frame Design," *Proceedings, 1991 Annual Technical Session*, Structural Stability Research Council, 65-76.

White, D.W. [1993], "Plastic Hinge Methods for Advanced Analysis of Steel Frames," *Journal of Constructional Steel Research*, 24(2), 121-152.

White, D.W., Liew, J.Y.R., and Chen, W.F. [1993], "Toward Advanced Analysis in LRFD," *Plastic Hinge Based Methods for Advanced Analysis and Design of Steel Frames: An Assessment of the State-of-Art*, D.W. White and W. F. Chen (ed.), Structural Stability Research Council, Bethlehem, PA, 95-173.

Wood, B.R., Beaulieu, D., and Adams, P.F. [1976], "Further Aspects of Column Design by P-Delta Method," *Journal of Structural Engineering*, ASCE, 102(ST3), 487-500.

Yura, J.A. [1988], "Elements for Teaching Load and Resistance Factor Design, Combined Bending and Axial Load," American Institute of Steel Construction, Chicago, IL.

Zaremba, C.J. [1988], "Strength of Steel Frames Using Partial Composite Girders," *Journal of Structural Engineering*, ASCE, 114(8), 1741-1760.

CHAPTER 2

EFFECTIVE LENGTH APPROACHES

2.1 Introduction

This chapter begins with an exposition of the fundamental assumptions and mechanisms contained in the use of effective length approaches for stability design of members and frames. In Section 2.2, the AISC LRFD column design equations are presented, and the tie between these equations and idealized buckling models is established. This is followed in Sections 2.2 and 2.3 by a categorization and discussion of the many types of models for determining buckling loads or effective lengths. One of the primary difficulties with the use of effective length approaches in practice is their maturity. The number of methods that have been presented in the literature for calculation of effective length is overwhelming. The fact of the matter is that, in many practical design situations, the structure is not "stability critical" at all, and an approximate determination of the effective length or even an effective length factor of $K = 1$ is sufficient[1]. However, in situations where the design is truly sensitive to stability effects, there are many complexities and subtleties associated with the proper calculation of buckling loads or column effective lengths.

Section 2.2 categorizes and discusses the different buckling models in general terms based on the type of material idealization employed, elastic or inelastic. Section 2.3 follows by organizing and explaining effective length methods in terms of the physical extent of the structure included in the underlying buckling model. The goal of these sections is to provide a thorough foundation for comparing and contrasting any of the approaches for determining effective length. This information is necessary for evaluating which methods are best suited for a given design situation.

Section 2.3 establishes "story-based" effective length methods, i.e., methods in which the underlying buckling model is composed of the entire story in which the column is contained, as the preferred approach for practically all "regular" building frames. This is followed by a lengthy Section 2.4 that presents two specific "story-based" procedures included in the AISC LRFD Specification Commentary [AISC, 1993]. The derivation, assumptions, accuracy, and usage of these procedures is outlined in detail.

---

[1] Criteria for use in determining when the design can be based on $K = 1$ are presented in Chapter 5.

2.2 Relationship of the AISC LRFD Column Strength Formulas to Idealized Buckling Models

The AISC LRFD Specification [1993] utilizes a single column curve. The rationale behind the development of this curve is explained in Tide [1985]. If the traditional allowable stress equations [AISC, 1989] (which include a variable factor of safety) are multiplied by $1.67\phi_c A_g$, the LRFD column design strength ($\phi_c P_n$) is closely approximated [Salmon and Johnson, 1996]. The term ϕ_c is the resistance factor for column strength in LRFD, equal to *0.85*, and A_g is the gross area of the column cross-section. For columns that fail or "buckle" inelastically, the LRFD column strength may be expressed as

$$\phi_c P_n = \phi_c \, 0.658^{[P_y/P_e]} P_y \tag{2.1a}$$

or as

$$\phi_c P_n = \phi_c \, 0.658^{\lambda_c^2} P_y \tag{2.1b}$$

where

$$\lambda_c = \sqrt{\frac{P_y}{P_e}} = \frac{KL}{r\pi}\sqrt{\frac{F_y}{E}} \tag{2.2}$$

In the above equations P_y is the yield load of the member, equal to $A_g F_y$, and P_e is an elastic buckling load based on the effective length *KL*. This load may be calculated as

$$P_e = \frac{\pi^2 EI}{(KL)^2} = \frac{\pi^2 EA_g}{(KL/r)^2} \tag{2.3}$$

where *K* is the member's effective length factor. In LRFD, a concentrically loaded column is assumed to fail inelastically and have its strength governed by Eqs. (2.1) whenever P_e is greater than or equal to $\frac{4}{9}P_y$, i.e., when $\lambda_c \leq 1.5$.

For columns that fail by "elastic buckling", the LRFD column curve takes the form

$$\phi_c P_n = \phi_c \, 0.877 \, P_e \tag{2.4a}$$

or

$$\phi_c P_n = \phi_c \left[\frac{0.877}{\lambda_c^2} \right] P_y \tag{2.4b}$$

That is, a column's strength is approximated as 0.877 of the elastic buckling load if the effective length is long enough such that the "perfect" member (with residual stresses) would buckle elastically. The nominal strength P_n is reduced below the elastic buckling load (by the constant 0.877) to account for the effects of geometric imperfections, such as out-of-straightness and accidental eccentricities. Equations (2.4) are applicable for columns in which P_e is less than or equal to $\frac{4}{9} P_y$, i.e., when $\lambda_c \geq 1.5$.

Any P_e and its corresponding effective length factor K are associated (either explicitly or implicitly) with the buckling of an idealized model of some sort. The use of either P_e or K in the column design equations is the only tie between the design equations and the selected buckling model. Many different types of buckling idealizations have been utilized or proposed in the engineering literature pertaining to column design. For problems that are truly stability critical, the selection of a proper buckling model, or in other words, the calculation of suitable effective length factors can involve many complexities and subtleties. The goal of the discussion that follows is to establish a general basis for comparing and contrasting the many different K factor procedures.

In simple terms, all buckling models or K factor procedures can be subdivided into two categories: elastic methods, in which case the effects of yielding within the structure are neglected, or inelastic methods, in which case the buckling or K factor calculations typically involve simple representations of the column inelasticity[2]. Although design calculations are often based on elastic effective lengths, significant economy may be gained by considering the column inelasticity in many building design situations [Yura, 1971; LeMessurier, 1977]. This is because the slenderness of the column members in typical steel buildings is often such that substantial yielding occurs before the design strength is achieved. If the columns are yielded at their design strengths, the relative restraint provided by the elastic girders is greater than if

---

[2] In addition to the categorization of K factor procedures as either elastic or inelastic, another important categorization is based on whether the buckling model is an isolated subassembly immediately surrounding the member, the full story in which the column is located, or the full structural system in which the member is contained. This second type of categorization is discussed in detail in Section 2.3.

the columns were still elastic. As a result, the associated effective length to be entered into Eqs. (2.1) (or more specifically, Eqs. (2.2) or (2.3)) may be reduced[3].

Tremendous insight into the physical basis of the column design formulas can be gained by posing Eqs. (2.1) and (2.4) in terms of a generic elastic/inelastic buckling model. If inelastic buckling is considered, ordinarily it is assumed that the yielding is uniform along the column length such that the inelastic members are in effect prismatic within the buckling model. Based on this idealization, Eqs. (2.1) and (2.4) can be written as one equation (applicable for either elastic or inelastic column failure) as

$$\phi_c P_n = \phi_c 0.877 \frac{\pi^2 \tau EI}{(KL)^2} = \phi_c 0.877 \frac{\pi^2 \tau E A_g}{(KL/r)^2} = \phi_c 0.877 P_{e\tau} \quad (2.5)$$

where τ is referred to as the inelastic stiffness reduction factor [AISC, 1994], and

$$P_{e\tau} = \frac{\pi^2 \tau EI}{(KL)^2} \quad (2.6)$$

is the column buckling load (elastic or inelastic). That is, if the equations for τ, presented below, are substituted into Eq. (2.5), Eqs. (2.1) (for $\tau < 1$) or (2.4) (for $\tau = 1$) are obtained after some algebraic manipulation. The term τ in these equations can be considered either as the ratio of the tangent modulus E_t to the elastic modulus of the column ($\tau = E_t / E$), or as the ratio of the rigidity of the "effective elastic core" at incipient buckling to the elastic rigidity of the full column cross-section ($\tau = EI_e/EI$, where EI_e is the rigidity of the effective elastic core).

The inelastic stiffness reduction factor is tabulated in the AISC LRFD manual [1994], and it may be expressed algebraically as either

[3] Specific procedures for calculating elastic and inelastic effective lengths are discussed in Sections 2.3 and 2.4. If overall story or system stability effects are considered, a column's K value from an inelastic procedure may in some cases be greater than the corresponding elastic value. This is a result of increased "leaning" effects from other framed columns that are loaded inelastically.

$$\tau = \begin{cases} 1.0 & \text{for } P_n \le \left[0.877\left(\frac{4}{9}P_y\right) = 0.39P_y\right] \\ -\dfrac{\frac{P_n}{P_y}}{0.877}\left[\dfrac{ln\left[\frac{P_n}{P_y}\right]}{ln(0.658)}\right] = -2.724\dfrac{P_n}{P_y}ln\left[\dfrac{P_n}{P_y}\right] & \text{for } P_n > 0.39P_y \end{cases} \quad (2.7a)$$

in terms of the nominal column strength P_n[4], or as

$$\tau = \begin{cases} 1.0 & \text{for } P_{e\tau} \le \frac{4}{9}P_y \\ -2.389\dfrac{P_{e\tau}}{P_y}ln\left[\dfrac{0.877P_{e\tau}}{P_y}\right] & \text{for } P_{e\tau} > \frac{4}{9}P_y \end{cases} \quad (2.7b)$$

in terms of the load in the idealized column at incipient elastic/inelastic buckling, $P_{e\tau}$.

It should be emphasized that, if the second of Eqs. (2.7a) is substituted into Eq. (2.5), Eqs. (2.1) are obtained after some algebraic manipulation. If $\tau = 1$ is substituted into Eq. (2.5), Eqs. (2.4) are obtained. Thus, Eq. (2.5) may be taken as a general statement of the AISC LRFD column strength for elastic and inelastic "buckling", with τ expressed by Eqs. (2.7a) or (2.7b).

The relationship between τ and P_n / P_y is plotted for P_n / P_y from *0.0* to *1.0* in Fig. 2.1. Also, the variation of τ versus P_n / P_y, determined from the conventional CRC column strength formula [SSRC, 1976], is plotted in the figure for purposes of comparison. The τ used in LRFD is generally smaller than that based on the conventional CRC column strength. This is due to the fact that the τ from the CRC equation does not include the effects of column geometric imperfections, whereas the value based on the LRFD column strength equations is influenced by these effects. Similar to the development of Eqs. (2.7), other equations for τ can be derived that correspond to other column strength equations (for example, see Section 2.3 of [ECCS, 1976]).

---

[4] Equation (2.7a) is obtained by substituting $\lambda_c^2 = 0.877\tau\, P_y / P_n$ into Eq. (2.1a) and then solving for τ. The basic steps of this procedure are outlined in Baker [1987].

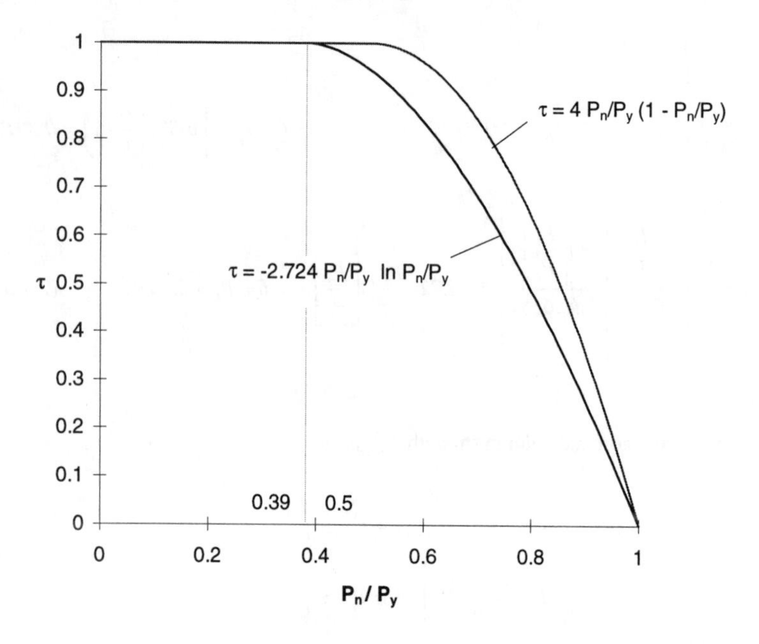

Figure 2.1. Comparison of the inelastic stiffness reduction factor τ as per AISC LRFD [AISC, 1994] versus τ implied by the CRC parabolic column strength formula [SSRC, 1976].

2.2.1 Elastic Versus Inelastic Models for Calculation of Effective Length

Equations (2.1) through (2.7) are all tied to an effective length factor K, either directly or through the member buckling loads, P_e or $P_{e\tau}$. In many cases, a sufficient but *conservative* design is obtained by ignoring column inelasticity in the effective length calculations, thus basing the design calculations on "elastic" effective lengths. However, within the context of AISC LRFD and ASD, a commonly accepted practice for columns designed by Eqs. (2.1) (or the comparable ASD equations) is to calculate effective lengths based on column inelastic stiffness properties [Yura, 1971; Disque, 1973; SSRC, 1976; AISC, 1994].

The "exact" effective length factors are associated with the column stiffnesses (elastic or inelastic) *at incipient buckling* of the idealized structure or subassembly. Calculation of the "exact" inelastic effective lengths generally requires an iterative procedure in which the applied loading is changed -- and the column inelastic stiffnesses are recomputed -- until buckling of the structure or subassembly being considered is achieved. At this point, the axial force in all the columns of the buckling model is by definition equal to $P_{e\tau}$ of Eq. (2.6). This value can be substituted into Eq. (2.5) to obtain the column design strength directly. The associated effective length factor may be obtained by solving Eq. (2.6) for K. This type of procedure was first discussed within the context of AISC Allowable Stress Design by Yura [1971]. Iteration is required because τ and the column axial forces at incipient buckling (equal to $P_{e\tau}$) are interdependent.

The use of an "exact" analysis as outlined above is rarely used in practical design. A non-iterative and often conservative approach to calculating inelastic effective lengths was first presented by Disque [1973]. Within the context of AISC LRFD, the logic behind Disque's approach can be summarized as follows:

(1) The maximum design force $P_{u,max}$ that can be supported by a column is $P_{u,max} = \phi_c P_n = \phi_c (0.877 P_{e\tau})$. For a beam-column member, P_u is always smaller than this value since a portion of the member's strength is "used up" by bending actions.

(2) Therefore, for determination of inelastic effective lengths, often it is conservative to calculate an approximate τ – denoted in this report as $\hat{\tau}$ – by substituting $\dfrac{P_u}{\phi_c}$ for P_n in Eq. (2.7a), or for $0.877P_{e\tau}$ in Eq. (2.7b) (use of an approximate $\hat{\tau} \geq \tau$ is unconservative if the column effective length is increased by column inelasticity; as explained in Footnote 3, this can happen if the "leaning" effects on a given column are increased due to inelasticity in other columns of the frame).

(3) Since P_u is always less than or equal to the column design strength $\phi_c P_n$, $\hat{\tau}$ tends to overestimate the "exact" value of τ associated with the state at

incipient buckling. As a result, the effective length of the inelastic column, restrained by elastic beams, tends to be overestimated—although to a lesser degree than if elastic column properties are employed in the calculations.

A table is provided in the second-edition AISC LRFD Manual [1994] for simple lookup of the above $\hat{\tau}$ values. A similar table was provided in the first edition; however, it should be noted that the inelastic stiffness reduction table in the first edition of the LRFD Manual [AISC, 1986] indicated that $\phi_c P_u$ should be used instead of $\frac{P_u}{\phi_c}$ in determining $\hat{\tau}$. This is not consistent with the basis of the inelastic stiffness reduction procedures, as outlined above, and erroneously resulted in $\hat{\tau}$ values that are larger than obtained using the "correct" load term. In the second edition of the Manual, the correct value $\frac{P_u}{\phi_c}$ is utilized in the derivation of this table (the $\hat{\tau}$ values are listed in terms of $\frac{P_u}{A_g}$ in the new table).

Unfortunately, if the design forces on the member being considered include substantial bending moment, the force P_u will be significantly smaller than the column's axial strength $\phi_c P_n$ for the required section. As a result, the approximate $\hat{\tau}$ may be much larger than the exact inelastic stiffness reduction factor, τ, associated with the member's design strength. An alternate non-iterative procedure that alleviates this problem is to estimate the value of $P_n \geq \frac{P_u}{\phi_c}$, and then calculate a $\hat{\tau}$ based on this estimate. If the estimate is smaller than the final calculated P_n, the computed value for $\phi_c P_n$ is conservative (assuming that the effective length is reduced by column yielding effects). *This procedure is recommended as the preferred non-iterative method for columns that are subjected to substantial bending actions.*

Specific procedures for calculation of elastic or inelastic effective lengths are discussed in Sections 2.3 and 2.4. Example computations are presented in Section 2.4 and in Chapter 6. Most of the procedures involve direct expressions for the effective length factor *K*. This factor may be substituted into Eqs. (2.2) or (2.3) to obtain the slenderness parameter λ_c, which is required for calculation of the design strength from Eqs. (2.1b) or (2.4b). Nevertheless, the *K* factor is always related to a member buckling load (approximate, based on an estimated $\hat{\tau}$, or "exact", based on the "exact" inelastic stiffness reduction at incipient buckling, in which case $\hat{\tau} = \tau$) by the equation

$$P_{e\hat{\tau}} = \frac{\pi^2 \hat{\tau} EI}{(KL)^2} \tag{2.8a}$$

where $\hat{\tau}$ is the inelastic stiffness reduction factor utilized in the calculation of the effective length factor K. The column design strength can always be obtained directly in terms of the buckling load $P_{e\hat{\tau}}$ by substituting

$$\lambda_c = \sqrt{\frac{\hat{\tau} P_y}{P_{e\hat{\tau}}}} \tag{2.8b}$$

into Eqs. (2.1b) or (2.4b)[5]. If $\hat{\tau} = 1$ is assumed for all the members in the frame, then the K values are the "elastic effective length factors" and $P_{e\hat{\tau}}$ is the "true" elastic buckling load. If the $\hat{\tau}$ values and their corresponding K factors are determined using Disque's non-iterative procedure (or the alternate non-iterative procedure outlined above), then $P_{e\hat{\tau}}$ is an approximate inelastic buckling load. This buckling load is implicitly based on the idealized distribution of yielding caused by the forces $\frac{P_u}{\phi_c}$ (or caused by the estimated $P_n \geq \frac{P_u}{\phi_c}$). If an "exact" inelastic buckling analysis is performed, $\hat{\tau}$ is equal to the exact inelastic stiffness reduction factor τ associated with the column strength ($0.877 P_{e\tau} \geq \frac{P_u}{\phi_c}$), and Eq. (2.8a) is identical to Eq. (2.6).

2.2.2 Mechanisms by which the Effects of Column Inelasticity are Captured in the AISC LRFD Column Equations, and Significance of $\hat{\tau}$ versus τ

Several important observations can be made from the above discussions. First, the careful reader may have noticed that the value P_e in Eqs. (2.1) through (2.4) is based on the elastic column rigidity, EI. However, the effective length utilized in determining P_e (see Eq. (2.3)) can be based either on elastic or inelastic member properties. The fact that an inelastic buckling model may be utilized to determine a K factor for calculation of the elastic member buckling load in Eq. (2.3) is not an anomaly. Algebraically, Eq. (2.1a), with P_e determined in this way, is identical to Eq. (2.5) for columns that fail inelastically (i.e., for $\tau < 1$). Therefore, it may be concluded that the effects of column inelasticity are captured most completely through a combination of:

[5] Equation (2.4) governs when $\lambda_c \geq 1.5$.

(1) the exponential form of Eq. (2.1a), and

(2) the consideration of column inelasticity in determining the effective length factor input to Eqs. (2.2) or (2.3).

Alternatively, it can be stated that the effects of column inelasticity are captured through the combined use of the τ and K factors in Eq. (2.5). If the effective length factor is determined based on the elastic column properties, then in many cases, a conservative estimate of the column design strength $\phi_c P_n$ is obtained[6]. However, the τ factor in the numerator of Eq. (2.5) must not be over-estimated. If it is, the resulting value for $\phi_c P_n$ will be unconservative.

It should be emphasized that, although $\hat{\tau} \geq \tau$ is commonly accepted and is often conservative for calculating the effective length factors for use in Eqs. (2.1) through (2.6), or for calculating λ_c from Eq. (2.8b) for use in Eqs. (2.1b) or (2.4b), *the value for τ in the numerator of Eq. (2.5) must always be the exact value for proper determination of the design strength $\phi_c P_n$ with this equation.* Therefore, usage of Eq. (2.5) is appropriate only with an exact iterative determination of the inelastic column buckling loads $P_{e\tau}$. Since Eqs. (2.1) and (2.4) do not require calculation of the exact τ values (i.e., the exact τ is implicitly contained within Eqs. (2.1)), these equations are better suited for ordinary design. Equations (2.5) through (2.8) are presented here to clarify the basis for calculating various elastic or inelastic effective lengths.

Section 2.3 presents and contrasts a number of the most widely accepted procedures for computing effective length based on a separate categorization, the extent of the structure included in the buckling model. Unless noted otherwise, all of the equations in Section 2.3 are presented in terms of the general parameter $\hat{\tau}$, where $\hat{\tau}$ is equal to one for elastic effective length calculations, or it is computed as a value less than or equal to one for any approximate or exact inelastic effective length procedure. *Once a value is determined for $\hat{\tau}$, the structure can be envisioned as being composed of "effective elastic" prismatic members with rigidities of $\hat{\tau} EI$ in any of the K factor calculation procedures.*

2.3 Subassembly, Story, and System Models for Calculation of Effective Length

A wide range of methods have been suggested in the engineering literature for the calculation of column effective length (K) factors. Section 2.2 has addressed the categorization of these procedures based on the degree to which column inelasticity is considered in the calculations. However, a separate but equally important

[6] Footnote 3 of this chapter describes when the elastic effective lengths may be unconservative.

categorization is obtained by observing the physical extent of the structure included in the buckling model used to obtain a member's effective length. Based on this organization, the various K factor procedures may be grouped according to the following types of models or idealizations: (1) the buckling of an idealized subassembly composed only of the column being investigated or designed and its immediately adjacent members, (2) the buckling of the story in which the column is contained, and (3) the buckling of the overall structural system, or some general subassembly of the structural system. Some of the methods are suitable for hand calculations. Others generally require a computer solution. All the methods involve certain assumptions and simplifications.

2.3.1 Isolated Subassembly Approach

Within the context of AISC LRFD and ASD practice, the most widely used method for calculating K factors is the nomograph (or alignment chart) method [SSRC, 1976; Kavanagh, 1962]. This approach is reviewed below for the separate cases of unbraced (sidesway uninhibited) and braced (sidesway inhibited) construction.

Unbraced Frames (Sidesway Uninhibited)

For an unbraced frame (sidesway uninhibited), the nomograph method is based on the buckling of the isolated idealized subassembly shown in Fig. 2.2a. A detailed step-by-step derivation of the equations governing the buckling of this subassembly is presented by Chen and Lui [1987]. This derivation involves the following assumptions:

1. All the members are prismatic and initially straight.

2. All the connections are fully-restrained. Joint size and deformations are neglected.

3. All the members are completely elastic. Distributed yielding in the columns, plastic hinging in the girders, nonlinearity of the connections, etc. are not considered.

4. Any axial forces in the girders are small enough such that their effect on the girder stiffnesses is negligible.

5. The stiffness parameter $L\sqrt{P/EI}$ is the same for all the columns in the subassembly: $c1$, $c2$, and $c3$. Therefore, the relative magnitude of the second-order stiffnesses for these columns (reduced by the effect of axial compression) is the same as that of their first-order elastic stiffnesses.

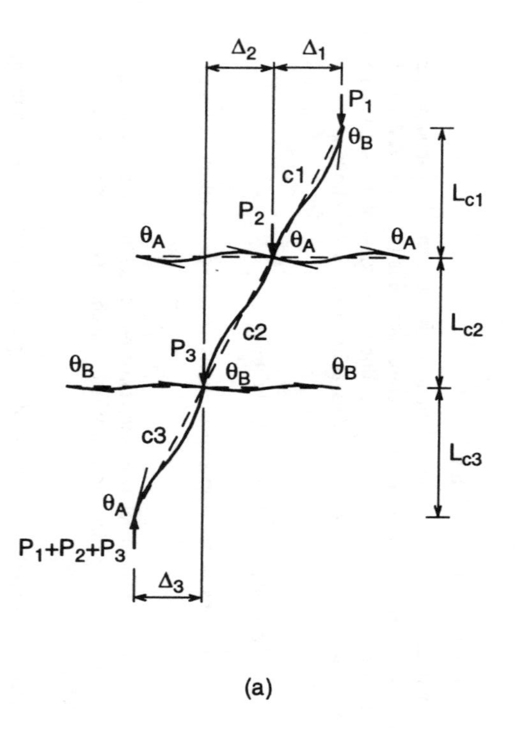

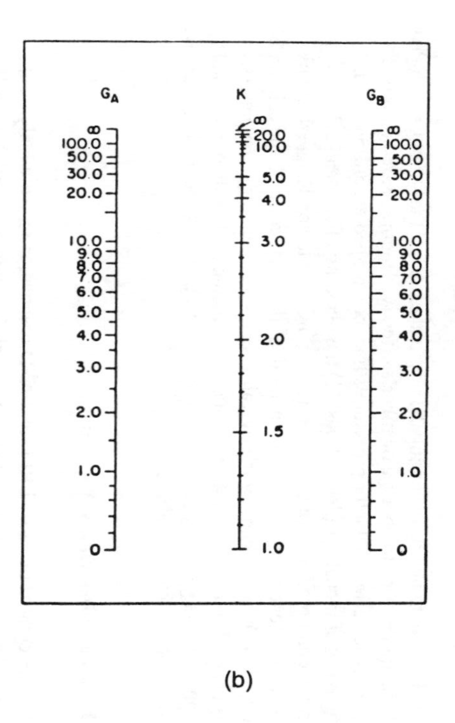

Figure 2.2. Subassembly model and alignment chart for unbraced (i.e. sidesway uninhibited) frame [Chen and Lui, 1987].

6. The buckling rotations at the joints in a given floor level (either θ_A or θ_B in the figure) are all equal and in the same direction at the ends of each of the girders, thus producing full reversed-curvature bending of the girders with an inflection point at their mid-lengths.

7. The joint rotations alternate from story to story within the subassembly, from θ_B at the top of column *c1*, to θ_A at the top of column *c2* (joint *A*), to θ_B again at the bottom of column *c2* (joint *B*), and to θ_A again at the bottom of column *c3*.

8. The column chord rotations associated with the buckling displacements, Δ/L_{c1}, Δ/L_{c2}, and Δ/L_{c3}, are all equal.

9. The subassembly is neither restrained nor loaded by the other portions of the three stories in which it is contained (and from which it is isolated). That is, at incipient buckling, none of the other portions of the frame provide any lateral restraint at the boundaries of the subassembly. Furthermore, the subassembly being considered is not required to restrain the sidesway movement of (or in other words, to provide stability to) any of the other columns within the frame. Zero shear force is transferred to this subassembly from the adjacent portions of the structure.

The only way that condition 9 can be satisfied is that the other framing within the stories associated with the subassembly shown in Fig. 2.2 must buckle simultaneously with the subassembly. Furthermore, the combination of conditions 5 through 8 leads to the restraint from the girders being distributed to the columns above and below the joints A and B in proportion to the I/L of the two columns at each joint.

If one invokes all of the above assumptions[7], the governing equation for the buckling of the subassembly in Fig. 2.2a may be written as

$$\frac{G_A G_B(\pi/K)^2 - 36}{6(G_A + G_B)} - \frac{(\pi/K)}{\tan(\pi/K)} = 0 \tag{2.9}$$

where

[7] These assumptions are seldom satisfied exactly in practice, although in many cases, the errors involved with their violation are small. However, a significant number of situations exist for which the error associated with these assumptions is significant and unconservative. Many of the examples illustrated in the report fall into this category, e.g., the example outlined in Section 1.4.

$$G_A = \frac{\sum_A (EI/L)_c}{\sum_A (EI/L)_g} = \frac{\Sigma \textit{ of stiffness of columns at joint A}}{\Sigma \textit{ of stiffness of girders at joint A}} \tag{2.10a}$$

$$G_B = \frac{\sum_B (EI/L)_c}{\sum_B (EI/L)_g} = \frac{\Sigma \textit{ of stiffness of columns at joint B}}{\Sigma \textit{ of stiffness of girders at joint B}} \tag{2.10b}$$

The solution of Eq. (2.9) can be expressed in the convenient form of the alignment chart shown in Fig. 2.2b.

As noted at the end of Section 2.2, the effect of column inelasticity can be included in any of the effective length procedures by simply envisioning the structure as being composed of "effective elastic" prismatic members with appropriately determined inelastic rigidities $\hat{\tau} EI$. Therefore, the assumption of elastic column behavior in the third item of the above list may be removed by rewriting Eqs. (2.10a) and (2.10b) as [Yura, 1971; Disque, 1973]:

$$G_A = \frac{\sum_A (\hat{\tau} EI/L)_c}{\sum_A (EI/L)_g} = \frac{\Sigma \textit{ of stiffness of columns at joint A}}{\Sigma \textit{ of stiffness of girders at joint A}} \tag{2.10c}$$

$$G_B = \frac{\sum_B (\hat{\tau} EI/L)_c}{\sum_B (EI/L)_g} = \frac{\Sigma \textit{ of stiffness of columns at joint B}}{\Sigma \textit{ of stiffness of girders at joint B}} \tag{2.10d}$$

Ordinarily it is assumed that the girders are fully elastic at incipient buckling (i.e., there are no distributed yielding effects associated with axial forces in the girders, and there is no significant yielding that has occurred due to bending actions). As previously discussed in Chapter 1, these assumptions are considered to be valid for elastic design of "rigidly-connected" or FR frames.

Since the above equations assume that the buckling rotations are equal and in the same direction at the ends of all the girders (assumption 6 in the above list), the stiffnesses of these members $(EI/L)_g$ need to be modified if the girders have different boundary conditions [AISC, 1993]. This may be handled conveniently by determining a modified length of the girder

$$L'_g = L_g \left[2 - \frac{M_F}{M_N} \right] \tag{2.11a}$$

or a modified girder stiffness parameter

$$(EI/L')_g = \frac{(EI/L)_g}{\left[2 - \frac{M_F}{M_N}\right]} \quad (2.11b)$$

where M_F is the moment at the far end of the girder under consideration and M_N is the moment at the near end, from a first-order lateral load analysis of the frame.

Equations (2.11) are based on the fact that the sidesway buckling mode of the structure often can be closely approximated by the deflected shape of the system under lateral loading. This modification is essential for effective length calculations in unsymmetrical frames. Equations (2.11) reduce the girder stiffness by a factor of 2 if the far end is pinned. They reduce the girder stiffness by a factor of 1.5 if the far end of the girder is completely fixed against rotation. If $\frac{M_F}{M_N}$ is less than one, the column effective length would be underestimated, and thus the buckling load of the member being considered would be overestimated if the nomograph were used directly without any modification of the girder stiffnesses. If $\frac{M_F}{M_N}$ is greater than 2, L'_g becomes negative. This is valid, but may in turn produce negative G factors, which are not addressed by the AISC nomograph [AISC, 1993]. In such cases, the "nomograph" effective length factor must be calculated directly from Eq. (2.9), which is valid for positive and negative G values. The committee is not aware of any design charts that directly provide the K factors associated with negative G values in sway frames.

If inelastic effective lengths are computed, the "most correct" L'_g in Eqs. (2.11) requires that the lateral load analysis to obtain these values be based on the inelastic column rigidities $\hat{\tau} EI$. Of course, approximations of the "exact" L'_g can certainly be obtained from an elastic lateral load analysis.

In a single story frame, any lateral load may be applied at the roof level to determine the ratio $\frac{M_F}{M_N}$. However, in multi-story frames, the parameter $\frac{M_F}{M_N}$ in general can vary significantly based on the type of lateral loading applied in the analysis. In many cases, the $\frac{M_F}{M_N}$ ratios can be determined with sufficient accuracy from a first-order lateral load analysis of the structure associated with the design wind loads. Nevertheless, it is well accepted that the most appropriate analysis for determining these ratios is a first-order analysis in which the total gravity load at each story (or some percentage thereof) is applied as a horizontal load at each of the corresponding floor levels [Horne, 1975; LeMessurier, 1993]. It is important to note that, if the frame being considered is unsymmetrical such that sway occurs under pure gravity load, the analysis used to determine the $\frac{M_F}{M_N}$ values should *not* include vertical loads.

Otherwise, the lateral deflections will not reflect solely the lateral load/lateral deflection characteristics of the structure.

Numerous adjustments are possible to accommodate other variations from the assumptions listed for the sidesway-uninhibited alignment chart. Variations from assumption 1 (prismatic members) are not addressed in this document. Chapter 3 addresses variations from assumption 2 (i.e., frames with PR connections). The story buckling approach discussed later in Section 2.4.1 involves an adjustment of the nomograph effective lengths that accounts for variations from assumption 9. However, violations of assumptions 5, 7, and 8 are generally not considered in alignment chart based procedures. The story buckling approach discussed in Section 2.4.2 accommodates variations from assumptions 6 through 9, and approximates conservatively the effects of axial force on the column stiffnesses at incipient buckling for most cases (assumption 5).

Lastly, a simple approximate adjustment that accounts for the effects of any axial force in the girders of a frame (i.e., the violation of assumption 4 in the previous list) is the multiplication of the $(EI/L)_g$ or $(EI/L')_g$ values in Eqs. (2.11) by

$$\left[1 - \frac{Q}{Q_e}\right] \tag{2.12}$$

where Q is the axial force in the girder (positive in compression) and Q_e is the elastic critical load of the girder, $\pi^2 EI_g/L_g^{\,2}$.

Braced Frames (Sidesway Inhibited)

For a braced frame (sidesway inhibited), the alignment chart procedure is based on buckling of the idealized isolated subassembly shown in Fig. 2.3a. The assumptions employed in this buckling idealization are the same as those employed in the development of the sidesway-uninhibited nomograph with the following exceptions:

6. The buckling rotations at all the joints in a given story (either θ_A or θ_B in the figure) are all equal and in opposite directions in each girder, causing uniform bending in each of these members.

8. There is zero sidesway, and thus the column chord rotations are zero.

9. The subassembly is completely restrained against sidesway by some means.

Similar to the unbraced case, the combination of conditions 5 through 8 leads to the restraint from the girders being distributed to the columns above and below joints A and B in proportion to the *I/L* of the two columns at each joint.

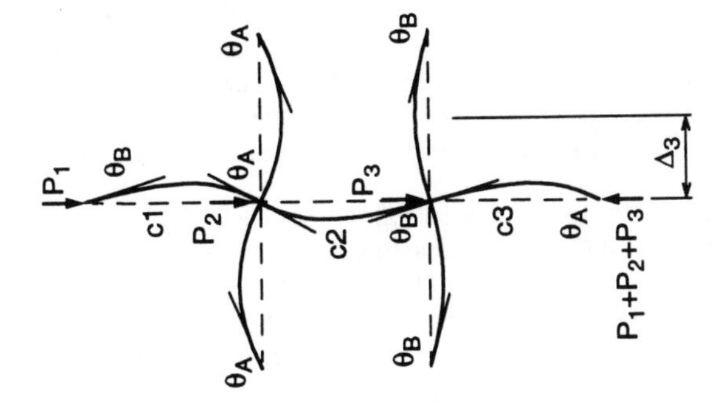

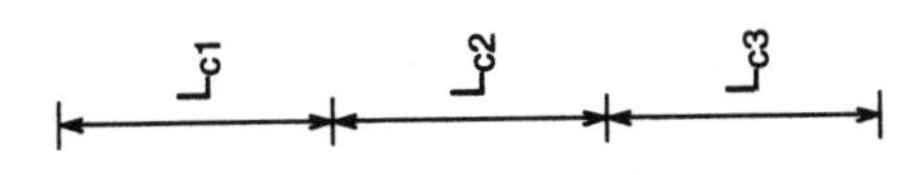

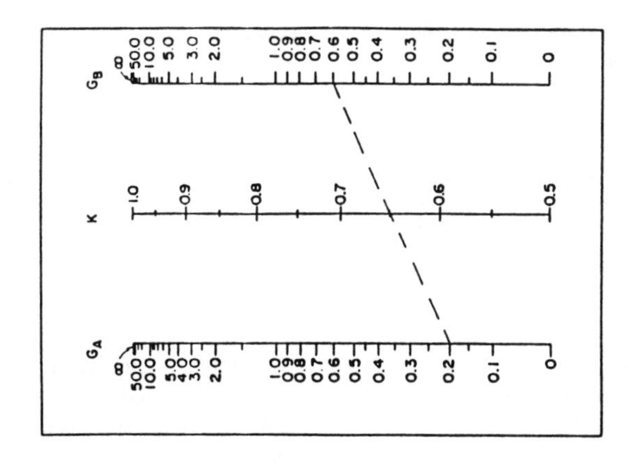

Figure 2.3. Subassembly model and alignment chart for braced (i.e. sidesway inhibited) frame [Chen and Lui, 1987].

Based on the full set of assumptions, the governing equation for buckling of the subassembly in Fig. 2.3a may be written as

$$\frac{G_A G_B}{4}\left(\frac{\pi}{K}\right)^2 + \left(\frac{G_A + G_B}{2}\right)\left[1 - \frac{\dfrac{\pi}{K}}{\tan\left(\dfrac{\pi}{K}\right)}\right] + \frac{2\tan\left(\dfrac{\pi}{2K}\right)}{\dfrac{\pi}{K}} - 1 = 0 \qquad (2.13)$$

The alignment chart developed from this equation is shown in Fig. 2.3b.

Equations (2.11) are not valid for use with the sidesway inhibited nomograph[8]. These equations are based on the assumption that the first-order deflected shape of the frame under some lateral loading produces the same distribution of moments in the girders as sidesway buckling of the frame. This type of simple assumption is not possible for braced buckling. For braced frames in which the actual girder end boundary conditions differ from those in the model of Fig. 2.3a, the following ideal cases exist:

i. If the far end of a girder is fixed, multiply the $(EI/L)_g$ of the member by 2.0.

ii. If the far end of the girder is pinned, multiply the $(EI/L)_g$ of the member by 1.5.

The restraint offered to the columns by the girders is smaller in the model assumed for derivation of the nomograph than in either of these ideal cases, since the columns are assumed to rotate by an equal and opposite amount at each of the girder ends in the nomograph model (assumption 6). In the actual frame, if the column at one of the girder ends tends to rotate by a greater angle than the column at the opposite end, then the effective restraint from the girder will be larger for the column with the larger buckling rotation. This trend is reflected in the above idealized cases where the opposite end of the girder is either fully fixed or pinned. Correspondingly, the column with the smaller buckling rotation receives lesser restraint from the girder.

The reader is referred to (Bridge and Fraser, 1987) for a development of equations for sidesway-inhibited K factors, accounting for general girder (or column) end-conditions, including cases with negative G factors. Of course, if a beam-to-column connection within a frame is ideally pinned, then the $(EI/L)_g$ contribution to the G_A or G_B equations for the column at this connection is zero. Furthermore, if the beam-to-column connections are semi-rigid or partially-restrained, adjustments must

---

[8] However, Eq. (2.12) is appropriate for braced and unbraced cases. Also, the inelastic stiffness reduction factor $\hat{\tau}$ is appropriate for braced or unbraced frames.

be made to the $(EI/L)_g$ values to account for the connection stiffness. These adjustments are addressed in Chapter 3.

The design examples in this report focus entirely on sidesway-uninhibited framing. This is in part due to the fact that the use of $K < 1$ for the design of column members has important ramifications on the second-order moments M_u in the beams. The beam bending moments may be amplified significantly in braced frames designed with $K < 1$. In the opinion of the committee, inclusion of braced-frame examples with $K < 1$ would require that the proper calculation of these second-order moments be addressed, and thus would require expansion of the scope of the report.

2.3.2 Story Buckling Approach

As noted in Section 2.3.1, the sidesway-uninhibited nomograph of the AISC LRFD Specification [AISC, 1993; AISC, 1994] assumes that zero shear force is transferred to the subassembly in Fig. 2.2 from the other columns in the stories involved with the sidesway buckling. Each of the columns in any story is assumed to buckle in a sidesway mode independent of the other columns in the story. Obviously, this is an idealization that, in some cases, may be significantly violated.

For example, consider the simple one-bay portal frame in Fig. 1.5. Column *C2* provides all the sidesway buckling resistance of the frame, and as illustrated in the figure, a significant destabilizing shear force acts on this column at buckling. The leaning column *C1* has zero sidesway buckling resistance, and it depends entirely on column *C2* for its lateral support. Yura [1971, 1972a, 1972b] observed that the elastic sidesway buckling of such a frame occurs approximately when the sum of the axial forces in all the columns of the story ($P_1 + P_2$ in this case) reaches the sum of the sidesway buckling resistances of the columns providing the lateral stiffness (this is simply the buckling resistance of column *C2* in this example).

In addition to moment frames that support leaning columns, other practical cases where there is significant "buckling interaction" between the columns in a story can include structural systems in which moment frames are used in two orthogonal directions. In this type of system, some of the columns within a given bent usually would be turned in weak-axis bending, and others in strong-axis bending. This type of situation is illustrated by a simple example in Fig. 2.4 (from LeMessurier [1991]). In this example, if the sidesway stability of the bent containing columns *C1* and *C2* is considered, the sidesway buckling capacity of column *C2* will tend to be much smaller than that of column *C1*. Therefore, column *C1* will tend to restrain the lateral movement of *C2* up to the load at which it can no longer resist sidesway buckling, including the destabilizing shear effects resulting from column *C2*. Furthermore, if the force in column *C2* is large compared to the force in column *C1*, column *C2* will depend to an even greater extent on column *C1* for its sidesway stability. In some practical cases, the weaker, more heavily-loaded column can in fact have an effective

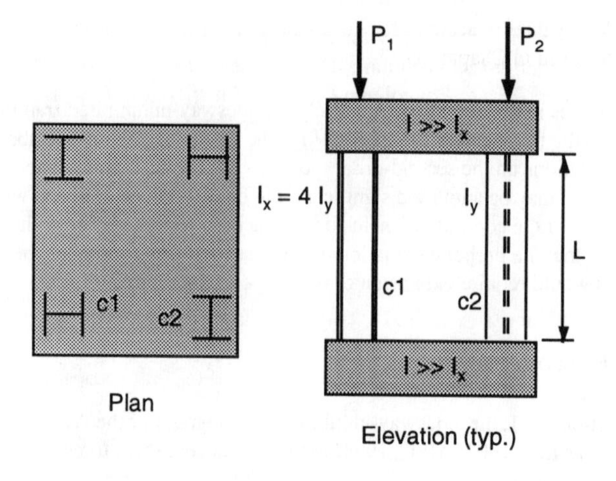

At incipient sidesway buckling,

$$P_1 + P_2 \cong \pi^2 E(I_x + I_y)/L^2$$

where $P_1 = \pi^2 EI_x/(K_1L)^2$

$$P_2 = \pi^2 EI_y/(K_2L)^2$$

assuming $K_2 \geq 0.5$

If $P_1 = P_2$, $K_1 \cong 1.27$ & $K_2 \cong 0.63$

Figure 2.4. Example case in which significant sidesway buckling interaction exists between the columns of a story [LeMessurier, 1991].

length factor less than one (i.e., the heavily-loaded column can be effectively "braced" by the stronger, more lightly-loaded column). This is the case for the example frame in the figure—the "exact" K factor for column $C2$ is 0.635 if $P_1 = P_2$. Conversely, due to the significant destabilizing effect from the weaker column, the load in column $C1$ at sidesway buckling of the story is reduced substantially from that associated with the alignment chart. In other words, the effective length of this member is significantly increased—the "exact" K factor for column $C1$ is 1.272 for equal column axial loads in LeMessurier's example.

To account for stories in which there is significant sidesway buckling interaction between the different columns, Yura [1971, 1972a, 1972b] proposed that each story in an unbraced frame should be considered as a unit. He based his philosophy on the earlier work of Higgins [1965], Zweig [1968], and Salem [1969], who had observed that:

(1) the stronger columns in a story will "brace" the weaker columns until an overall story buckling load is reached, and

(2) at the story buckling load, the story as a whole buckles in a sidesway mode.

By considering the buckling capacity of each story as a unit, Yura established a more general basis for determining effective length.

Yura's work formed the basis for the studies by LeMessurier several years later [LeMessurier, 1976; LeMessurier, 1977]. LeMessurier provided a comprehensive discussion of story stability behavior, and he developed a refined practical approach for computing effective length to account for this behavior. With the exception of system-based approaches for computing effective length (such as the use of an eigenvalue buckling analysis, as discussed in Section 2.3.3), most of the "comprehensive" methods currently available for computing effective length, and thus the buckling capacity of a column, are in essence derived from the above papers by Yura and LeMessurier. Some of the more recent papers include [Wu, 1985], [Baker, 1987], [Lui, 1992], [Aristizabel-Ochoa, 1994], [Hajjar and White, 1994], and [Geschwindner, 1994].

For the case of stories with equal column lengths, all the story-based procedures for assessing structural stability can be formulated based on the following equation:

$$\lambda_{story} \sum_{all} P_u = \sum_{non-leaner} P_{cr(story)} \tag{2.14a}$$

where $P_{cr(story)}$ represents the contribution from each column to the story sidesway buckling resistance, and λ_{story} is the load magnifier or buckling parameter that the factored loads P_u must be scaled by to achieve story sidesway buckling. The symbol

$\sum_{all}$ indicates summation over all of the columns in the story, and $\sum_{non-leaner}$ represents summation only over the columns contributing to the sidesway resistance (the leaning columns, which are assumed to be ideally pinned-pinned in the model of the structure, provide zero contribution to the sidesway stiffness of the story). The value of the story buckling parameter is obtained from Eq. (2.14a) as:

$$\lambda_{story} = \frac{\sum_{non-leaner} P_{cr(story)}}{\sum_{all} P_u} \tag{2.14b}$$

All the various story-based formulations that have been proposed in the literature differ primarily in only one aspect: the manner in which they compute $\sum_{non-leaner} P_{cr(story)}$, the resistance of the story as a whole to buckling in a sidesway mode. Two specific procedures for computing this resistance are discussed in Section 2.4.

Since it is assumed that all the design axial forces are increased proportionally by λ_{story} to achieve buckling, the force in a given column at incipient buckling is

$$\lambda_{story} P_u = \frac{\sum_{non-leaner} P_{cr(story)}}{\sum_{all} P_u} P_u \tag{2.15}$$

A story-based effective length factor may be calculated by equating this force to the column buckling load, $P_{e\hat{\tau}}$, as defined by Eq. (2.8). That is,

$$\left[P_{e\hat{\tau}(story)} = \frac{\pi^2 \hat{\tau} EI}{(K_{story} L)^2} \right] = \left[\lambda_{story} P_u = \frac{\sum_{non-leaner} P_{cr(story)}}{\sum_{all} P_u} P_u \right] \tag{2.16a}$$

where the subscript "*(story)*" has been added to $P_{e\hat{\tau}}$ and K to distinguish these terms from the column buckling load and effective length factor determined by other procedures. Equation (2.14a) establishes a condition that is satisfied at incipient buckling of the story. Equation (2.16a) restates the definition of the member buckling load, as given by Eq. (2.8), and equates this load to the axial force in the column at story buckling. By considering the equality of the second and fourth expressions in Eq. (2.16a), and solving for the effective length factor, one obtains:

$$K_{story} = \sqrt{\frac{1}{P_u} \frac{\pi^2 \hat{\tau} EI}{L^2} \frac{\sum_{all} P_u}{\sum_{non-leaner} P_{cr(story)}}} \tag{2.16b}$$

Alternatively, the term $P_{e\hat{\tau}(story)}$ can be calculated directly as

$$P_{e\hat{\tau}(story)} = \frac{\sum\limits_{non-leaner} P_{cr(story)}}{\sum\limits_{all} P_u} P_u \tag{2.16c}$$

based on the equality of the first and fourth expressions in Eq. (2.16a). Either Eq. (2.16b) or Eq. (2.16c) may be used to determine a column's design strength $\phi_c P_n$. The result from Eq. (2.16b) may be substituted either into Eq. (2.2) to determine λ_c or into Eq. (2.3) to determine P_e, and subsequently, these values may be used either in Eqs. (2.1b) and (2.4b), or Eqs. (2.1a) and (2.4a) respectively. Alternately, the value from Eq. (2.16c) may be substituted into Eq. (2.8b) to determine λ_c for use in Eqs. (2.1b) and (2.4b).

The inelastic stiffness reduction factor, $\hat{\tau}$ in Eqs. (2.16a) and (2.16b), accounts for distributed plasticity in the column due to the combination of axial force and initial residual stresses. As discussed in Section 2.2, often it is conservative to use $\hat{\tau} = 1$ for calculation of the effective length. However, a simple approximate value for $\hat{\tau} < 1$ may be computed for columns that fail inelastically by substituting an under-estimated value for P_n into Eq. (2.7a) (see Section 2.2 for a discussion of this procedure). It is important to note that the $\hat{\tau}$ values for each of the columns within the story, which are substituted into the numerator of Eq. (2.16b) to obtain the individual K_{story} values, also must be utilized in evaluating the $P_{cr(story)}$ contribution from each of the columns in Eqs. (2.14) through (2.16).

Further clarification of some of the terms introduced in the above discussion is appropriate at this point. It should be emphasized that $P_{e\hat{\tau}(story)}$ is the column buckling load, including all the lateral resistance and destabilizing effects within the story. In other words, the story being considered will buckle when the applied axial forces in all of the columns reach their corresponding $P_{e\hat{\tau}(story)}$ values. This should be distinguished from $P_{cr(story)}$, which is the resistance of an *isolated* column to overall sidesway buckling, neglecting any second-order interactions with the other columns within the story. For $\hat{\tau} = 1$, $P_{e\hat{\tau}(story)}$ is the story-based *elastic* column buckling load, whereas $P_{cr(story)}$ is the elastic resistance of the column to sidesway buckling. The load $P_{e\hat{\tau}(story)}$, obtained in this way, is identical to the term P'_{ei} in the commentary of the AISC LRFD Specification [AISC, 1993]. Similarly, $P_{cr(story)}$ is identical to the term P_{e2} in the LRFD commentary.

In general, $P_{cr(story)}$ is not equal to $\lambda_{story} P_u$, and therefore by Eq. (2.16a) it is also not equal to $P_{e\hat{\tau}(story)}$[9]. This is because $P_{cr(story)}$ is the resistance of an isolated

[9] Specific procedures for determining $P_{cr(story)}$ are discussed in Section 2.4

column to story sidesway buckling. The parameters λ_{story} (Eq. (2.14b)) and $P_{\hat{e\tau}(story)}$ (Eq. (2.16c)) account for the distribution of the vertical loads among the column members, and thus they account for the interaction of all the columns in the story in resisting sidesway buckling.

It is instructive to write Eq. (2.16c) in the form

$$\frac{P_{\hat{e\tau}(story)}}{\sum\limits_{non-leaner} P_{cr}} = \frac{P_u}{\sum\limits_{all} P_u} \tag{2.17}$$

This equation indicates that the ratio of a column's buckling capacity $P_{\hat{e\tau}(story)}$ to the full story sidesway buckling resistance $\sum\limits_{non-leaner} P_{cr(story)}$ is always equal to the ratio of the member's design axial force P_u to the full factored gravity load supported by the story $\sum\limits_{all} P_u$. Obviously, this must be true since the loads P_u are simply scaled by λ_{story} as per Eq. (2.15) to obtain the column buckling loads. However, all the story-based procedures either assume that $\sum\limits_{non-leaner} P_{cr(story)}$ is a constant, independent of the distribution of the column axial loads across the story, or they only approximate the influence of the axial load distribution on $\sum\limits_{non-leaner} P_{cr(story)}$. Section 2.4.5 discusses limits that must be imposed on two specific story-based procedures to prevent their application in situations where the story strength may be significantly smaller than estimated using the story-based equations.

Implications of the Story-Buckling Concept on the Axial Strength Term of the Beam-Column Interaction Equations

By calculating $P_{e\tau(story)}$ from Eq. (2.16c) using the exact inelastic stiffness reduction factors (τ) in determining $\sum\limits_{non-leaner} P_{cr(story)}$, and then substituting this expression into Eq. (2.5) to obtain $\phi_c P_n$, the axial strength ratio of the AISC-LRFD beam-column interaction equations, $\frac{P_u}{\phi_c P_n}$, can be written as

$$\frac{P_u}{\phi_c P_n} = \frac{P_u}{\phi_c\,(0.877\,P_{e\tau(story)})} = \frac{\sum\limits_{all} P_u}{0.877\phi_c \sum\limits_{non-leaner} P_{cr(story)}} = \frac{1}{0.877\phi_c \lambda_{story}} \tag{2.18}$$

for all the columns within a given story. That is, all the columns within the story will have the same value for $\frac{P_u}{\phi_c P_n}$. The last expression in Eq. (2.18) is obtained by substituting Eq. (2.14b).

Furthermore, since the term $\sum_{non-leaner} P_{cr(story)}$ in this equation is based on the "exact" τ values associated with incipient elastic or inelastic buckling, this term can be replaced by $\frac{1}{0.877} \sum_{non-leaner} P_n$ such that the axial strength term can be expressed as a constant value,

$$\frac{P_u}{\phi_c P_n} = \frac{\sum_{all} P_u}{\phi_c \sum_{non-leaner} P_n} \tag{2.19}$$

for all the columns within the story. It should be emphasized that this relationship is true only if the "exact" τ is used consistently throughout all the column/story stability calculations. If Eqs. (2.1) and (2.4) are utilized for calculating $\phi_c P_n$, the exact τ shown in the numerator of Eq. (2.6) is implicitly included within these equations. However, if $P_{cr(story)}$ for the individual columns is not based on the exact τ values in the calculation of K_{story} (Eq. (2.16b)) or $P_{e\tau(story)}$ (Eq. (2.16c)), $\frac{P_u}{\phi_c P_n}$ will not be constant over all the columns within the story[10].

If an approximate inelastic stiffness reduction factor, $\hat{\tau} \leq 1$, is used in determining the $\sum_{non-leaner} P_{cr(story)}$ value input to Eq. (2.16b), and the resulting K_{story} value is used in determining $\phi_c P_n$ through either Eqs. (2.1) or (2.4), the resulting $\phi_c P_n$ often will be smaller than the column design strength obtained by using the exact τ in the equation for K_{story}. This is because the K_{story} based on an approximate $\hat{\tau}$ is often larger than the exact inelastic K_{story}. However, as noted in Section 2.2, K_{story} is increased due to inelasticity when the increased "leaning" effects from other framed columns that are loaded inelastically outweigh any beneficial effects of increased stiffness of the elastic girders relative to the elastic or inelastic column. This type of situation would occur for example if the column being considered is lightly loaded such that it remains elastic and has an exact τ equal to one, while other columns

[10] Nevertheless, for the story-based procedures with an approximate $\hat{\tau}$, the variation of $\frac{P_u}{\phi_c P_n}$ among the columns of a story is small in many practical cases.

within the story are subjected to loads large enough such that their $\hat{\tau}$ values are less than one.

Due to the relationship between the computed values of $\phi_c P_n$ based on an approximate $\hat{\tau}$ versus the exact τ, as discussed above, it can be concluded that if $\frac{P_u}{\phi_c P_n}$ is evaluated at a column for which K_{story} is reduced by the inelastic stiffness reduction, this value of $\frac{P_u}{\phi_c P_n}$ can be used conservatively for all the other columns of the story. Furthermore, the column that has the smallest $\hat{\tau}$ (i.e., the largest axial stress, or the largest $\frac{P_u}{P_y}$) is ensured of having its K_{story} reduced by the inelastic effects), and it will generally produce the most accurate estimate of the "exact" $\frac{P_u}{\phi_c P_n} = \frac{\sum\limits_{all} P_u}{\phi_c \sum\limits_{non-leaner} P_n}$. Conversely, the value of $\frac{P_u}{\phi_c P_n}$ for the columns that are most lightly loaded (i.e., the columns that have the smallest axial stress or values of $\frac{P_u}{P_y}$), obtained based on an approximate $\hat{\tau} > \tau$, generally will be smaller than the "exact" value for the story's $\frac{P_u}{\phi_c P_n}$. *Therefore, as a general design procedure, it is recommended that* $\frac{P_u}{\phi_c P_n}$ *should be calculated using the story-buckling equations with the column of the lateral-resisting system having the largest value of* $\frac{P_u}{P_y}$, *and that this value of the axial strength ratio should be used in the beam-column interaction check for all the lateral-resisting columns of the story.* This procedure streamlines the design calculations while also avoiding any potential unconservatism associated with a separate calculation of the $\frac{P_u}{\phi_c P_n}$ for the more "lightly-loaded" columns. The accuracy of this approach in estimating the "exact" uniform $\frac{P_u}{\phi_c P_n}$ is reasonably good in most cases, and is equal to the accuracy of the $\frac{P_u}{\phi_c P_n}$ calculated for the single most heavily loaded column.

Story Buckling and the Effective Length of Leaning Columns

The sidesway-buckling resistance, $P_{cr(story)}$, is of course zero for any leaning (i.e., "pin-connected") columns in the story. These members are effectively braced by the story's lateral-resisting system. Therefore, the buckling load $P_{e\tau}$ and/or the design strength $\phi_c P_n$ of any leaning column should be computed based on $K = 1$, rather than Eq. (2.16b) or (2.16c). Usage of these equations for leaning columns is one area of much confusion and mis-application of the story-buckling concept. If either of these equations is applied, any value of K (greater or less than one) can be obtained depending on the amount of axial force in the leaning column at incipient buckling of the story. In actuality, the buckling of the leaning column by itself and of the story as a whole are two separate phenomena. The story buckling load is affected by the destabilizing P-Δ shear forces from the leaning columns, and these effects are included in Eqs. (2.16b) and (2.16c). Either of these formulas may be substituted into the appropriate equations of Section 2.2 to compute the design strength of the columns within the lateral-resisting system. However, the actual stability behavior of the leaning columns is that of a "pinned-pinned" column, braced by the lateral-resisting system of the story (i.e., $K = 1$). *That is, a sidesway buckling failure of the leaning columns is actually not a buckling failure of the leaning columns at all. It is a buckling failure of the story's lateral resisting system.*

Anomolies occur in the calculation of story-based K factors when PR connections are considered and the connection stiffness is assumed to change from zero to a small finite value. This problem has been referred to by Cheong-Siat-Moy and others as the "K factor paradox" (e.g., see Cheong-Siat-Moy [1986]). This and other related problems are addressed in Section 2.3.4 and in Chapter 3, Section 3.6.

2.3.3 System Buckling Approach

Effective length factors also may be calculated from an elastic or inelastic buckling analysis of the entire structural system. Some of the most common procedures for system buckling analysis are outlined in Appendix A. The term "system buckling analysis" often implies an eigenvalue analysis of the entire structural system. Of course, direct eigenvalue analysis also can be applied on a member or subassembly level, if so desired. The "system" K factors associated with this type of analysis are obtained by equating the axial force in a member at incipient buckling of the entire frame, $\lambda_{system} P_u$, to the buckling load $P_{e\hat{\tau}(system)}$ determined from Eq. (2.8) with an effective length $K_{system} L$:

$$\lambda_{system} P_u = P_{e\hat{\tau}(system)} = \frac{\pi^2 \hat{\tau} EI}{(K_{system} L)^2} \tag{2.20}$$

In this equation, the member axial force P_u is based on a linear elastic analysis associated with the factored design load combination being considered, and λ_{system} is the multiple of P_u at which system buckling occurs. It is assumed that any "pre-buckling" displacements have a negligible effect on the forces within the frame. The effective length factor is then determined by solving Eq. (2.20) for K_{system}:

$$K_{system} = \sqrt{\frac{\pi^2 \hat{\tau} EI / L^2}{P_{e\hat{\tau}(system)}}} \tag{2.21}$$

where $\hat{\tau}$ is equal to one for "elastic" K factors (based on an elastic system buckling analysis), and $\hat{\tau}$ may be less than one for calculation of "inelastic" K factors (based on an inelastic system buckling analysis). The K factor obtained from Eq. (2.21) inherently accounts for interactions within and between stories at incipient buckling of the system model. Similar to the procedures discussed for calculating inelastic K_{story} values in the previous section, any approximate $\hat{\tau}$ may be used in Eq. (2.21). However, the same $\hat{\tau}$ must be utilized in the system buckling analysis to obtain $P_{e\hat{\tau}(system)} = \lambda_{system} P_u$.

If an exact inelastic system buckling analysis is conducted (see Appendix A for a discussion of what is entailed by such an analysis), the inelastic stiffness reduction factors in each of the members of the frame will be equal to the "exact" values, τ. In this case, the axial strength ratio $\dfrac{P_u}{\phi_c P_n}$ will be equal to a constant value for all the columns within the frame. This can be observed from Eq. (2.20) by noting that based on the first two expressions in that equation

$$\frac{P_u}{P_{e\hat{\tau}(system)}} = \frac{1}{\lambda_{system}} \tag{2.22a}$$

is a constant value for all the members within the frame. If the exact τ values are computed in the analysis, $P_{e\hat{\tau}(system)} = P_{e\tau(system)}$, and based on Eq. (2.5),

$$\frac{P_u}{\phi_c P_n} = \frac{1}{0.877 \phi_c \lambda_{system}} \tag{2.22b}$$

is a constant value for all the members within the frame.

For some types of framing, a system buckling analysis may be necessary for accurate assessment of the frame stability behavior. For example, slender frames may have significant buckling interactions between stories due to the fact that a large percentage of the lateral deflection may come from "cantilever action" of the structure and the associated axial deformation of the columns. Therefore, modeling of the stability behavior and determining column effective lengths on a story-by-story basis may not be sufficiently accurate in certain cases [CTBUH, 1979]. Also, the design of innovative structural systems, such as mega-braced tall frames may require consideration of complex system buckling modes that are not captured properly by isolated subassembly or story-based models [Abdelrazaq et al., 1993]. In Section 2.4.5 of this report, system buckling analyses are used to obtain the "exact" solutions to assess the limits of applicability of several story-based methods.

Nevertheless, for most ordinary rectangular "shear-type" building frames, the story-based procedures tend to produce answers for design that are at least as good as that obtained from a system buckling analysis. In fact, as will be discussed in Section 2.3.4, in some cases there are anomolies associated with the effective lengths obtained based on Eq. (2.21) such that the story-based K factors are considered to be more representative of the stability behavior for design. These anomolies are related to the fact that, if a system buckling analysis is utilized, all the members in the frame will have the same value for $\frac{P_u}{\phi_c P_n}$ by Eq. (2.22b) (if based on the exact τ), or approximately the same value otherwise.

2.3.4 Interpretation of Extremely Large Effective Length Factors

The use of a story- or system-based buckling model can in some cases lead to "unexpectedly large" effective length factors in columns having small axial forces at the buckling of the story or structural system. In fact, based on Eqs. (2.16b) and (2.21), it can be observed that the K_{story} or K_{system} values approach infinity as the axial force in a member approaches zero relative to $\frac{\pi^2 EI}{L^2}$. However, this does not necessarily mean that the member's strength is "used up". In these situations, it can be shown that the $P_u / \phi_c P_n$ term in Eqs. (1.1) will be equal to $1/(0.877\, \phi_c \lambda_{story})$ or $1/(0.877 \phi_c\, \lambda_{system})$ as $\lambda_{story} P_u$ or $\lambda_{system} P_u$ approach zero (and as K_{story} or K_{system} approach infinity), regardless of whether the exact τ is computed throughout the model or not[11]. This is because, as K_{story} or K_{system} approach infinity for a member, the column strength is based on elastic buckling and therefore $\tau = 1$ and $P_{e\tau} = P_e = \frac{P_n}{0.877}$. Based on Eq. (2.5) and equating $P_{e\tau} = P_e = \frac{P_n}{0.877}$ to either $\lambda_{story} P_u$ or $\lambda_{system} P_u$, the ratio of the axial load to the column design strength may be written as

[11] The term $\lambda_{system}\ P_u$ is the load in a member at incipient buckling of the structural system, as calculated from a system buckling analysis. Similarly, $\lambda_{story}\, P_u$ is the corresponding load in the member from a story-based approach.

$$\frac{P_u}{\phi_c P_n} = \frac{P_u}{\phi_c(0.877\lambda_{story} P_u)} = \frac{1}{0.877\phi_c\lambda_{story}} \tag{2.23a}$$

for the story-based approach or

$$\frac{P_u}{\phi_c P_n} = \frac{P_u}{\phi_c(0.877\lambda_{system} P_u)} = \frac{1}{0.877\phi_c\lambda_{system}} \tag{2.23b}$$

for the system-based approach in cases where the axial force in a member approaches zero [Bridge, 1989] [12].

Any situation in which the axial force is small compared to $\frac{\pi^2 EI}{L^2}$ will always be associated with one of the following cases:

(1) The axial loading effects in the member are negligible, and the member is for all practical purposes acting as a "beam". As a beam, the member may or may not be contributing substantially to the buckling resistance of the frame (by providing restraint to the column members).

(2) The member is a column (i.e., the member's axial force is non-negligible); however, the member *is not* contributing significantly to the computed story or system buckling resistance. Nevertheless, if the size of the member is reduced by a large enough amount, the buckling strength of the structure (or story) might be controlled by a different buckling mode that *does* depend significantly on this member.

(3) The member is a column that *is* contributing significantly to the computed story or system buckling resistance. In this case, if the size of the member is changed by even a small amount, the buckling strength of the structure or story will be changed significantly.

In the first case, the axial strength term in the beam-column interaction equations (i.e., Eqs. (2.23)) should be taken as zero. If the member is significantly influencing the buckling strength of the story or system by restraining other members that are subjected to significant compression, this restraint has already been accounted for within the buckling analysis. Since the buckling model is based on the EI that has been specified for this member, the buckling strength must of course be reevaluated if the member is significantly restraining the buckling mode and if its cross-section is changed. Conversely, in the third case, the member must be designed for the non-

---

[12] As discussed in Sections 2.3.2 and 2.3.3, if the buckling calculations are based on the exact τ values, Eqs. (2.23) are satisfied for all columns within the story or the structural system, regardless of the value of P_u.

negligible compression force, and the axial strength term (Eqs. (2.23)) most definitely should be computed and included in evaluating the member's interaction equation for axial compression plus bending.

Nevertheless, in many situations where some of the column axial forces at incipient buckling of the story or system are small compared to $\frac{\pi^2 EI}{L^2}$, the members that have small axial forces (and thus large effective length factors) are actually not participating significantly in the buckling of the story or system at all. These members correspond to case 2 in the above list. In this situation, the effective length factor obtained from the story or system buckling analysis is not the most appropriate K value to use for design. In other words, these members are being "penalized" by the fact that $\frac{P_u}{\phi_c P_n}$ is constant or near-constant for all the members within the buckling model. A more appropriate K value can be computed by a "story-buckling" procedure, if the K_{system} value is high, or by a "nomograph" procedure, if the K_{story} value is high.

Without the execution of time-consuming buckling sensitivity studies, the determination of whether a column member belongs to case 2 or to case 3 can require significant insight and judgment by the engineer. Nevertheless, if the engineer decides that a member belongs to case 2, then the buckling resistance of this member should not be included in the calculation of K_{system} (if large K_{system} values are being "disregarded"), or it should not be included in the calculation of K_{story} (if large K_{story} values are "over-ruled").

To understand the scope of the above problems, it is helpful to consider a few specific examples. Two situations where extremely large K factors can occur are discussed below:

System Buckling Analysis of Multi-Story Frames

One instance where the above problem can happen is with the system buckling analysis of multi-story frames [Bridge, 1989]. Due to usage of the same column section throughout several stories, the lower stories within a particular column lift may be "more critical" than the upper stories. Furthermore, the top few stories within a tall moment-frame may have small axial forces relative to their column strength due to bending actions significantly influencing their design, whereas the columns in the bottom stories may be more "stability critical" (i.e., the contribution from the axial load term in the beam-column interaction equations may be larger). In these situations, the upper stories may "ride along" essentially in a rigid-body mode at the buckling of the full structural system. Since the axial loads in the columns of the upper stories may be relatively small, "unexpectedly large" effective length factors may be obtained within these stories.

In these cases, one can argue that the column effective lengths should be based on a story buckling analysis, since:

(1) the interaction effects between stories in influencing the buckling of the structural system may be negligible,

(2) the system buckling is then essentially a "critical story" phenomenon, with the most critical story governing the buckling capacity of the frame, and

(3) the system stability is meaningless to the design of the individual stories for all except the critical story level.

The effective length factors of the lightly-loaded columns (e.g., the columns in the upper stories) will be smaller in general if calculated in this way. However, in some situations, a member might have a large K_{system} factor from the system buckling analysis because it is restraining a "more critical" member or several members in another story at incipient buckling of the full structure. If this restraint is significant and counted upon in the design of the more critical members, and if the member's axial force is non-negligible, then the large K_{system} value must be used.

Mixing of K_{story} and K_{system} factors is difficult and requires subtle interpretations of the behavior. The best approach is that, if significant interactions are expected among a number of stories, a system-buckling analysis might be conducted for a subassembly composed of these stories, while a "standard" story-buckling analysis might be employed for other stories of the building. An example where this approach might be utilized would be a building frame in which several columns have moment connections at their ends and extend unsupported for more than one story.

Buckling Analysis of a Story Containing a Lightly-Loaded "Near-Leaning" Column

An anomaly similar to the one discussed above can occur in the calculation of K_{story} if the contribution of a member to the story sidesway buckling capacity is small. For example, suppose that a column is lightly loaded and is connected to its girders by a very flexible connection, while the other columns within the story are loaded with relatively large gravity loads and are framed with "rigid" moment connections. If the connection to the first column is made more and more flexible, the contribution of this column to the story sidesway stiffness, $\sum_{non-leaner} P_{cr(story)}$ in Eq. (2.16b), of course becomes smaller and smaller. If the connection is made flexible enough, $\sum_{non-leaner} P_{cr(story)}$ effectively becomes independent of the contribution from this column. Nevertheless, if $\dfrac{\sum_{all} P_u}{P_u}$ is large, this member's effective length factor, K_{story}, will be large relative to that for the other columns of the story, although the member is for all practical purposes a leaning column, braced against sidesway by the other columns, with $K = 1$.

Suppose that the engineer decides to select a section with a larger EI for the above column due to this member failing a design check based on K_{story}. Because of the small connection flexibility, the only parameter that significantly changes in Eq. (2.16b) would be the term EI. As a result, K_{story} actually increases, and consequently $P_{e\tau(story)}$ in Eqs. (2.16a) remains essentially constant. The fact that $P_{e\tau(story)}$ remains essentially constant for this case can be seen also from Eq. (2.16c), since each of the parameters in this equation are essentially unchanged in the above scenario. Since the increase in EI does not result in any significant increase in the buckling capacity of the "near-leaning" column, the engineer may be led to increase the sizes of the other columns in the story to increase $\sum_{non-leaner} P_{cr(story)}$, and thus to increase the $P_{e\tau(story)}$ for this column as per Eq. (2.16c). However, if the contribution from the small connection stiffness is neglected and the column is:

(1) not included in $\sum_{non-leaner} P_{cr(story)}$, and

(2) considered as a leaning column (with $K = 1$),

this column as well as the other columns in the story may be found to be quite adequate. The problem here is that the column actually is not contributing significantly to the story sidesway buckling resistance. This situation is discussed in the context of a specific frame example in Section 3.6.

It should be noted that, if the hypothetical column discussed above is framed with a stiff enough connection such that it does provide a large contribution to $\sum_{non-leaner} P_{cr(story)}$, then this member actually fits under the third case outlined at the beginning of this section. In this case, it is probably most economical to design the member based on the large K_{story} value (caused by the small P_u and the fact that this member is providing substantial assistance to the story sidesway stability).

Conversely, if $\dfrac{\sum_{all} P_u}{P_u}$ is small in the above examples (i.e., if P_u on the column is relatively large for either of the cases of a flexible or a relatively stiff connection), the K_{story} for this member will be small compared to the other columns of the story. The member's K_{story} can even be smaller than 0.5, which is not physically possible, and in which case the "braced buckling" of this member would control. The limits developed in Section 2.4.5 prevent this type of situation from occurring.

At the present time (1995), the story-buckling concept is well accepted for regular, rectangular "shear-building" frames in which significant interaction effects may exist between the columns of a story. Inter-story interaction effects on the column buckling loads are treated in a very idealized way in the typical story-buckling equations. Several of the predominant story-buckling based approaches are considered in detail in Section 2.4.

2.4 Calculation of Elastic and Inelastic Critical Story Load in Unbraced Frames

This section presents two story-based effective length procedures that are included in the AISC LRFD Specification Commentary to Chapter C [AISC, 1993][13], and discusses in detail their derivation, assumptions, and use for practical design. Both procedures are based on Eq. (2.16b) and (2.16c), and they differ only in the manner in which they estimate the story sidesway buckling capacity $\sum\limits_{non-leaner} P_{cr(story)}$. The first approach, labeled K_{K_n} and outlined in Section 2.4.1, bases this computation on the AISC sidesway-uninhibited nomograph [AISC, 1993]. The second approach, outlined in Section 2.4.2 and labeled K_{R_L}, does not require the use of the nomograph, but instead presumes that the columns of the lateral-resisting system are subjected to a relatively large, destabilizing P-δ effect. Because of this simplifying assumption, K_{R_L} is efficient to compute, while maintaining reasonably good accuracy.

All the equations in this section are written presuming that the columns in a story have the same length. Appendix B derives these two effective length factors for the more general situation involving columns of unequal length.

2.4.1 Nomograph-Based Effective Length Factor, K_{K_n}

One story-based effective length factor that has received some attention in practice [Wu, 1985; Baker, 1989; Liew et al., 1991; AISC, 1993; Hajjar and White, 1994] assumes that the buckling capacity of the story is equal to the sum of the column buckling loads computed using an effective length factor (labeled K_n) based upon the AISC sidesway-uninhibited nomograph [AISC, 1993]. That is,

$$\sum_{non-leaner} P_{cr(story)} = \sum_{non-leaner} \frac{\pi^2 \hat{\tau} EI}{(K_n L)^2} \tag{2.24}$$

In calculating the K_n values that are entered into Eq. (2.24), all the appropriate modifications to the basic alignment chart procedure discussed in Section 2.3.1 should

---

[13] As noted in Chapter 1, the story-based effective length and notional load approaches are equally applicable and beneficial for frames designed according to the AISC Allowable Stress Design Specification [AISC, 1989].

be made. This includes modification of the girder length from L_g to L'_g according to Eq. (2.11a) and adjustment of the G factors to account for the inelastic reduction factor, $\hat{\tau}$, according to Eqs. (2.10c) and (2.10d). However, the sidesway buckling interaction between the different columns of the story is not considered in determining K_n. This adjustment is accomplished through Eqs. (2.16).

Upon substituting Eq. (2.24) into Eq. (2.16a), the following equation is obtained:

$$\left[P_{e\hat{\tau}(K_n)} = \frac{\pi^2 \hat{\tau} EI}{(K_{K_n} L)^2} \right] = \left[\lambda_{K_n} P_u = \frac{\sum\limits_{non-leaner} \frac{\pi^2 \hat{\tau} EI}{(K_n L)^2}}{\sum\limits_{all} P_u} P_u \right] \tag{2.25a}$$

By equating the second and fourth terms of this equation, and solving for K_{K_n}, the story-based effective length factor may be expressed as

$$K_{K_n} = \sqrt{\frac{1}{P_u} \frac{\pi^2 \hat{\tau} EI}{L^2} \frac{\sum\limits_{all} P_u}{\sum\limits_{non-leaner} \frac{\pi^2 \hat{\tau} EI}{(K_n L)^2}}} = \sqrt{\frac{\sum\limits_{all} P_u}{P_u} \frac{\hat{\tau} I}{\sum\limits_{non-leaner} (\frac{\hat{\tau} I}{K_n^2})}} \tag{2.25b}$$

If $\hat{\tau}$ is assumed equal to one, this equation is identical to Eq. (C-C2-4a) in the LRFD Commentary Chapter C [AISC, 1993]. Alternatively, the buckling load $P_{e\hat{\tau}(story)}$ can be calculated directly as

$$P_{e\hat{\tau}(K_n)} = \frac{\sum\limits_{non-leaner} \frac{\pi^2 \hat{\tau} EI}{(K_n L)^2}}{\sum\limits_{all} P_u} P_u \tag{2.25c}$$

based on the equality of the first and fourth expressions in Eq. (2.25a). Equations (2.25b) and (2.25c) are simply Eqs. (2.16b) and (2.16c) with Eq. (2.24) substituted for $\sum\limits_{non-leaner} P_{cr(story)}$. Either of these formulas may be substituted in Eq. (2.8b) to determine λ_c for use in the LRFD column strength equations. Furthermore, the load magnifier that the factored loads P_u must be scaled by to achieve story sidesway buckling may be written as

$$\lambda_{K_n} = \frac{\sum\limits_{non-leaner} \frac{\pi^2 \hat{\tau} EI}{(K_n L)^2}}{\sum\limits_{all} P_u} \tag{2.25d}$$

Implicit in Eqs. (2.25) is the assumption that the story buckling resistance is unique and independent of the distribution of the vertical loads on the story. It is presumed that a column's contribution to the story sidesway buckling resistance, $P_{cr(story)}$ (Eq. (2.24)), is that of a member which buckles independently of the other columns in the story. If substantial buckling interaction exists between a story's columns such that some of the framed columns have a relatively large stiffness parameter $L\sqrt{P/\hat{\tau}EI}$ and are heavily leaned against other columns, and if the mode shape for story sidesway buckling involves substantial curvature in these weaker columns, then Eq. (2.24) may significantly overestimate the $P_{cr(story)}$ contribution from the weaker columns. Section 2.4.5 derives a limit on the use of Eqs. (2.25) which ensures that any associated unconservative errors are small. A step-by-step procedure for computing K_{K_n} is presented in Section 2.4.6 using the portal frame from Fig. 1.6. Additional examples are given in Chapter 6.

2.4.2 Practical Story-Based Effective Length Factor, K_{R_L}

The effective length factor obtained from the sidesway-uninhibited nomograph, K_n, provides a theoretically exact assessment of the buckling capacity of a column if all the assumptions embedded in the nomograph derivation are satisfied for the equivalent problem obtained after the factors G_A and G_B are modified (see Section 2.3.1). If there is substantial buckling interaction between the column members of the story, or if there are significant destabilizing effects from leaning columns, Eqs. (2.25) accurately account for these effects in most cases.

Nevertheless, there are several assumptions embedded in K_n, and thus in K_{K_n}, that are not accounted for by the modifications discussed in Sections 2.3.1 and 2.4.1. These are the fifth, seventh, and eighth assumptions of the sidesway-uninhibited alignment chart, as outlined in Section 2.3.1. For convenience, these assumptions are repeated again below:

1. The stiffness parameter $L\sqrt{P/EI}$ is the same for all the columns in the subassembly: *c1*, *c2*, and *c3*. Therefore, the relative magnitude of the second-order stiffnesses for these columns (reduced by the effect of axial compression) is the same as that of their first-order elastic stiffnesses.

7. The joint rotations alternate from story to story within the subassembly, from θ_B at the top of column *c1*, to θ_A at the top of column *c2* (joint *A*), to θ_B again at the bottom of column *c2* (joint *B*), and to θ_A again at the bottom of column *c3*.

8. The column chord rotations associated with the buckling displacements, Δ/L_{c1}, Δ/L_{c2}, and Δ/L_{c3}, are all equal.

Assumption 5 is tied to the effect of the axial forces on the stiffnesses of the columns in each of the story levels (i.e., the P-δ effects). Assumptions 7 and 8 are simplifying assumptions about the regularity of the deformations from story to story throughout the height of the frame. If these assumptions are significantly violated, there may be significant buckling interactions between the adjacent stories that are not captured by the K_{K_n} model. For example, if some of the columns in one story are relatively weak compared to the columns immediately above or below these members in an adjacent story, the stronger columns may help restrain the weaker adjacent columns at incipient buckling of the frame. As will be seen in Section 2.3.7 and in Chapter 6, K_{K_n} is often sufficiently accurate and conservative, but in some situations, the above assumptions can be violated to an extent such that it can be unconservative.

Furthermore, the calculation of the modified G factors, the lookup of the associated K_n values, and the subsequent calculation of the K_{K_n} values is a relatively laborious process. If a simpler story-based K factor procedure is available that approximates conservatively the effects of axial force on the column stiffnesses at incipient buckling for most cases, then such a method deserves consideration. LeMessurier [1993, 199?] has developed such a procedure which also has the benefit that it accounts to some extent for variations from assumptions 7 and 8.

LeMessurier's approach does not require usage of the alignment charts. Instead, it is based on the results from a first-order lateral load analysis of the structural system. This approach represents the P-Δ effects on the story buckling capacity exactly, and as noted above, it accounts conservatively for the P-δ effects within the columns in most practical cases. By accounting conservatively for the P-δ effects, the resulting effective length factor, labeled K_{R_L}, is relatively simple to calculate. This method accounts for story interaction effects associated with the violation of assumptions 7 and 8 to the extent that any irregularities in the buckling deformations from story to story are approximated by the deformations in the first-order lateral load analysis. This K factor procedure, which is derived in Sections 2.4.2.1 through 2.4.2.4, is based on approximation of the story sidesway buckling resistance as

$$\sum_{non-leaner} P_{cr(story)} = \frac{\sum_{non-leaner} HL}{\Delta_{oh}}(0.85 + 0.15R_L) \tag{2.26}$$

where

$$R_L = \frac{\sum_{leaner} P_u}{\sum_{all} P_u} \tag{2.27}$$

and $\sum_{leaner} P_u$ represents the summation of the axial forces in the leaning columns of the story. The symbol Δ_{oh} represents the first-order interstory drift due to the story shear $\sum_{non-leaner} H$.

Upon substituting Eq. (2.26) into (2.16a), the following equation is obtained:

$$\left[P_{e\hat{\tau}(R_L)} = \frac{\pi^2 \hat{\tau} EI}{(K_{R_L} L)^2} \right] = \left[\lambda_{R_L} P_u = \frac{\dfrac{\sum_{non-leaner} HL}{\Delta_{oh}}(0.85 + 0.15 R_L)}{\sum_{all} P_u} P_u \right] \quad (2.28a)$$

By equating the second and fourth terms of this equation and solving for K_{R_L}, the story-based effective length factor may be written as

$$K_{R_L} = \sqrt{\frac{1}{P_u}\frac{\pi^2 \hat{\tau} EI}{L^2} \frac{\sum_{all} P_u}{\dfrac{\sum_{non-leaner} HL}{\Delta_{oh}}(0.85 + 0.15 R_L)}} \quad (2.28b)$$

Alternatively, the column buckling strength can be calculated directly as

$$P_{e\hat{\tau}(R_L)} = \frac{\dfrac{\sum_{non-leaner} HL}{\Delta_{oh}}(0.85 + 0.15 R_L)}{\sum_{all} P_u} P_u \quad (2.28c)$$

If $\hat{\tau}$ is assumed equal to one, Eq. (2.28c) is identical to Eq. (C-C2-5a) of the LRFD Specification [AISC, 1993]. Equations (2.28) are most commonly employed with $\hat{\tau} = 1$ to obtain elastic buckling loads and effective lengths. However, an inelastic K_{R_L} may be obtained by using a $\hat{\tau} \leq 1$ in the numerator of this equation, and by using the corresponding $\hat{\tau}$ values for all the column stiffnesses in the first-order lateral-load analysis of the frame. Either of Eqs. (2.28b) or (2.28c) may be substituted into Eq. (2.8) to determine λ_c for use in the LRFD column strength equations. The corresponding story buckling parameter for this procedure may be expressed as

$$\lambda_{R_L} = \frac{\dfrac{\sum\limits_{non-leaner} HL}{\Delta_{oh}}(0.85 + 0.15R_L)}{\sum\limits_{all} P_u} \tag{2.28d}$$

As will be seen in the next section, $\dfrac{\sum\limits_{non-leaner} HL}{\Delta_{oh}}$ is the first-order sidesway resistance of the story. The K_{R_L} approach thus reduces this sidesway resistance by a factor $(0.85 + 0.15\, R_L)$, which ranges from 0.85 to 1.0, depending on the percentage of the story gravity load supported by leaning columns. This reduction factor accounts for the P-δ effects within the columns of the story.

Similar to that for Eqs. (2.25), the derivation of Eqs.(2.28) is based on the assumption that the story buckling load is unique and independent of the distribution of axial forces among the columns of a story. Section 2.4.4 derives a limit on the use of Eqs. (2.28) that serves the same purpose as the limit placed upon Eqs. (2.25) -- limiting the unconservative error in the calculated buckling resistances. A step-by-step procedure for computing K_{R_L} is presented in Section 2.4.6 for the portal frame of Fig. 1.6. Further examples of its calculation and use are given in Chapter 6. First, however, to best understand the qualities of this approach, it is necessary to take a closer look at how the story sidesway buckling resistance, including P-Δ and P-δ effects, is approximated by Eq. (2.26). This is addressed in the next section.

To highlight the implications of the approaches that have been presented in Sections 2.4.1 and 2.4.2, the assumptions invoked in the development and use of the K_{K_n} and K_{R_L} formulas are summarized below:

1. All the members are initially straight and prismatic.
2. All the connections are fully-restrained (partially-restrained connections are addressed in Chapter 3).
3. All the girders are completely elastic. If desired, distributed yielding due to axial loading may be considered in the columns by use of the inelastic stiffness reduction factor $\hat{\tau}$. Any inelasticity along the column lengths is assumed to be uniformly distributed such that the inelastic columns are effectively prismatic.
4. Axial force effects in the girders are neglected entirely in the K_{R_L} equations. They are commonly neglected, but may be accounted for approximately through Eq. (2.12), in the K_{K_n} approach.
5. The story buckles as a unit in a sidesway mode, with equal lateral deflection of all columns in the story. Braced buckling of a "weak column", restrained against

sidesway by the other story framing, is not captured by the equations presented thus far (this mode of failure is accommodated by the equations developed in Section 2.4.5).

6. The factored design loads P_u are increased proportionally until story sidesway buckling is achieved. The multiple of P_u corresponding to incipient buckling of the story is given by λ_{K_n} (Eq. (2.25d)) and λ_{R_L} (Eq. (2.28d)) for the two approaches presented.

7. The end restraint provided by the girders to the columns is modeled based on the results of a first-order lateral load analysis in the K_{R_L} approach. In the K_{K_n} method, the column end restraint is modeled through use of the G_A and G_B parameters (Eqs. (2.10)), which may be adjusted to account for the lateral deflected shape of the specific structure being investigated based on Eqs. (2.11). A repeated pattern of deformations from story to story is assumed (see assumptions 7 and 8 in Section 2.3.1).

8. The P-Δ effects are modeled exactly in both approaches.

9. The P-δ effects are only approximated in both approaches. These approximations allow for the development of reasonably simple equations for the effective lengths and buckling resistances, and the fact that the story buckling resistance is independent of the load distribution within the story in both the K_{K_n} and K_{R_L} approaches is a primary result. In the K_{K_n} approach, the P-δ effects are modeled as per the assumptions of the AISC nomograph. Specifically, the accuracy of this approach is limited by the accuracy of assumptions 5 (equal $L\sqrt{P/\hat{\tau}EI}$ in all the columns), 7 (the repeated pattern of column end rotations shown in Fig. 3), and 8 (equal drift ratio Δ/L of each of the stories) of the nomograph buckling model. Each of these assumptions affect the buckling curvature in the column being evaluated, and thus affect the approximation of the P-δ effects. In the K_{R_L} approach, the second-order stiffnesses of the columns in a given column stack are implicitly assumed to be proportional to the first-order stiffnesses (i.e., equal $L\sqrt{P/\hat{\tau}EI}$ is assumed within each column stack). Furthermore, the P-δ effects on the story sidesway stability are assumed to depend only upon the ratio R_L. The term $(0.85 + 0.15R_L)$ gives the reduction in buckling resistance associated with these effects. For most practical cases, this term accounts conservatively for the P-δ reduction. However, for some extreme cases, it is unconservative. The equations developed in Section 2.4.5 specify limits such that the unconservative errors associated with modeling of the P-δ effects in both the K_{K_n} and the K_{R_L} approaches are acceptable.

10. All columns in a story have equal length. Appendix B rederives the K_{K_n} and K_{R_L} equations for stories having columns of unequal length.

11. Shear deformations within the structural members are neglected in the K_{K_n} equations, i.e., Euler-Bernoulli bending theory is assumed (plane cross-sections remain plane and normal to the axis of the members during flexure). In the K_{R_L} approach, the effects of shear deformations within the members, or other effects such as panel zone deformations, may be included to the extent that they are modeled in the linear elastic analysis used to obtain the lateral story deflections Δ_{oh}. However, Euler-Bernoulli assumptions are involed in the calculation of K_{R_L} (Eq. (2.28b)).

2.4.2.1 First-Order Sidesway Story Resistance

In 1977, LeMessurier introduced two related story-based procedures for computing the effective length of a column, as well as its critical load. The key to the derivation of these formulas lies in the accurate assessment of $\sum_{non-leaner} P_{cr(story)}$ in Eq. (2.16). In LeMessurier's formulation [LeMessurier, 1976; LeMessurier, 1977], the buckling resistance of the story is obtained quite accurately by writing simple approximations of the member's buckled shape and considering all significant geometrically nonlinear effects, including the well known P-Δ and P-δ effects, as well as column inelasticity. LeMessurier termed this procedure "a practical approach to second-order analysis" [LeMessurier, 1976; LeMessurier, 1977]. The manner in which LeMessurier's approach accounts for geometric nonlinearity is described in Sections 2.4.2.2 and 2.4.2.3. The effects of column inelasticity on the story's stability are taken into account by including an inelastic stiffness reduction factor (see Sections 2.1 and 2.2). This is described further in Section 2.4.5.

Before investigating the nonlinear behavior of a story, first consider the cantilever column shown in Fig. 2.5a[14]. The first-order moment diagram for this cantilever, assuming elastic behavior, is shown in Fig. 2.5b. From the moment-area theorem, the *first-order* deflection, Δ_{oh}, shown in Fig. 2.5c, is given by:

$$\Delta_{oh} = \frac{HL^3}{3EI} \tag{2.29}$$

By rearranging terms:

$$H = \frac{3EI}{L^3}\Delta_{oh} \tag{2.30}$$

The first-order sidesway stiffness of this cantilever (k) is equal to the transverse force required to displace the tip of this cantilever by a unit displacement $(\Delta_{oh} = 1)$:

---

[14] Several of the figures in Section 2.4.2 and Appendix B are patterned after those presented in [LeMessurier, 1977]; their notation has been changed to coordinate the discussion with the remainder of this report.

$$k = \frac{3EI}{L^3}(1) = \frac{3EI}{L^3} \tag{2.31}$$

As an alternative, LeMessurier preferred to consider the sidesway resistance as the force required to cause the cantilever to undergo a "unit rotation" (i.e., $\Delta_{oh}/L = 1$, or $\Delta_{oh} = L$, from a first-order elastic analysis). This force, which he defined as P_L, is obtained from Eq. (2.30) as:

$$P_L = H\big|_{\Delta_{oh}=L} = \frac{3EI}{L^3}\Delta_{oh}\Big|_{\Delta_{oh}=L} = \frac{3EI}{L^3}L = \frac{3EI}{L^2} \quad \Rightarrow \quad P_L = \frac{3EI}{L^2} \tag{2.32}$$

Also, by definition (comparing Figs. 2.5c and 2.5d):

$$\frac{H}{\Delta_{oh}} = \frac{P_L}{L} \tag{2.33}$$

or, rearranging terms:

$$P_L = \frac{H}{\Delta_{oh}/L} = \frac{HL}{\Delta_{oh}} \tag{2.34}$$

Thus, for this cantilever, it may be seen that the elastic *first-order sidesway resistance*, P_L, may be represented either in terms of the geometry, end restraint, and section rigidity of the member, as in Eq. (2.32), or directly in terms of the ratio of the shear force, H, to the interstory rotation, Δ_{oh}/L, caused by that shear, as in Eq. (2.34). Since P_L is the first-order sidesway stiffness of the story, it is also equal to the critical buckling load of the story if P-δ effects are neglected. This fact is demonstrated in the next section. Also, as will be shown in Section 2.4.2.4, it is for this reason that LeMessurier utilized this quantity in the development of K_{R_L}.

2.4.2.2 P-Δ Effect on Story Buckling Load

This section illustrates how the P-Δ effect is accounted for in story-based effective length procedures. Consider the structure shown in Fig. 2.6, in which the leaning column j on the right is loaded with a gravity load P_{uj}, and column i on the left is subjected to zero gravity load. Both columns are to have the same length.

As both Yura and LeMessurier showed (and as discussed in Section 2.3.2), column i must resist the shear forces created by the leaning column as the structure sways. When the structure is subjected to sidesway due to a lateral load H, it displaces an amount Δ_{ph} to its final equilibrated position, where Δ_{ph} is the elastic *second-order* displacement of the structure. This displacement equals the first-order sidesway displacement, Δ_{oh}, plus any deflection due to geometric

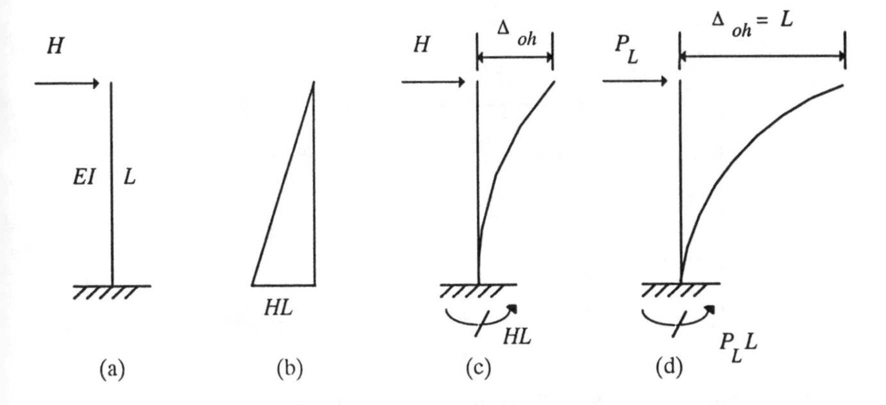

Figure 2.5 First-Order Flexural Behavior of a Cantilever Column without Axial Load

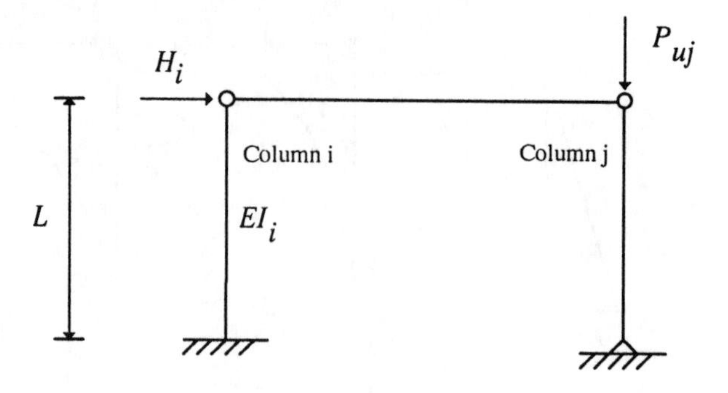

Figure 2.6 One-Bay, One-Story Frame with Axial Load Applied to Leaning Column

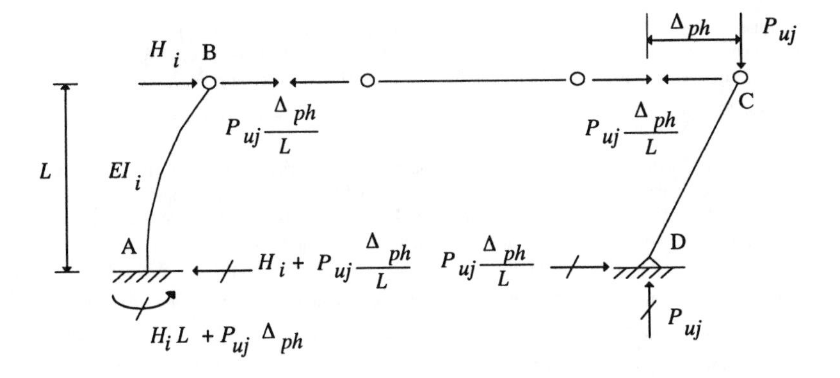

Figure 2.7 Transfer of P-Δ Effect from Leaning Column to Rigidly-Connected Column

nonlinearity in the system. In this specific case, the only geometric nonlinearity is due to the P-Δ effect caused by the axial force in the leaning column acting through its lateral displacement. Therefore, as shown in Fig. 2.7, the axial force P_{uj}, displaced an amount Δ_{ph}, creates a shear $P_{uj}\Delta_{ph}/L$ in the leaning column to balance the second-order couple, $P_{uj}\Delta_{ph}$.

This shear is transferred into the girder as an axial force, to maintain equilibrium at joint C, and this force is then transferred into the cantilever column as a shear to maintain equilibrium at joint B (note that it is assumed here that the lateral displacements at both ends of the girder are identical). Therefore, the shear at the top and bottom of the cantilever column is equal to $H_i + P_{uj}\Delta_{ph}/L$, which equals the first-order shear plus the shear due to the P-Δ effect. The corresponding moment at the bottom of the rigidly-connected column equals the first-order moment plus the frame's P-Δ moment $(H_i + P_{uj}\Delta_{ph}/L)L = H_i L + P_{uj}\Delta_{ph}$. Fig. 2.8b shows the first-order bending moment diagram of column i, and Fig. 2.8c the additional second-order bending moment. Fig. 2.8d shows the total bending moment in the cantilever column.

As long as neither column buckles, and assuming elastic behavior, the final, equilibrated lateral deflection of the structure may be found from moment-area principles. For this simple case, this second-order lateral deflection also equals the shear in the rigidly-connected column divided by the first-order sidesway stiffness of the cantilever column. Since this column is not subjected to any axial force, its elastic sidesway stiffness is equal to the elastic first-order stiffness from Eq. (2.31), $3EI_i/L^3$. Therefore, presuming the rigidly-connected column and the leaning column are of equal length:

$$\Delta_{ph} = \left(H_i + \frac{P_{uj}}{L}\Delta_{ph}\right)\frac{L^3}{3EI_i} = \left(H_i + \frac{P_{uj}}{L}\Delta_{ph}\right)\frac{1}{P_{Li}/L} = \frac{H_i L}{P_{Li}} + \frac{P_{uj}\Delta_{ph}}{P_{Li}} \tag{2.35}$$

Solving for Δ_{ph} in Eq. (2.35), we obtain:

$$\Delta_{ph} = \frac{H_i L}{P_{Li} - P_{uj}} \tag{2.36}$$

Eq. (2.36) is an expression for the *second-order* elastic displacement of this simple structure, including its P-Δ effect. As long as the standard assumptions embedded in this analysis (i.e., cross sections which are initially plane and normal to the centroidal axis of the member remain so during loading, etc.) are satisfied, this is the correct second-order displacement is exact for this structure. This is because the cantilever column is not subjected to any direct axial force (i.e., its stiffness is not affected by the loading).

As LeMessurier noted, from Eq. (2.36) it may be seen that Δ_{ph} approaches infinity when P_{uj} is increased by a factor λ_{story} until (assuming both columns are of equal length):

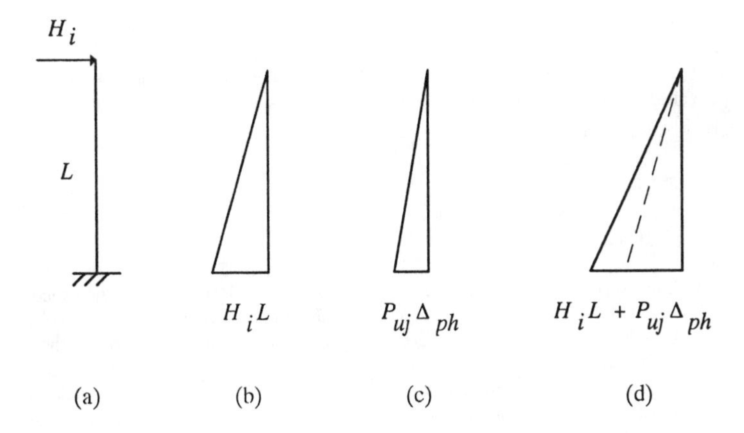

Figure 2.8 Flexural Behavior of Rigidly-Connected Column Due to P-Δ Effect from Leaner Column

$$\lambda_{story} P_{uj} = P_{Li} \tag{2.37}$$

This represents the load at which this structure buckles into its fundamental sidesway mode, and λ_{story} is called the "buckling parameter" of the story. Therefore, the contribution of column i to the story buckling strength is:

$$P_{cr_i} = \lambda_{story} P_{uj} = P_{Li} \tag{2.38}$$

for this simple structure.

The buckling load for this unbraced frame has been determined by considering the structure's deflected shape, and then expressing the second-order effects associated with the axial forces moving through these displacements. In this structure, the first-order displaced shape is the same as the shape of the second-order geometry, and it is also the same as the first-mode buckled shape of the structure. Therefore, the above analysis yields the exact solution for sidesway buckling. This is because the lateral-resisting column i is not subjected to any P-δ effect. Consequently, an accurate buckling load for this unbraced frame has been determined by considering the structure's *first-order sidesway story stiffness* and its *second-order sidesway displaced shape*. This is the essence of LeMessurier's "practical approach to second-order analysis." The next section explains how LeMessurier approximates the buckled shape when the lateral-resisting column is subjected to P-δ effects.

2.4.2.3 Combined P-Δ and P-δ Effects on Story Buckling Load

Consider next the second-order behavior of the same structure, but with the cantilever column loaded by axial force instead of the leaning column (Fig. 2.9). The leaning column now contributes nothing to the behavior of the structure. Consequently, the second-order behavior of the cantilever may be considered individually. Figure 2.10a shows a free-body diagram of the deformed cantilever, Fig. 2.10b shows the first-order bending moments for the cantilever, and Fig. 2.10c shows the additional second-order moments.

Note the difference between the second-order moment diagrams of Figs. 2.8c and 2.10c. The axially loaded cantilever column has a second-order moment not only due to the axial force acting through the rotation of its chord (i.e., $P_{ui}\Delta_{ph}$), but also due to the axial force acting through the curvature of the element. This moment corresponds to the shaded region in Figs. 2.10c through 2.10d. The curvature in the column gives rise to the displacements δ from the member's chord, shown in Fig. 2.10a, and this extra moment corresponds to the P-δ effect. The total moment on the member is shown in Fig. 2.10d.

It is necessary to determine the first-mode sidesway buckled shape of this structure, after which its buckling load may be calculated. In keeping with LeMessurier's basic approach, the buckled shape of the columns may be approximated conservatively by a sine function which is equal to the second-order displaced shape at incipient buckling. The derivation is lengthy and is

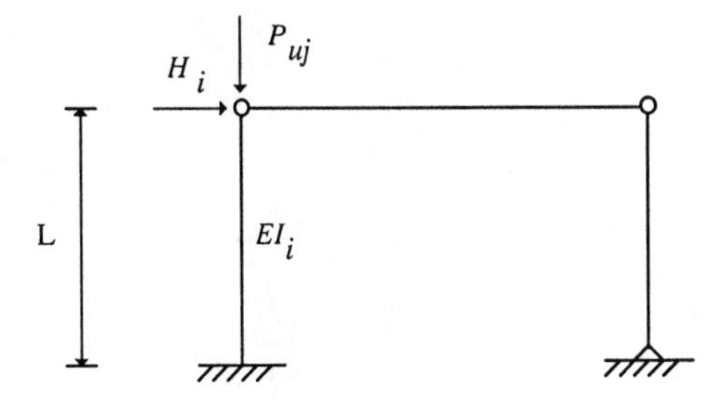

Figure 2.9 One-Bay, One-Story Frame with Axial Load Applied to Rigidly-Connected Column

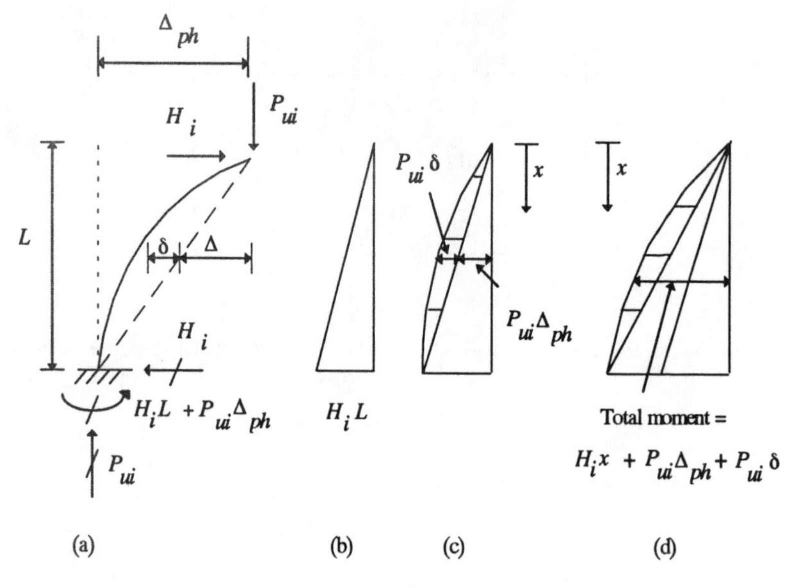

Figure 2.10 Combined P-Δ and P-δ Effects in Axially Loaded Column

detailed in Appendix B. LeMessurier [1977] showed that the second-order deflection equals:

$$\Delta_{ph} = \frac{H_i L_i}{P_{Li} - P_{ui} - C_{Li} P_{ui}} \tag{2.39}$$

where he called C_L the *Clarification Factor*. The third term of the denominator accounts for the P-δ effect on the cantilever column. This term is directly proportional to the second term, which accounts for the P-Δ effect. The third term did not occur in Eq. (2.36), since the rigidly-connected column of Fig. 2.6 experienced no P-δ effect (it was subjected to zero axial force).

As with Eq. (2.36), the buckling capacity of this column equals the load at which the denominator of Eq. (2.39) equals zero (i.e., at which the lateral displacement becomes infinity). Thus, buckling occurs when P_{ui} is increased by a factor λ_i until:

$$\lambda_i P_{ui} = P_{Li} - \lambda_i C_{Li} P_{ui} \tag{2.40}$$

and the elastic critical load, P_{cr_i}, of column i, and therefore of this structure, is:

$$P_{cr_i} = \lambda_i P_{ui} = \frac{P_{Li}}{1 + C_{Li}} \tag{2.41}$$

Comparing Eqs. (2.38) and (2.41), one may observe that the destabilizing effect of the P-δ moments on the buckling capacity of this structure take the form of $1 + C_{Li}$ in the denominator.

As will be seen in the next section, an engineer need not ever calculate C_L to use the story-based effective length factor, K_{R_L}. However, to better understand the derivation of this K factor, it is useful to explore further the typical range of values taken by C_L. LeMessurier derived a general expression for C_L by considering a column with rotational springs at its ends and free to translate at its top (Fig. 2.11). Using a *stiffness factor* for the column, β, which accounts for the effects of the column's elastic end restraints, he showed that a column's elastic sidesway resistance may be written as:

$$P_L = \beta \frac{EI}{L^2} \tag{2.42}$$

This is an alternate form of Eq. (2.34), and the generalized form of Eq. (2.32). Then, substituting the elastic buckling capacity, $P_{e(n)}$, of the column as predicted by the sidesway uninhibited nomograph:

$$P_{e(n)} = \frac{\pi^2 EI}{(K_n L)^2} \tag{2.43}$$

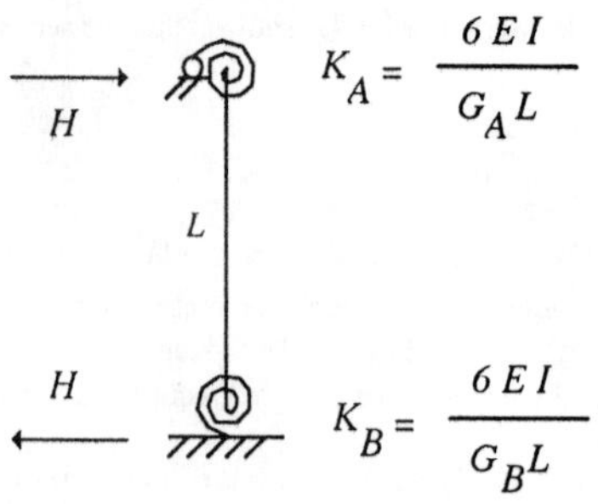
H
L
H
$K_A = \frac{6EI}{G_A L}$
$K_B = \frac{6EI}{G_B L}$

Figure 2.11 Representation of Elastic Restraint at the Ends of a Column

an expression for C_L may be obtained as:

$$C_L = \frac{P_L}{P_{e(n)}} - 1 = \frac{\beta K_n^2}{\pi^2} - 1 \tag{2.44}$$

where K_n is the sidesway uninhibited nomograph effective length factor [AISC, 1993] for the column.

To calculate C_L, one must determine legitimate values for β and K_n, which are at a minimum dependent upon the end conditions of the member. LeMessurier proposed that β be derived in a manner consistent with the assumptions used for the derivation of K_n, using properly determined end restraints. He also indicated that the use of these values of β and K_n provide a sufficiently accurate value for C_L [LeMessurier, 1977]. The column in Fig. 2.11 is used for the derivation of β. As may be seen from the values of the spring stiffnesses given in that figure, a sidesway mode of failure is assumed and the spring stiffnesses are associated with full reverse curvature loading of the beams. From an elastic first-order analysis of this system (see [LeMessurier, 1977] for details of this derivation), it may be determined that:

$$\beta = \frac{6(G_A + G_B) + 36}{2(G_A + G_B) + G_A G_B + 3} \tag{2.45}$$

The G factors may be computed from Eq. (2.10).

The value of C_L computed using Eq. (2.44) does not depend upon the amount of axial force present in the member. This is an approximation made by LeMessurier [1977] to simplify the computation of the P-δ effect on a column. As shown in Appendix B, the true value of C_L for any column in a story actually depends upon the axial force in that column. This value generally ranges from 0.0 to 0.216 [LeMessurier, 1977], although C_L may be higher for extreme cases [LeMessurier, 1993] (C_L is always equal to zero for a leaning column since a leaning column does not exhibit any P-δ effect).

An important attribute of C_L is that it captures approximately the varying P-δ effect that causes the buckling capacity of any story in an unbraced frame to vary depending upon how the gravity forces are distributed among the columns in that story. To clarify this variable P-δ contribution further, consider the effects of inducing axial force in both the rigidly-connected and the leaning columns simultaneously, as shown in Fig. 2.12. The P-Δ effect for this frame remains constant (it equals the sum of the axial forces in the two columns times Δ_{ph}). If one keeps the total gravity load on the frame at a constant value, but then decreases the ratio P_{ui}/P_{uj} (i.e., shifts load from the cantilever column to the leaning column), the P-δ effect on the lateral stiffness of the frame changes (it approaches zero as the leaning column takes the full load). By multiplying C_L by the rigidly-connected column's axial force, as in Eq. (2.40), this changing P-δ

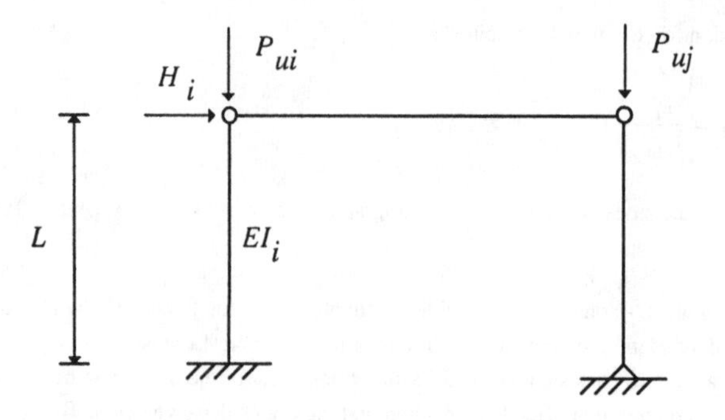

Figure 2.12 One-Bay, One-Story Frame with Axial Load Distributed Between Rigidly-Connected Column and Leaner Column

effect is largely taken into account: as the cantilever column's axial force decreases, so too does the P-δ effect on the frame. LeMessurier [1977] showed that this approach provides an accurate assessment of the buckling capacity of a frame.

2.4.2.4 Derivation of K_{R_L}

For a frame having multiple non-leaning and/or leaning columns, Eq. (2.40), which expresses the buckling capacity of the story having only one column providing lateral resistance, generalizes to:

$$\lambda_{story} \sum_{all} P_u = \sum_{non-leaner} P_L - \lambda_{story} \sum_{non-leaner} C_L P_u \qquad (2.46)$$

where, as stated in Section 2.3.2, it is assumed that all columns in the story buckle when the axial forces in the story are factored by a single buckling factor (i.e., $\lambda = \lambda_{story}$ for all columns). Also, as a generalization of the derivation presented in the previous three sections, the right hand side of the equation above may be computed assuming either elastic behavior, or assuming an effective elastic cross section for each column (i.e., accounting for an inelastic stiffness reduction factor, $\hat{\tau}$ when computing P_L and C_L for each non-leaning column, as discussed in Sections 2.1 and 2.2).

Rearranging terms in Eq. (2.46), one obtains an expression for λ_{story} based upon the use of LeMessurier's Clarification Factor:

$$\lambda_{story} \sum_{all} P_u + \lambda_{story} \sum_{non-leaner} C_L P_u = \sum_{non-leaner} P_L \Rightarrow \lambda_{story} = \frac{\sum\limits_{non-leaner} P_L}{\sum\limits_{all} P_u + \sum\limits_{non-leaner} C_L P_u} \qquad (2.47)$$

Therefore, Eq. (2.16) yields:

$$P_{e\hat{\tau}(story)} = \frac{\pi^2 \hat{\tau} EI}{\left(K_{story} L\right)^2} = \lambda_{story} P_u = \frac{\sum\limits_{non-leaner} P_L}{\sum\limits_{all} P_u + \sum\limits_{non-leaner} C_L P_u} \left(P_u\right) \qquad (2.48)$$

Recently, LeMessurier [1993, 199?] proposed the second story-based effective length factor presented in the AISC LRFD Commentary Chapter C [1993]. It may be formulated based upon the derivation of Eqs. (2.46) through (2.48). As stated in the previous section, C_L generally varies from 0 to 0.216, although it can be larger than 0.216 in more unusual circumstances (for an example of this, see Section 6.2) [LeMessurier, 1993]. To simplify the calculation of Eq. (2.48), one may assume any constant, average value for C_L for all the columns in the frame (C_L is taken as zero for all leaning columns in the frame). This yields:

$$\sum_{non-leaner} C_L P_u \Rightarrow C_{L_{average}} \sum_{all} P_u \tag{2.49a}$$

Consequently:

$$P_{e\hat{\tau}(story)} = \frac{\pi^2 \hat{\tau} EI}{(K_{story} L)^2} = \lambda_{story} P_u = \frac{\sum_{non-leaner} P_L}{\left(1 + C_{L_{average}}\right) \sum_{all} P_u} (P_u) \tag{2.49b}$$

If a conservative, average value of 0.216 is assumed for C_L for all columns in a story, Eq..(2.46) becomes:

$$(1.216)\lambda_{story} \sum_{all} P_u = \sum_{non-leaner} P_L \Rightarrow \lambda_{story} \sum_{all} P_u = 0.82 \sum_{non-leaner} P_L \tag{2.50}$$

On the other hand, in the limit of all the columns in the story being leaning columns, there is no debilitating P-δ effect, and:

$$\lambda_{story} \sum_{all} P_u = \sum_{non-leaner} P_L \tag{2.51}$$

Therefore, since $\lambda_{story} \sum_{all} P_u = \sum_{non-leaner} P_{cr(story)}$ from Eq. (2.14a), the amount of axial force which a story may withstand prior to story buckling usually falls within the following range:

$$(0.82) \sum_{non-leaner} P_L \leq \sum_{non-leaner} P_{cr(story)} < \sum_{non-leaner} P_L \tag{2.52}$$

(a more specific version of this equation appears in the LRFD Commentary Chapter C [AISC, 1993]). The term on the left controls in the limit that none of the loaded columns in the story are leaning columns, and all are framed rigidly into girders at one or both ends, resulting in large P-δ effects. The term on the right controls in the limit that all of the loaded columns in the story are leaning columns, so that the story essentially feels no P-δ effect.

The effective length factor is obtained by solving for K_{story} in Eq. (2.49) and incorporating Eq. (2.52) through the following definition for $\sum_{non-leaner} P_{cr(story)}$:

$$\sum_{non-leaner} P_{cr(story)} = \frac{\sum_{non-leaner} HL}{\Delta_{oh}} (0.85 + 0.15 R_L) \tag{2.26}$$

where Eq. (2.34) has been generalized to a story having several non-leaning columns:

$$\sum_{non-leaner} P_L = \frac{\sum_{non-leaner} HL}{\Delta_{oh}} \tag{2.53}$$

and R_L is defined as:

$$R_L = \frac{\sum_{leaner} P_u}{\sum_{all} P_u} \tag{2.27}$$

where $\sum_{leaner} P_u$ represents the summation of axial force in the leaning columns of the story. This effective length factor, labeled K_{R_L}, has the following form:

$$K_{R_L} = \sqrt{\frac{\hat{\tau} I}{P_u} \frac{\pi^2 E}{L^2} \frac{\sum_{all} P_u}{\sum_{non-leaner} P_L (0.85 + 0.15 R_L)}}$$

$$= \sqrt{\frac{\hat{\tau} I}{P_u} \frac{\pi^2 E}{L^2} \frac{\sum_{all} P_u}{\frac{\sum_{non-leaner} HL}{\Delta_{oh}} (0.85 + 0.15 R_L)}} \tag{2.28b}$$

This effective length factor takes into account the range of behavior encompassed within Eq. (2.52), although to avoid the unlikely occurrence of having a story in which the P-δ effect is extreme in all columns, the value of 0.82 has been rounded to 0.85. Thus, while K_{R_L} does not explicitly require a calculation of C_L for each column, or, for that matter, the calculation of a nomograph K factor, it accounts approximately for the P-δ effect by reducing the first-order sidesway stiffness of the story by up to 17.6% $\left([1-0.85]/0.85 = 0.176\right)$. Also, by using R_L (the ratio of axial force in the leaning columns to the total axial force taken by the story), it accounts for the fact that the P-δ effect in a story diminishes as the relative amount of axial force supported by the leaning columns increases. In addition, by making the simplifying assumptions with respect to the manner in which the P-δ effect is accommodated, K_{R_L} is considerably easier to calculate than K_{K_n}, primarily since no G factors need to be calculated.

The LRFD Commentary Chapter C expresses this approach in the form of a column buckling capacity in Eq. (C-C2-5a) [AISC, 1993]. This equation is obtained by substituting K_{R_L} from Eq. (2.28b) into K_{story} in Eq. (2.16):

$$P_{e\hat{\tau}(R_L)} = \frac{P_u}{\sum_{all} P_u} \sum_{non-leaner} P_L (0.85 + 0.15 R_L) = \frac{P_u}{\sum_{all} P_u} \frac{\sum_{non-leaner} HL}{\Delta_{oh}} (0.85 + 0.15 R_L) \tag{2.28a}$$

where

$$\lambda_{R_L} = \frac{\dfrac{\sum_{non-leaner} HL}{\Delta_{oh}} (0.85 + 0.15 R_L)}{\sum_{all} P_u} \tag{2.54}$$

is the story buckling parameter for this approach. As outlined in Section 2.1, it is convenient to use Eq. (2.30a) directly to compute λ_c in Eq. (2.8b). Section 2.4.2 identifies the key assumptions of the story-based effective length factor, K_{R_L}.

2.4.3 Analysis of the Accuracy of Story-Based Effective Length Computations

The manner in which the story-based effective length factors are derived is critical to their ability to contribute properly to the prediction of the story's buckling capacity. In this section we consider these derivations further (see also [Hajjar, 1994]). To begin the analysis of the accuracy of these effective length factors, the "exact" values of effective length predicted from an eigenvalue buckling analysis, as discussed in Section 2.3.3, will be established as a benchmark. Consider first the following equation, obtained from Eqs. (2.18) and (2.19) in Section 2.3.3, which expresses the buckling capacity of a column in terms of the buckling parameter (or effective length factor) determined from an eigenvalue buckling analysis:

$$P_{e\hat{\tau}(system)} = \lambda_{system} P_u = \frac{\pi^2 \hat{\tau} EI}{(K_{system} L)^2} \tag{2.55}$$

This equation may be rewritten as:

$$\lambda_{system} K_{system}^2 = \frac{\pi^2 \hat{\tau} EI}{L^2 P_u} \tag{2.56}$$

Summing Eq. (2.55) over all columns in the story, and realizing that all columns have an identical value of λ_{system}, and that $P_{e\hat{\tau}(system)}$ equals zero for leaning columns, we obtain:

$$\lambda_{system} \sum_{all} P_u = \sum_{non-leaner} P_{e\hat{\tau}(system)} \Rightarrow \lambda_{system} = \frac{\sum_{non-leaner} P_{e\hat{\tau}(system)}}{\sum_{all} P_u} \tag{2.57}$$

This equation is analogous to Eq. (2.14a) and is an expression of the buckling capacity of the story, obtained based upon effective length factors computed from a system (eigenvalue) buckling analysis.

The story-based effective length factors, in turn, are derived based upon the constraints expressed in a general form by Eq. (2.16):

$$P_{e\hat{\tau}(story)} = \frac{\pi^2 \hat{\tau} EI}{(K_{story} L)^2} = \lambda_{story} P_u = \frac{\sum_{non-leaner} P_{cr(story)}}{\sum_{all} P_u} (P_u)$$

and therefore:

$$\lambda_{story} K_{story}^2 = \frac{\pi^2 \hat{\tau} EI}{L^2 P_{ui}} \tag{2.58}$$

where here K_{story} may be taken as any story-based effective length factor (note that the value of λ_{story} would necessarily be different for each different approach to computing K_{story}).

The right hand sides of Eqs. (2.56) and (2.58) are equal (presuming the same value for $\hat{\tau}$ is used for the two approaches, which is a reasonable assumption). Therefore, for any given column:

$$\lambda_{system} K_{system}^2 = \lambda_{story} K_{story}^2 \tag{2.59}$$

or

$$\frac{K_{story}^2}{K_{system}^2} = \frac{\lambda_{system}}{\lambda_{story}} = (1+\varepsilon)^2 \tag{2.60}$$

where ε is the percent error in the calculation of the individual column story-based effective length factor, as compared to the column's eigenvalue (system) buckling effective length factor:

$$\varepsilon = \frac{K_{story} - K_{system}}{K_{system}} \tag{2.61}$$

Therefore, from Eq. (2.60), if K_{story} is in error by +5%, then the ratio $\lambda_{system}/\lambda_{story}$ equals 1.1025 (i.e., 10.25% error). While an engineer never actually calculates λ_{story} for the story-based procedures, its value is important: the closer it is to the structure's first mode buckling eigenvalue, λ_{system}, the more accurate the effective length factors K_{story} are for that story, and vice versa. Thus, ideally, λ_{story} for each story will approximately equal λ_{system}. Of course, the errors in the story-based procedures grow as their assumptions are breached by a particular frame.

Recall that the general expression for λ_{story} in Eq. (2.14b) assumes that its value is the same for all columns in the story:

$$\lambda_{story} = \frac{\sum\limits_{non-leaner} P_{cr(story)}}{\sum\limits_{all} P_u} \tag{2.14b}$$

Because Eq. (2.14b) is used inherently in the calculation of K_{story} and $P_{e\hat{t}(story)}$ for all non-leaning columns in the story, Eq. (2.61) indicates that the error in the story-based effective length factors (as compared to system buckling K factors) is the same for all non-leaning columns in the story (since $\lambda_{system}/\lambda_{story}$ in Eq. (2.60) is constant across the story). *Other columns in the story do not, for example, "compensate" for one column's unconservative effective length factor by having conservative effective length factors.* Eqs. (2.14b), (2.60), and (2.61) reemphasize a key point made in Section 2.3.2: any story-based buckling load computation is only as accurate as its assessment of $\sum\limits_{non-leaner} P_{cr(story)}$. Section 2.4.5 further investigates the applicability of the two effective length procedures presented in this chatper, and Hajjar [1994] discusses the implications of Eq. (2.60) on the accuracy of these approaches for assessing frame stability.

2.4.4 Imposed Limits on Story-Based Effective Length Computations

Story-based effective length approaches have two primary limitations from the standpoint of accuracy and comprehensiveness. First, these approaches are based on the fundamental assumption that the value of λ_{story} that is computed (or, more to the point, implicitly *presumed*) for a given story in a multi-story frame is within an acceptable unconservative error of the "exact" eigenvalue buckling parameter, λ_{system}, of the frame. This error was discussed in the previous section (see also [Bridge, 1989; Liew, 1991; Hajjar, 1994] for related discussions). Second, these approaches cannot represent a braced buckling mode of a non-leaning column in the story, even if this is the governing buckling mode of the story. A significant portion of these errors are due to the fact that these story-based approaches only approximately account for the distribution of the P-δ effect in the story (recall Section 2.4.2.3).

Both of these deficiencies may be alleviated by imposing a minimum limit on the value of the effective length factor (or, alternately, a maximum limit on the value of the column's buckling capacity) computed using these approaches. By imposing a limit on an individual column's buckling load, a braced buckling mode may be approximately and conservatively captured, and overall errors in the use of story-based effective length approaches may be limited to acceptable values [LeMessurier, 1993; LeMessurier, 199?]. The sections that follow derive and discuss the limits on the values of effective length, or their corresponding expressions for buckling load, that are outlined in LRFD Commentary Chapter C [AISC, 1993]. These limits apply equally to elastic structures, and to structures whose columns have an effective elastic core with a moment of inertia, $\hat{\tau}I$.

2.4.4.1 Derivation of Imposed Limits on Effective Length: One-Bay, One-Story Frame

The two column bents of Fig. 2.13 will be used in the derivation of the imposed limits. A system (eigenvalue) buckling analysis may be used to determine accurately the buckling capacity of these frames. In Fig. 2.14, the total buckling load (i.e., $\lambda_{system}(P_{uj} + P_{uk})$) of the frame in Fig. 2.13a is plotted (for the member properties shown in the inset) versus the distribution of axial force in the two columns. Axial deformation is neglected in this computation, and the elastic modulus is taken as 1[15].

For simplicity, the concept of a "moving load" (introduced by LeMessurier [1993; 199?]) is used schematically to depict this distribution of axial force, as shown in the inset in Fig. 2.14. The moving load is not actually an intraspan load; it generates no moment in the girder. In the buckling analysis, gravity loads are only applied directly to the tops of the two columns. The moving load in the figure simply indicates that each column's proportion of the constant total applied gravity load is varied. As this proportion is varied in fixed increments (i.e., as x is varied from 0 to 1, since the length of the girder in these examples equals 1), a new eigenvalue buckling analysis is performed to determine the buckling capacity of the frame.

One may see from Fig. 2.14 that the actual total buckling load of the frame, as predicted by the eigenvalue analysis, changes as the gravity load distribution varies from the extremes of having all of the gravity load applied completely to one column (e.g., $x = 0$ or $x = 1$) to having the load divided equally among the two columns (e.g., $x = 0.5$). For this symmetric frame, the P-δ effect is largest when all of the load is applied to a single column; the capacity of the frame is weakest at this point. As the load is distributed more evenly between the columns, the P-Δ and P-δ effects, as a percentage of the frame's stiffness, decrease, thus leading to a rise in the frame's buckling capacity as x approaches 0.5. However, for this symmetric frame, there is no distribution of loading at which one of the columns buckles in a braced mode (this is true for the two girders having any value of stiffness).

[15] An elastic modulus of 1.0 may be used for the examples presented in Section 2.4.4 since it is only the relative value of EI/L between the members of the frame that dictate the behavior being illustrated in this section.

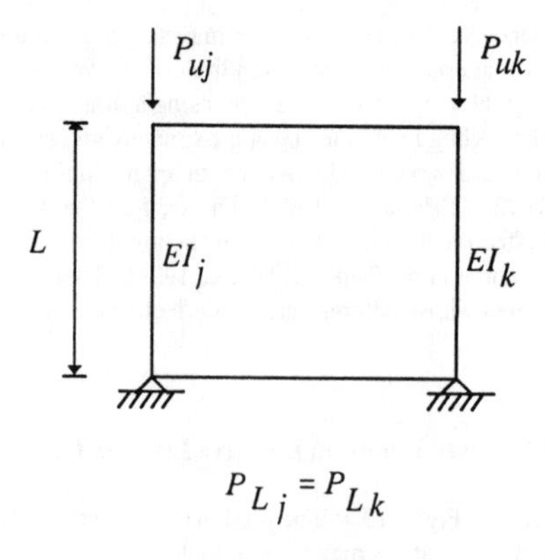

$P_{L_j} = P_{L_k}$

a) Symmetric Frame

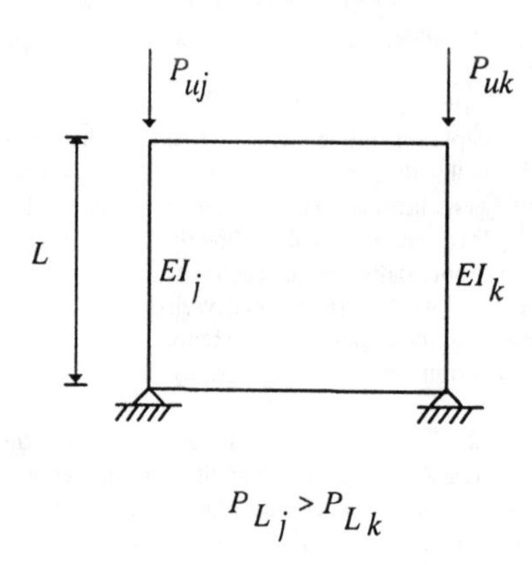

$P_{L_j} > P_{L_k}$

b) Unsymmetric Frame

Figure 2.13 One-Bay, One-Story Frame with Axial Load Distributed Between Two Rigidly-Connected Columns

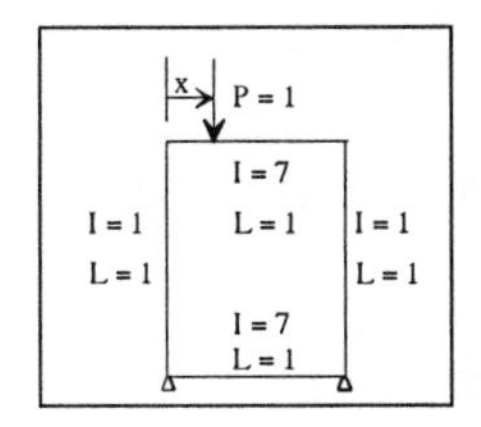

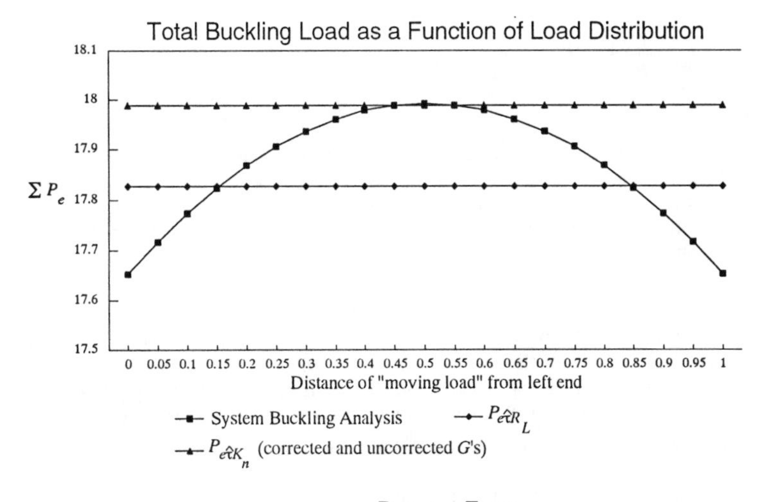

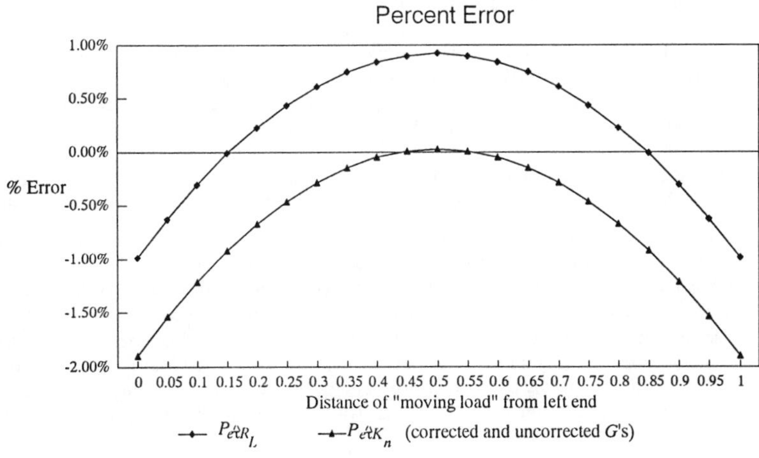

Figure 2.14 Buckling Load of a Symmetric One-Bay, One-Story Frame

The total buckling loads for the story computed using the effective length factors K_{K_n} and K_{R_L} are also shown in Fig. 2.14. The corresponding buckling capacities of each column, $P_{e\hat{\tau}(K_n)}$ and $P_{e\hat{\tau}(R_L)}$, are expressed in Eqs. (2.25a) and (2.28a), respectively:

$$P_{e\hat{\tau}(K_n)} = \frac{P_u}{\sum\limits_{all} P_u} \sum_{non-leaner} \frac{\pi^2 \hat{\tau} EI}{(K_n L)^2} \tag{2.27a}$$

$$P_{e\hat{\tau}(R_L)} = \frac{P_u}{\sum\limits_{all} P_u} \sum_{non-leaner} P_L (0.85 + 0.15 R_L) \tag{2.30a}$$

In this example, for a given approach to computing buckling load, the total buckling load of the story is obtained simply by summing the computed capacities of the two columns in the frame. In the figure, negative values of percent error indicate that the story-based buckling load is larger than the capacity obtained from a system buckling analysis, thus indicating that the story-based computation is unconservative.

For K_{K_n}, the nomograph effective length factor, K_n, is computed using G factors calculated at each end of the column according to Eq. (2.10). In the examples presented in this section, K_n is computed using both the uncorrected and corrected girder lengths (the corresponding G factors will be referred to as "uncorrected" and "corrected" G factors, respectively, referring to the correction of Eq. (2.11)). In Fig. 2.14, the frame capacity computed using K_{K_n} with either uncorrected or corrected G factors is identical due to the symmetry of the frame.

One may observe in Fig. 2.14 that both $\sum\limits_{non-leaner} P_{e\hat{\tau}(K_n)}$ and $\sum\limits_{non-leaner} P_{e\hat{\tau}(R_L)}$ yield a constant total buckling load for this frame. In general, while the buckling loads of the individual columns vary with the distribution of axial force, the total buckling load predicted by these approaches remains constant (if all columns in the story are of equal length). This is a significant assumption of story-based effective length procedures, as outlined in Section 2.3.2.

Fig. 2.15 plots the total buckling capacity of a frame, first investigated by LeMessurier [1993], in which the right column is weaker than the left ($P_{L_j} > P_{L_k}$ in Fig. 2.13b). Referring to the results of the system buckling analysis, one can see that, as the weaker column is increasingly loaded, the frames' buckling capacity steadily decreases. This is contrast to the behavior seen in the symmetric frame of Fig. 2.14. For a constant total gravity load on the frame, this steady decrease in capacity is due to the change in the P-δ effect. As the load on the weak right column increases, its P-δ increases more quickly than the P-δ effect of the left column decreases. In addition, as more load is shifted to the weak column, the frame's buckled shape deviates from a simple sidesway mode towards a combined sidesway/braced mode of buckling. When nearly all of the load is applied to the right column (i.e., when x equals approximately 0.85), a sudden

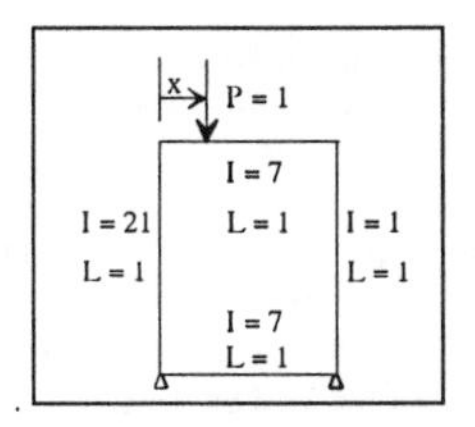

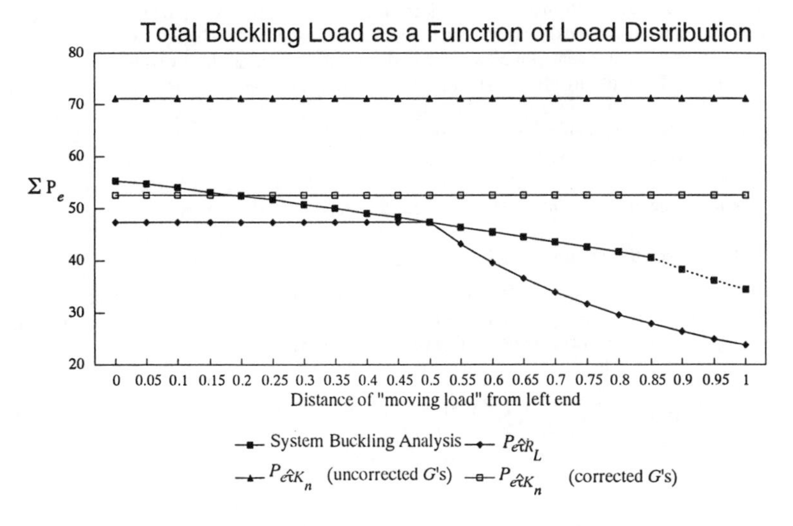

Figure 2.15 Buckling Load of an Unsymmetric One-Bay, One Story Frame

deviation occurs in the system buckling capacity: this corresponds to the right column buckling in a sidesway inhibited (e.g., braced) mode (shown dotted in the figure), rather than the frame as a whole buckling in a sidesway mode.

As the left column's stiffness approaches that of the right (i.e., as in the frame shown in Figs. 2.13a and 2.14), the right column must resist an increasing percentage of the gravity load (i.e., x must be increasingly closer to a value of 1) to achieve a braced mode of buckling. Consequently, one may conclude that placing all of the load on the right column in the symmetric two-column frame (Fig. 2.13a) results in a critical buckling load that is potentially just on the cusp between a sidesway uninhibited buckling mode and a sidesway inhibited buckling mode (it is actually just on the cusp when the moment of inertia of the girder equals infinity; as the girder moment of inertia is decreased from infinity, the frame must be increasingly unsymmetric before any braced buckling occurs). The capacity of the fully loaded right column of Fig. 2.13a may thus conservatively be assumed to be the maximum load the member can carry before a braced mode of buckling can occur. LeMessurier established the limit on the column buckling capacity, computed by the story-based approaches described earlier in this section, to be this value, and so assured that the braced buckling mode of a column is captured either accurately or conservatively.

Based upon this philosophy, a limit for $P_{e\hat{\tau}(R_L)}$ may be derived as follows. If one uses the approach outlined in Sec. 2.3.2 for computing a column's buckling capacity, an expression for the buckling capacity of the right column of Fig. 2.13a may be obtained from Eq. (2.30a) (note that R_L equals zero since there are no leaning columns in this frame):

$$P_{e\hat{\tau}(R_L)k} = \frac{P_{uk}}{\sum_{all} P_u} \sum_{non-leaner} P_L (0.85) \tag{2.62}$$

where the right column is labeled as k (and the left as j) in both the figure and the above equation. For the symmetric frame of Fig. 2.13a, $\sum_{non-leaner} P_L = 2P_{Lk}$. Therefore:

$$P_{e\hat{\tau}(R_L)k} = \frac{P_{uk}}{\sum_{all} P_u} \left[2P_{Lk}(0.85)\right] \tag{2.63}$$

Placing all of the load on the right column of this symmetric frame gives:

$$P_{uk} \bigg/ \left(\sum_{all} P_u \right) = P_{uk} / P_{uk} = 1 \tag{2.64}$$

which then establishes the maximum value which the right column may resist before buckling in a braced mode:

$$P_{e\hat{\tau}(R_L)k_{max}} = \frac{P_{uk}}{\sum_{all} P_u}\left[2P_{Lk}(0.85)\right] = (1)\left[2P_{Lk}(0.85)\right] = 1.7P_{Lk} \qquad (2.65)$$

This limit equation may be applied to any column in any story of a structure. Therefore, the limiting capacity for any general column in a story is:

$$P_{e\hat{\tau}(R_L)_{max}} = 1.7P_L \qquad (2.66)$$

where P_L was expressed in Eq. (2.34):

$$P_L = \frac{HL}{\Delta_{oh}} \qquad (2.34)$$

Therefore

$$P_{e\hat{\tau}(R_L)} \leq 1.7\frac{HL}{\Delta_{oh}} \qquad (2.67)$$

This is Eq. (C-C2-5b) in the LRFD Commentary Chapter C [AISC, 1993]. Alternately, this limit may be rewritten in terms of K_{R_L}:

$$P_{e\hat{\tau}(R_L)} = \frac{\pi^2\hat{\tau}EI}{\left(K_{R_L}L\right)^2} \leq 1.7\frac{HL}{\Delta_{oh}}$$

Therefore:

$$K_{R_L} \geq \sqrt{\frac{\pi^2}{1.7}\frac{\hat{\tau}EI}{L^2}\frac{\Delta_{oh}}{HL}} \qquad (2.68)$$

Note that these equations were derived presuming all columns in the story are of equal length. Once a column breaches this limit, it should almost always be redesigned [Squarzini, 1993], as it is carrying too much load.

If one now considers the unsymmetric frame of Fig. 2.13b, the above argument demands that the right column must breach the limit of Eq. (2.67) at some ratio of $P_{uk}\Big/\sum_{all} P_u$ that is less than one, since this column is weaker than the left column. If more load is shifted to the right column (while holding the total axial force in the story constant), thus loading the right column in excess of $P_{uk_{(limit)}}$, the right column's capacity should be considered to be constant at a value of $1.7P_{Lk}$. This is because beyond $P_{uk_{(limit)}}$, the right column is assumed to be buckling in a braced mode, and thus it is presumed to have reached its maximum capacity. Any additional load applied to the right column does not affect its capacity.

Appendix C derives an expression for the buckling capacity of the frame in Fig. 2.13b once the weaker column is loaded such that it has breached its limit (Eq. (C.10)). For the unsymmetric frame shown in Fig. 2.15, this occurs when x is larger than approximately 0.5. Note that the envelope created by the constant total capacity predicted by Eq. (2.30a) (before the weak right column breaches its limit) and the total capacity computed after the limit is breached (Eq. (C.10)) remains conservative for all values of x. If the limit were not imposed, the constant value of frame capacity predicted by Eq. (2.30a) would overestimate the capacity of this frame as x approaches 1.

2.4.4.2 Accuracy of Story Buckling Prediction for Stories Subjected to a Large P-δ Effect

Using the approach outlined in Section 2.4.4.1, a limit corresponding to Eq. (2.67) may be established for the approach to computing buckling capacity described in Section 2.4.1 ($P_{e\hat{\ell}(K_n)}$, from Eq. (2.27a)). Considering Fig. 2.13 again, the buckling capacity of column k as predicted by Eq. (2.27a) is:

$$P_{e\hat{\ell}(K_n)k} = \frac{P_{uk}}{\sum_{all} P_u} \sum_{non-leaner} P_{e\hat{\ell}(nomo)} \tag{2.69}$$

where the right column is labeled as k and the left as j in both the figure and the above equation. For the symmetric frame of Fig. 2.13a, $\sum_{non-leaner} P_{e\hat{\ell}(nomo)} = 2P_{e\hat{\ell}(nomo)k}$. Therefore:

$$P_{e\hat{\ell}(K_n)k} = \frac{P_{uk}}{\sum_{all} P_u}\left[2P_{e\hat{\ell}(nomo)k}\right] \tag{2.70}$$

Placing all of the load on the right column then establishes the maximum value which the right column may resist before buckling in a braced mode:

$$P_{e\hat{\ell}(K_n)k_{\max}} = \frac{P_{uk}}{\sum_{all} P_u}\left[2P_{e\hat{\ell}(nomo)k}\right] = (1)\left[2P_{e\hat{\ell}(nomo)k}\right] = 2.0P_{e\hat{\ell}(nomo)k} \tag{2.71}$$

This limit equation may be applied to any column in any story of a structure. Therefore, the limiting capacity for any general column in a story is:

$$P_{e\hat{\ell}(K_n)} \le 2.0P_{e\hat{\ell}(nomo)} \tag{2.72}$$

Note that this equation was derived presuming all columns in the story have equal length.

In addition to insuring that a braced mode of buckling is predicted conservatively, LeMessurier's limits also seek to insure that any unconservative error associated with a frame having a large P-δ effect has a maximum value of 5% [AISC, 1993]. Recall, for example, that Eq. (2.27a) accounts for P-δ effects only to the extent that they are incorporated into K_n (see Sections 2.4.1 and 2.4.2). Consequently, if a frame is subjected to a larger P-δ effect than is presumed in Eq. (2.27a), then $P_{e\hat{\tau}(K_n)}$ will exhibit an unconservative error. This is shown in Fig. 2.15, particularly when uncorrected G factors are used. The total story buckling capacity predicted by Eq. (2.27a) remains constant regardless of the distribution of loading on the story. Note that when corrected G factors are used, this predicted capacity equals the capacity of the frame predicted by a system buckling analysis at the distribution of loading which causes the stiffness parameter $L\sqrt{P_u/EI}$ to be equal for both columns. However, once the weaker column on the right becomes heavily loaded, the capacity computed using Eq. (2.27a) becomes unconservative. In Fig. 2.16, the limit of Eq. (2.72) is imposed and the capacity of the frame recomputed according to the approach outlined in Appendix C. While this limit insures that any braced buckling of the weak right column is captured conservatively (as is evident in the figure), it does not insure that the capacity of the story is predicted conservatively prior to the potential braced buckling behavior. Even when corrected G factors are used, the unconservative error approaches 10%.

To restrict this unconservative error for this approach, LeMessurier changed the limit of Eq. (2.72) to:

$$P_{e\hat{\tau}(K_n)} \le 1.6 P_{e\hat{\tau}(nomo)} \tag{2.73}$$

Fig. 2.16 also shows the effect of imposing this more restrictive limit: if (and only if) corrected G factors are used, $\sum_{non-leaner} P_{e\hat{\tau}(K_n)}$ now consistently provides a conservative estimate of the frame capacity. This is Eq. (C-C2-3b) of the LRFD Commentary [AISC, 1993]. The Commentary also expresses this limit in terms of the effective length factor K_{K_n} of Eq. (2.27b):

$$K_{K_n} \ge \sqrt{\frac{1}{1.6}} K_n = \sqrt{\frac{5}{8}} K_n \tag{2.74}$$

This is Eq. (C-C2-4b) of the LRFD Commentary Chapter C [AISC, 1993].

It has been shown through rigorous testing [LeMessurier, 1993; Squarzini, 1993; LeMessurier, 199?] that the approach to computing column buckling outlined in Section 2.4.2, along with the corresponding limit established in Eq. (2.67), almost always insures an unconservative error of less than 5%. Because the value of the P-δ effect (i.e., the 0.85 factor)

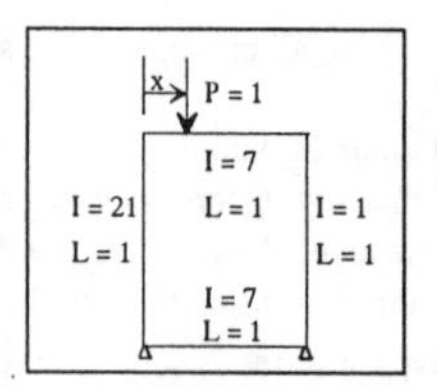

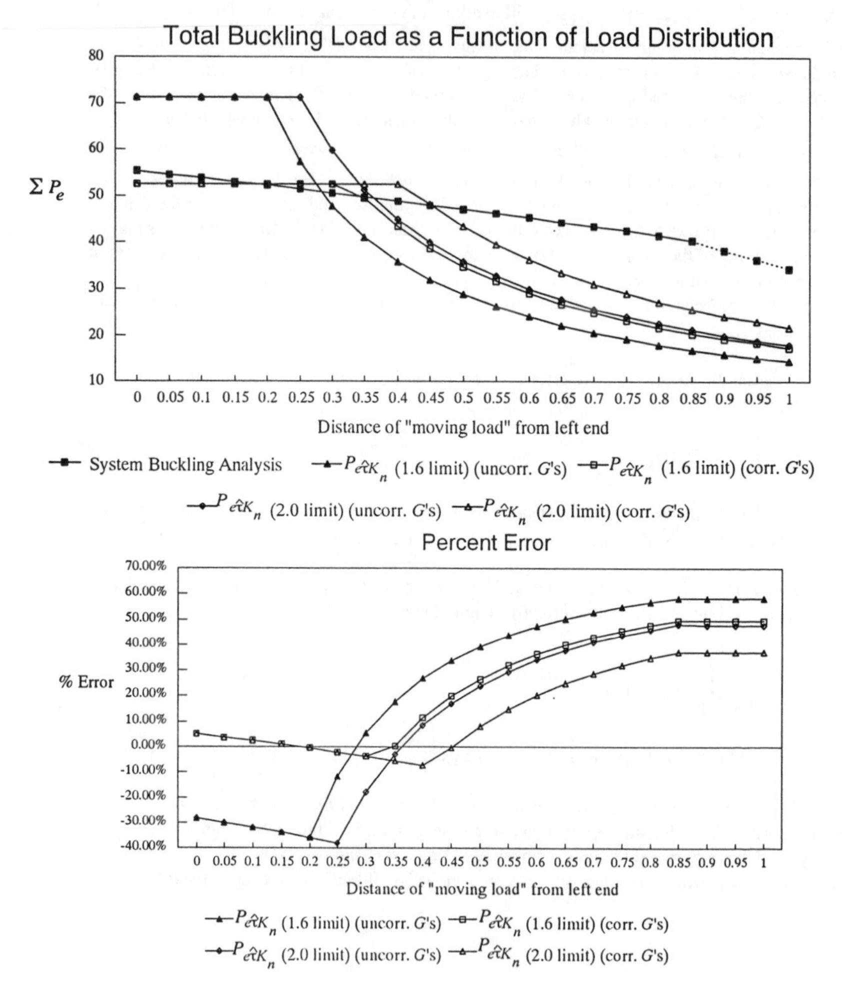

Figure 2.16 Comparison of Limits Imposed on Story-Based Buckling Load Computation

which is inherently assumed in the derivation of $P_{e\hat{\tau}(R_L)}$ in Eq. (2.30a) is sufficiently conservative, no further reduction in the limit of Eq. (2.67) is required to insure accurate results for this method (the accuracy of $\sum_{non-leaner} P_{e\hat{\tau}(R_L)}$ is evident in Fig. 2.15 and in the studies presented in Section 2.4.6).

2.4.4.3 Derivation of Imposed Limits on Effective Length: Two-Bay, One-Story Frame

The justification for the limit of Eqs. (2.67) and (2.73) insuring conservative prediction of the buckling load of a story was outlined in Sections 2.4.4.2 only for two-column systems. However, this limit is sufficiently comprehensive to insure a conservative prediction of buckling, including sidesway inhibited buckling, for stories having more than two non-leaning columns.

Consider the limit of Eq. (2.67). The philosophy behind the derivation of this limit presumes that no braced buckling in a column (referred to here as column A) occurs if the combination of stiffnesses of the other non-leaning columns in its story is less than or equal to the stiffness of column A, regardless of the distribution of axial force in the story's columns. This is a reasonable philosophy since, if all of the other non-leaning columns in story have a stiffness that sums only to the stiffness of column A, then the stiffnesses of these other columns must each be weaker than that of column A. While these weaker columns may breach the limit of Eq. (2.67), column A is in fact the stronger column in the story and will thus buckle in a sidesway mode. If, on the other hand, the stiffnesses of the other columns in the story have a combined stiffness that is at all greater than that of column A, then the limit for column A from Eq. (2.67) will be breached if column A is required to resist more than twice its own buckling load. Thus, the limit of Eq. (2.67) must be conservative for any multi-bay framing system. A similar argument may be made for Eq. (2.72), and, by extension, for Eq. (2.73), which remains conservative compared to Eq. (2.72).

To investigate multi-column stories further, the frame of Fig. 2.17 is used. When the right column's moment of inertia equals that of the left column (i.e., when $\gamma = 1$), Fig. 2.17 is symmetric, with respect to both loading and geometry. Analogous to the frame of Fig. 2.13a, no member should buckle in a braced mode under this loading configuration. However, if γ is increased to 1.05, the frame becomes unsymmetric and a weak left column exists. Using the capacity limit of Eq. (2.67), the left column exceeds its limit when it carries 49.5% of the load, the middle exceeds when it carries 99.6% of the load, and the right column exceeds its limit when it carries 50.9% of the total load. Consequently, the limit is reached on the left column when the gravity load is concentrated on the two outside columns, and the limit is reached on the middle column when the gravity load is concentrated at the center of the frame. LeMessurier [199?] and Squarzini [1993] provide examples showing that these approaches for computing story capacity, coupled with appropriate limits, model these multi-bay structures well. Chapter 6 provides further examples of multi-bay structures.

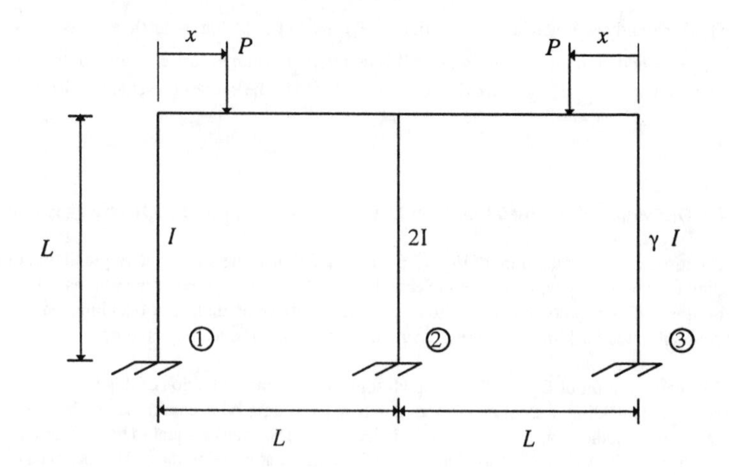

Figure 2.17 Two-Bay, One-Story Frame with Axial Load Distributed Between Three Rigidly-Connected Columns

2.4.5 Practical Design Considerations in the Use of Story-Based Effective Length Factors

This section illustrates the step-by-step computation of K_{K_n} and K_{R_L}. The frame of Fig. 1.6 is used as an example (in particular, the structure having a girder framing into both the top and bottom of the rigidly-connected column is used). Actual loading, wide-flange section sizes and member lengths are identified below for clarity (the member sizes and lengths are chosen to approximate closely the parameters of the problem specified in Section 1.4). As explained in Section 1.4, this frame elucidates the potential benefits of assessing stability using a story-based procedure. As will be seen below, any elastic or inelastic effective length factor computed using the nomograph, without accounting properly for story or system stability, yields an interaction equation value less than 1.0, while the more accurate and appropriate procedures for assessing the frame's stability show the rigidly-connected column to be clearly *overstressed.*

The procedure to compute $P_{e\hat{\tau}(K_n)}$ (or K_{K_n}), is outlined first, followed by the computations for $P_{e\hat{\tau}(R_L)}$ (or K_{R_L}). It is presumed that the loads applied to the frame as shown are factored, and are thus referred to here as P_u and H_u. The following loading is used for these calculations:

$$P_u = 430 \text{ kips}$$
$$H_u = 9.2 \text{ kips}$$

The following assumptions are made in the analysis and design of this frame [Kanchanalai, 1977]:

- All members are assumed to be compact and therefore not susceptible to local buckling.
- Strong axis bending of all members is assumed.
- The out-of-plane (weak) axis of the members is braced and does not control the calculation of P_n
- Shear deformation effects are ignored in the analytical computations.
- All beam-to-column connections are either perfectly rigid or perfectly pinned.
- The column L/r equals 40.
- A W14x90 wide-flange section is used for the rigidly-connected column; therefore: $A_g = 26.5 \text{ in}^2$, $I_x = 999 \text{ in}^4$, $Z_x = 157 \text{ in}^3$, $r_x = 6.14''$ and $L_c = 20.46' = 245.5''$ (the column size deviates from the W8x31 used in Chapter 1 so as to achieve more practical dimensions; the length is selected to achieve the appropriate L/r).
- A W18x106 wide-flange section is used for the girder; therefore: $I_x = 1910 \text{ in}^4$ and $L_g = 39.11' = 469.4''$ (the length is selected to achieve the appropriate G factors).
- The residual stress pattern of this column is assumed to vary linearly through the flange width, with a peak compressive residual stress of $0.3F_y$, while it retains a constant residual tension in the web.
- Uncorrected G factors: $G_A = 1$; $G_B = 1$
- G factors corrected to account for the pin at the far end of each girder: $G_A = 2$; $G_B = 2$. Only G factors corrected by use of Eq. (2.11) will be used in the computations below.
- $F_y = 36$ ksi.

- $E = 29000$ ksi.

As the W14x90 is sufficiently braced out-of-plane:

$$M_{nx} = M_{px} = Z_x F_y = (157)(36) = 5652 \text{ kip-feet.}$$

The required second-order elastic bending moment may be found by amplifying the first-order elastic bending moment by the appropriate amplification factors. The AISC LRFD B_1 / B_2 approach [AISC, 1993] will be used here. For this simple frame, the column experiences negligible bending moment due to gravity load, and only a B_2 factor is required (AISC LRFD Eq. C1-4 [AISC, 1993] is used here). As explained below (step 2 in Section 2.4.5.2), the first-order elastic analysis required to obtain Δ_{oh} uses an applied lateral load at each story having a magnitude equal to the total gravity load applied to the story (note that it is only the relative magnitude of H compared to Δ_{oh} that is important; the actual magnitude of Δ_{oh} is not relevant). This analysis yields:

$$\sum_{non-leaner} \frac{HL}{\Delta_{oh}} = \frac{(860)(245.5)}{109.9} = 1922 \text{ kips} \tag{2.75}$$

$$B_2 = \frac{1}{1 - \dfrac{\sum_{all} P_u}{\sum_{non-leaner} HL / \Delta_{oh}}} = \frac{1}{1 - \dfrac{860}{1922}} = 1.81 \tag{2.76}$$

The maximum first-order elastic bending moment in the rigidly-connected column equals:

$$M_x = H_u L_c / 2 = (9.2)(245.5)/2 = 1130 \text{ kips inches}$$

Therefore, it's second-order elastic required moment is:

$$M_{ux} = (1.81)(1130) = 2045 \text{ kip-inches}$$

Note that this slender frame has a substantial, but not unrealistic [Springfield, 1991], moment amplification.

To illustrate the full potential of accounting for column inelasticity when assessing column stability, inelastic stiffness reduction is accounted for in this example through the use of an iteratively-computed τ factor (Section 2.2). However, the procedure below is described presuming the use of either an iteratively-computed τ factor, or a more practical, non-iterative inelastic reduction factor, $\hat{\tau}$. Chapter 6 presents examples in which a non-iterative $\hat{\tau}$ factor is used to account for inelasticity in the columns.

Section 1.4 indicates that the rigidly-connected column's "exact" inelastic effective length factor computed using a system buckling analysis (see Section 2.3.3) equals 1.95. For the W14x90, this yields a value of P_n equal to 692.7 kips, and, correspondingly an iteratively-computed τ factor equal to 0.635 from Eq. (2.7) (of course, normally one must iterate as described below to compute τ, since a system buckling effective length factor would not be known a priori). Consequently, the corrected inelastic G factors at the top and bottom of the column equal $0.635 * 2.0 = 1.27$, which yields an inelastic effective length factor computed from the sidesway inhibited nomograph of $K_n = 1.40$. Recall also from Section 1.4 that the elastic effective length factor using uncorrected G factors equals 1.32, and the elastic effective length factor using corrected G factors equals 1.59. These, in turn, yield the following values for the interaction equation (also shown are the results of using the elastic and inelastic effective length factors computed using a system buckling analysis):

Table 2.2 Interaction Equation Values for Nomograph and System Effective Length Calculation Procedures

Effective Length Procedure	Effective Length Factor	P_n (kips)	Interaction Equation Value
Nomograph, elastic, uncorrected G factors	1.32	823.8	0.971
Nomograph, inelastic, corrected G factors	1.40	808.9	0.982
Nomograph, elastic, corrected G factors	1.59	771.1	1.01
System buckling analysis, inelastic	1.95	692.7	1.08
System buckling analysis, elastic	2.24	626.5	1.16

One can see the substantial difference between using an effective length factor computed using an accurate system buckling procedure, versus that computed using the AISC nomograph, which has no means of accounting for the destabilizing load on the leaning column in this frame. The story-based procedures presented below provide an effective compromise between accuracy and efficiency of computation.

2.4.5.1 Design Procedure Using K_{K_n}

1. *If iterative inelastic effective length factors are to be calculated, use an initial estimate of P_n for each non-leaning column in the story, including an estimate of an effective length factor, to compute an initial value of τ from Eq. (2.7). After an initial value of P_n is obtained in Step 6, return to this step to recompute the inelastic stiffness reduction factors as described in Sections 2.1 and 2.2. If non-iterative $\hat{\tau}$ factors are to be used, conduct a*

first-order elastic analysis to determine a force distribution in the structure to provide a value for P_u/ϕ_c *for use in Eq. (2.7).*

The force distribution for this non-iterative calculation is often best obtained from an appropriate factored gravity load combination of the frame. For the frame of Fig. 1.6, the force distribution that would be used for the computation of $\hat{\tau}$ for the rigidly-connected column equals the factored gravity load, $P_u = 430$ kips, on the rigidly-connected column.

For this example, as stated above, it was determined iteratively that $\tau = \hat{\tau} = 0.635$. If an elastic K_{K_n} is to be used, $\tau = \hat{\tau} = 1.0$ in Eq. (2.25).

2. *Compute either elastic or inelastic G factors for every non-leaning column in the frame using Eq. (2.10). If the frame is so irregular that the assumptions embedded in the derivation of* K_{K_n} *are inappropriate, modify the G factors using Eq. (2.11).*

   If corrected G factors are used, it is necessary to determine appropriate bending moments for use in Eq. (2.11). To generate these bending moments, an appropriate first-order sidesway analysis must be run using the approach described in Section 2.3.1. If inelastic effective length factors are being used, multiply each non-leaning column's moment of inertia (in the plane of bending) and, if desired, cross sectional area by its $\hat{\tau}$ factor during the analysis.

   As explained in Section 2.3.1, for this analysis, the applied lateral load at each story should have a magnitude equal to the total gravity load *applied* to the story (not the axial force in the columns themselves, due to all stories above, but the load actually applied to the structure solely at that story). While this is only an issue for a multi-story frame, for the frame in Fig. 1.6 this lateral load would equal $2P_u$.

3. *For each non-leaning column, compute* K_n *from the sidesway uninhibited nomograph.*

   Given $G_A = 1.27$ and $G_B = 1.27$, $K_n = 1.40$ for the rigidly-connected column.

4. *Compute the two summations of Eq. (2.25). All columns contribute to the calculation of* $\sum_{all} P_u$, *while only the non-leaning columns contribute to the calculation of* $\sum_{non-leaner} \frac{\hat{\tau} I}{K_n^2}$.

   $$\sum_{all} P_u = 2P_u = 860 \text{ kips}$$

   $$I = I_x = 999 \text{ in}^4$$

$$\sum_{non-leaner} \frac{\hat{\tau} I}{K_n^2} = \frac{(0.635)(999)}{(1.40)^2} = 323.7$$

5. *Compute K_{K_n} for each non-leaning column using Eq. (2.25b).*

$$K_{K_n} = \sqrt{\frac{\pi^2 \hat{\tau} EI}{P_u L^2} \frac{\sum_{all} P_u}{\sum_{non-leaner} \frac{\pi^2 \hat{\tau} EI}{(K_n L)^2}}} = \sqrt{\frac{\hat{\tau} I}{P_u} \frac{\sum_{all} P_u}{\sum_{non-leaner} \frac{\hat{\tau} I}{K_n^2}}} = \sqrt{\frac{(0.635)(999)}{(430)} \frac{(860)}{(323.7)}} = 1.98$$

Alternately, one may compute $P_{e\hat{\tau}(K_n)}$ from Eq. (2.25a):

$$P_{e\hat{\tau}(K_n)} = \frac{\sum_{non-leaner} \frac{\pi^2 \hat{\tau} EI}{(K_n L)^2}}{\sum_{all} P_u} (P_u) = \frac{\frac{\pi^2 (0.635)(29000)(999)}{(1.40)^2 (245.5)^2}}{860} (430) = 768.4 \text{ kips}$$

and

$$P_{e\hat{\tau}(K_n)} = 768.4 \quad < \quad 1.6 \frac{\pi^2 \hat{\tau} EI}{(K_n L)^2} = 1.6 \frac{\pi^2 (0.635)(29000)(999)}{(1.40)^2 (245.5)^2} = 2459 \text{ kips}$$

so the limit is not breached.

6. *Use this effective length factor (or, alternately, $P_{e\hat{\tau}(K_n)}$) to compute P_n for use in the first term of the LRFD interaction equation (see Section 2.1). Leaning columns may be designed with an effective length factor of 1.0. If iteratively-computed inelastic effective lengths are to be used, return to step 1 to recompute τ based upon the value of P_n computed in this step. Iterate on P_n using steps 1 through 6 until convergence is reached.*

$$\lambda_c = \frac{K_{K_n} L_c}{\pi r_x} \sqrt{\frac{F_y}{E}} = \frac{(1.98)(245.5)}{\pi (6.14)} \sqrt{\frac{36}{29000}} = 0.888 \quad < \quad 1.5$$

or, alternately:

$$A_g = 26.5 \text{ in}^2$$

$$P_y = A_g F_y = (26.5)(36) = 954 \text{ kips}$$

$$\lambda_c = \sqrt{\frac{\hat{\tau} P_y}{P_{e\hat{\tau}(K_n)}}} = \sqrt{\frac{(0.635)(954)}{768.4}} = 0.888 \; < \; 1.5$$

As this value is less than 1.5, the column would fail by inelastic buckling. The nominal axial strength of the column by Eq. (2.1) is thus:

$$P_n = 0.658^{\lambda_c^2} P_y = 0.658^{(0.888)^2}(954) = 685.8 \text{ kips}$$

7. *Compute the value of the LRFD interaction equation.*

$$\frac{P_u}{\phi_c P_n} = \frac{430}{(0.85)(685.8)} = 0.738 > 0.2$$

and AISC LRFD Eq. H1-1a (Eq. 1.1a) applies:

$$\frac{P_u}{\phi_c P_n} + \frac{8}{9}\frac{M_{ux}}{\phi_b M_{nx}} = 0.738 + \frac{8}{9}\frac{(2045)}{(0.9)(5652)} = 1.09 > 1.0 \qquad \text{No Good}$$

From Table 2.2, one can see that this procedure produces an inelastic story-based effective length factor almost identical to that yielded by an inelastic system buckling analysis. Similarly, if K_{K_n} is based on an elastic nomograph effective length factor ($K_n = 1.59$), using a $\hat{\tau} = 1$, then $K_{K_n} = 2.25$, and the interaction equation yields 1.17. From Table 2.2, one can see that this is close to the value of 1.16 achieved using the elastic system buckling effective length factor.

2.4.5.2 Design Procedure Using K_{R_L}

1. *If iterative inelastic effective length factors are to be calculated, use an initial estimate of P_n for each non-leaning column in the story, including an estimate of an effective length factor, to compute an initial value of τ from Eq. (2.7). After an initial value of P_n is obtained in Step 5, return to this step to recompute the inelastic stiffness reduction factors as described in Sections 2.1 and 2.2. If non-iterative $\hat{\tau}$ factors are to be used, conduct a first-order elastic analysis to determine a force distribution in the structure to provide a value for P_u/ϕ_c for use in Eq. (2.7).*

The force distribution for this non-iterative calculation is often best obtained from an appropriate factored gravity load combination of the frame. For the frame of Fig. 1.6, the force distribution that would be used for the computation of $\hat{\tau}$ for the rigidly-connected column equals the factored gravity load, $P_u = 430$ kips, on the rigidly-connected column.

For this example, as stated above, it was determined iteratively that $\tau = \hat{\tau} = 0.635$. If an elastic K_{R_L} is to be used, $\tau = \hat{\tau} = 1.0$ in Eq. (2.28).

2. *Run an appropriate first-order sidesway analysis to determine Δ_{oh}, as well as H for each non-leaning column in the story.*

   The approach described in Section 2.3.1 to compute corrected *G* factors should be used to run this analysis. If inelastic effective length factors are being used, multiply each non-leaning column's moment of inertia (in the plane of bending) and, if desired, cross sectional area by its $\hat{\tau}$ factor during the analysis. As explained in Section 2.3.1, for this analysis, the applied lateral load at each story should have a magnitude equal to the total gravity load *applied* to the story (not the axial force in the columns themselves, due to all stories above, but the load actually applied to the structure solely at that story). While this is only an issue for a multi-story frame, for the frame in Fig. 1.6 this lateral load would equal $2P_u$.

   For this frame, this analysis, using a moment of inertia for the rigidly-connected column equal to $\hat{\tau}I = (0.635)(999) = 634$ in^4, yields $\Delta_{oh}^{inelastic} = 130.9$ inches based upon a horizontal load of $2P_u = 860$ kips.

3. *Compute the two summations of Eq. (2.28). All columns contribute to the calculation of $\sum_{all} P_u$, while only the non-leaning columns contribute to the calculation of $\sum_{non-leaner} HL \Big/ \Delta_{oh}^{inelastic}$. For this last summation, H represents the shear in the non-leaning columns due to the lateral load (i.e., from the analysis in step 2).*

$$\sum_{all} P_u = 2P_u = 860 \text{ kips}$$

$$\sum_{non-leaner} \frac{HL}{\Delta_{oh}^{inelastic}} = \frac{(860)(245.5)}{130.9} = 1613 \text{ kips}$$

4. *Compute K_{R_L} for each non-leaning column using Eq. (2.28b).*

$$R_L = \frac{\sum_{leaner} P_u}{\sum_{all} P_u} = \frac{430}{860} = 0.50$$

$$K_{R_L} = \sqrt{\frac{\hat{\tau} I}{P_u} \frac{\pi^2 E}{L^2} \frac{\sum\limits_{all} P_u}{\sum\limits_{non-leaner} \frac{HL}{\Delta_{oh}^{inelastic}} (0.85 + 0.15 R_L)}}$$

$$= \sqrt{\frac{(0.635)(999)}{(430)} \frac{\pi^2 (29000)}{(245.5)^2} \frac{860}{1613[0.85 + (0.15)(0.50)]}} = 2.01$$

Alternately, one may compute $P_{e\hat{\tau}(R_L)}$ from Eq. (2.28a):

$$P_{e\hat{\tau}(R_L)} = \frac{\sum\limits_{non-leaner} \frac{HL}{\Delta_{oh}^{inelastic}} (0.85 + 0.15 R_L)}{\sum\limits_{all} P_u} (P_u) = \frac{1613[0.85 + (0.15)(0.5)]}{860} (430) = 746.0$$

and

$$P_{e\hat{\tau}(R_L)} = 746.0 \quad < \quad 1.7 \frac{HL}{\Delta_{oh}^{inelastic}} = 1.7(1613) = 2742 \text{ kips}$$

so the limit is not breached.

5. *Use this effective length factor (or, alternately, $P_{e\hat{\tau}(R_L)}$) to compute P_n for use in the first term of the LRFD interaction equation (see Section 2.1). Leaning columns may be designed with an effective length factor of 1.0. If inelastic effective lengths are to be used, return to step 1 to recompute τ based upon the value of P_n computed in this step. Iterate on P_n using steps 1 through 5 until convergence is reached.*

$$\lambda_c = \frac{K_{K_n} L_c}{\pi r_x} \sqrt{\frac{F_y}{E}} = \frac{(2.01)(245.5)}{\pi (6.14)} \sqrt{\frac{36}{29000}} = 0.901 \quad < \quad 1.5$$

or, alternately:

$$A_g = 26.5 \text{ in}^2$$

$$P_y = A_g F_y = (26.5)(36) = 954 \text{ kips}$$

$$\lambda_c = \sqrt{\frac{\hat{\tau} P_y}{P_{e\hat{\tau}(R_L)}}} = \sqrt{\frac{(0.635)(954)}{746.0}} = 0.901 \; < \; 1.5$$

As this value is less than 1.5, the column would fail by inelastic buckling. The nominal axial strength of the column by Eq. (2.1) is thus:

$$P_n = 0.658^{\lambda_c^2} P_y = 0.658^{(0.901)^2}(954) = 679.2 \text{ kips}$$

6. *Compute the value of the LRFD interaction equation.*

$$\frac{P_u}{\phi_c P_n} = \frac{430}{(0.85)(679.2)} = 0.745 > 0.2$$

and AISC LRFD Eq. H1-1a (Eq. 1.1a) applies:

$$\frac{P_u}{\phi_c P_n} + \frac{8}{9}\frac{M_{ux}}{\phi_b M_{nx}} = 0.745 + \frac{8}{9}\frac{(2045)}{(0.9)(5652)} = 1.10 > 1.0 \qquad \text{No Good}$$

From Table 2.2, one can see that this procedure produces an inelastic story-based effective length factor almost identical to that yielded by an inelastic system buckling analysis. Similarly, if K_{R_L} is computed using a $\hat{\tau} = 1$, then $K_{R_L} = 2.31$, and the interaction equation yields 1.18. From Table 2.2, one can see that this is close to the value of 1.16 achieved using the elastic system buckling effective length factor.

As is expected, K_{R_L} in general is slightly more conservative than K_{K_n}. In addition, both the story-based procedures and the system buckling procedures show the rigidly-connected column to be clearly overstressed, while the nomograph effective length factors show the column to be satisfactory.

2.4.6 Examples: Portal Bent Studies of Effective Length Approaches K_{K_n} and K_{R_L}

Squarzini and Hajjar [1993] studied a series of portal bents to elucidate the basic characteristics of the two procedures for computing story-based effective length, K_{K_n} and K_{R_L}, which are included in the LRFD Commentary Chapter C [AISC, 1993]. The frames are shown in Fig. 2.18, and include:

- One-bay, one-story unsymmetric portal bents.
- One-bay, one-story portal bents having columns of varying length
- One-bay, one-story portal bents which including a leaning column.
- Two-bay, one-story frames, such as the one shown in Fig. 2.16.

To create a comprehensive suite of test frames in these studies, the left column's and the girders' moments of inertia were varied methodically between 1 and 35 times the moment of inertia of the right column. The cross sectional area of these frames was taken as an arbitrarily large value so as to neglect the effects of axial deformation in these studies. The primary conclusions and a sampling of the results for the first set of rigidly-connected portal bents (Frames 1 through 6 in Fig. 2.18, constituting forty-one frames in total) were presented in Section 2.4.4.

2.4.6.1 Studies of Portal Bents with Columns of Unequal Length

Frames 7 through 9 in Fig. 2.18 include bents having columns of unequal lengths. In Appendix B, it is shown that the second-order deflection of these frames may be expressed by:

$$\Delta_{ph} = \frac{H}{P_L/L} + \frac{P_u \Delta_{ph}/L}{P_L/L} + \frac{P_u \Delta_{ph}/L}{P_L/L} C_L \tag{2.77}$$

The second term of this equation indicates that the P-Δ effect decreases as a greater proportion of the total load is shifted towards the longer column. A decreasing P-Δ effect in turn results in a decreasing P-δ effect, as seen in the third term of the equation. This in turn implies that total capacity of a frame such as that shown in Fig. 2.19 will increase as more load is shifted to the longer column.

Figure 2.19 illustrates a frame with different length columns having identical moments of inertia. As the load translates from the shorter left column towards the longer right column, the total frame capacity, as predicted by a system buckling analysis, increases. The story-based capacities approximate this trend with great accuracy.

Of course, the limit on an individual member's buckling load, Eq. (2.68) or Eq. (2.74), must still be imposed on a story having columns of unequal length. One may see in Fig. 2.19 that the story-based approach to computing buckling load detects braced buckling of the right column

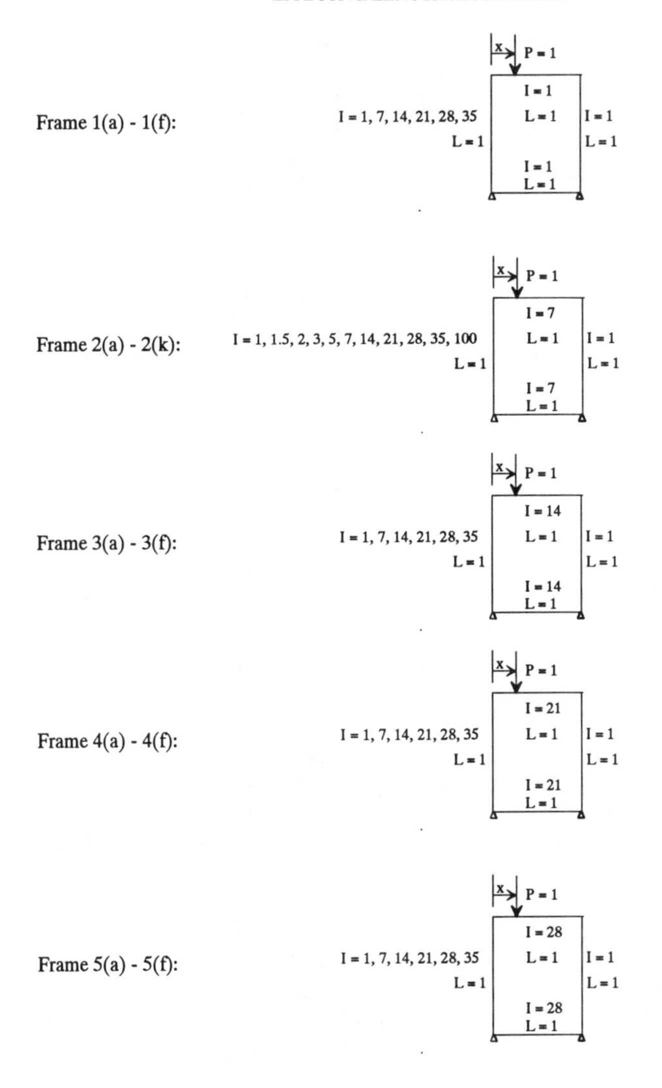

Figure 2.18 Frames Used to Study Characteristics of Effective Length Procedures

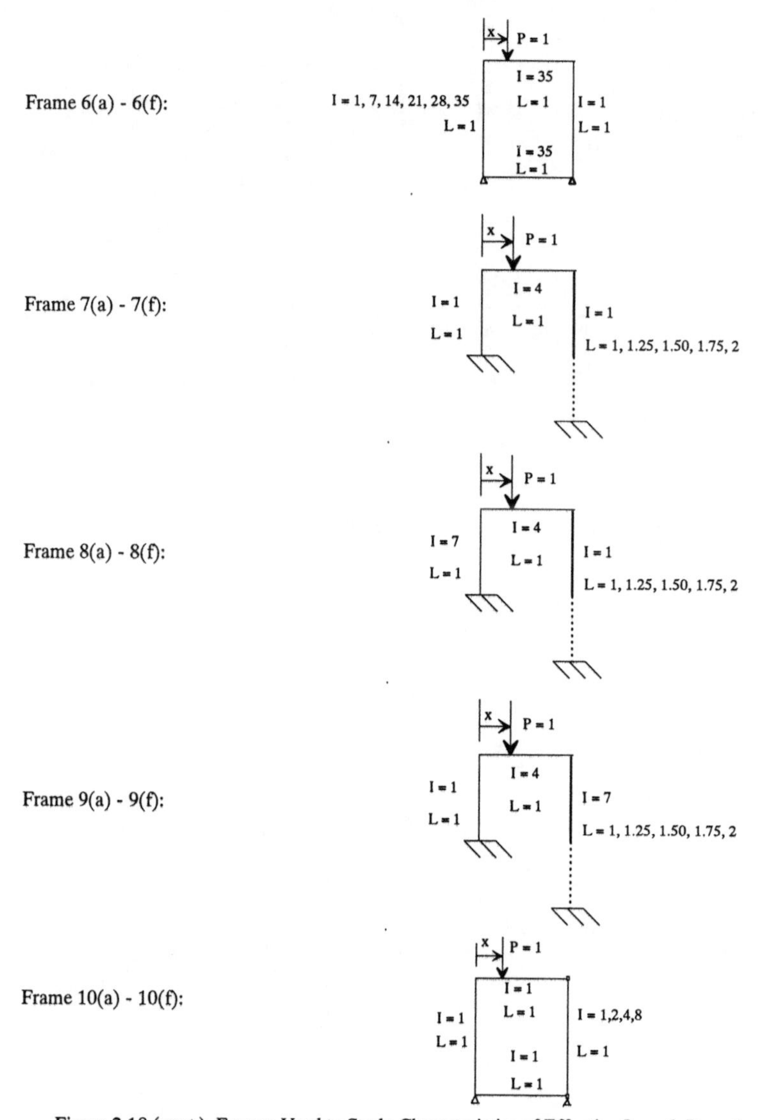

Figure 2.18 (cont.) Frames Used to Study Characteristics of Effective Length Procedures

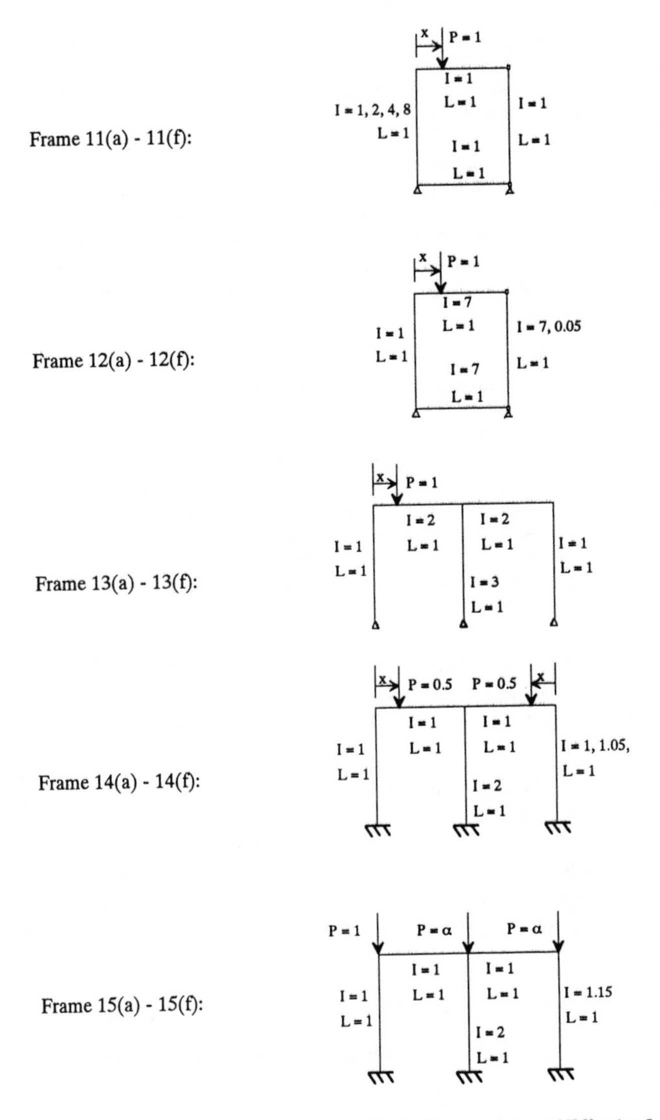

Figure 2.18 (cont.) Frames Used to Study Characteristics of Effective Length Procedures

sooner than the system buckling solution, as might be expected from previous discussions. As the right column becomes increasingly braced by the left column (i.e., as the left column inertia increases), the imposed limits detect braced buckling of the right column with improved accuracy. This trend is indicated by the frame in Fig. 2.20.

Figures 2.19 and 2.20 illustrate a redistribution of loading that increasingly loads the weaker column. On the other hand, if the right column is given an inertia of seven, as in Fig. 2.21, the limits of Eqs. (2.68) and (2.74) detect the left column as potentially buckling in a braced mode when most or all of the applied load is on that column (i.e., x is near 0). However, as the length of the right column is increased (Frame 9 in Fig. 2.18), that column becomes the weaker of the two, and the limits of Eqs. (2.68) or (2.74) are breached as the right column is increasingly loaded (Fig. 2.22).

2.4.6.2 Studies of Portal Bents Incorporating Leaning Columns

Frames 10 through 12 of Fig. 2.18 were studied by Squarzini and Hajjar [1993] to investigate further the accuracy of story-based effective length factors when leaning columns are included in the story. Fig. 2.23 illustrates a typical frame from this set of studies, designating the right column as a pin-ended, leaning column. For these studies, both the left and right column moments of inertia are progressively increased (Frames 10 and 11 in Fig. 2.18). Additionally, frames with either a strong or weak leaning column, and braced by relatively strong girders, were also analyzed (Frame 12, Fig. 2.18). The stronger restraining girders were analyzed to illustrate their relation to a frame's P-δ effect.

For these frames, $P_{e\hat{t}(K_n)}$ and $P_{e\hat{t}(R_L)}$ equal zero for the leaning column, since that column does not contribute to the sidesway resistance of the frame. This is illustrated in Fig. 2.23, which indicates that capacities computed from both the system buckling calculation and from the story-based procedures do not change as the stiffness of the leaning column changes (so long as the leaning column is sufficiently strong to resist buckling in a braced mode). This lends credence to the common design practice in which a leaning column is designed with an effective length factor equal to one (assuming ideal pins at its top and bottom).

Since $P_{e\hat{t}(R_L)}$ assumes a constant, relatively conservative value for the P-δ effect for all non-leaning columns, this approach usually yields conservative results when the load is placed entirely on the non-leaning column. Fig. 2.23 confirms this trend. Alternately, $\sum_{non-leaner} P_{e\hat{t}(R_L)}$ exactly matches the result from the system buckling analysis when the load is placed entirely on the leaning column. The figure also reaffirms that, for a frame of this sort, $\sum_{non-leaner} P_{e\hat{t}(R_L)}$ is not constant as the load is redistributed. This behavior of $P_{e\hat{t}(R_L)}$ becomes more pronounced if the girder sizes are increased [Squarzini and Hajjar, 1993].

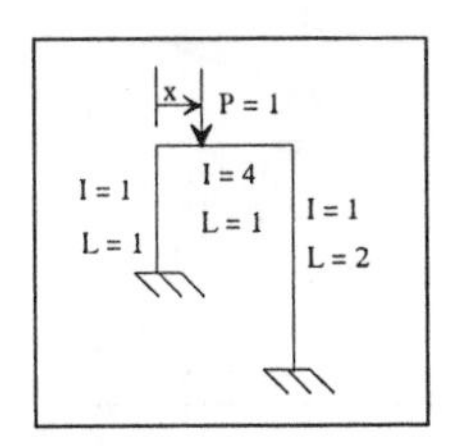

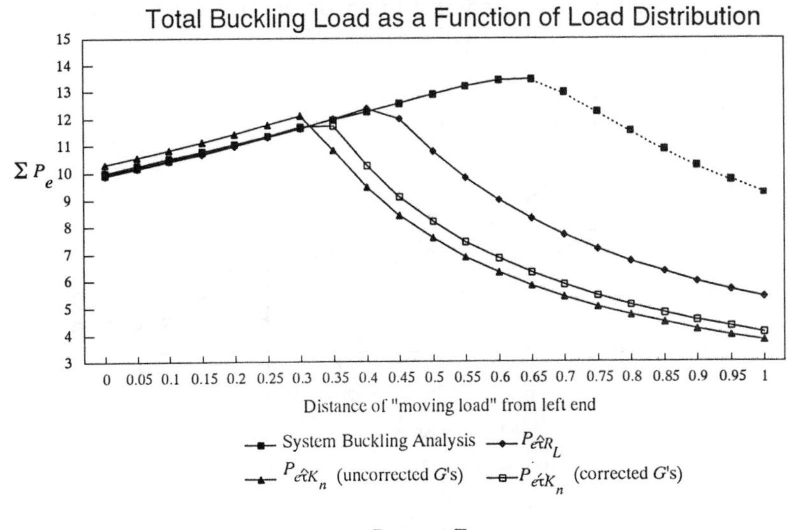

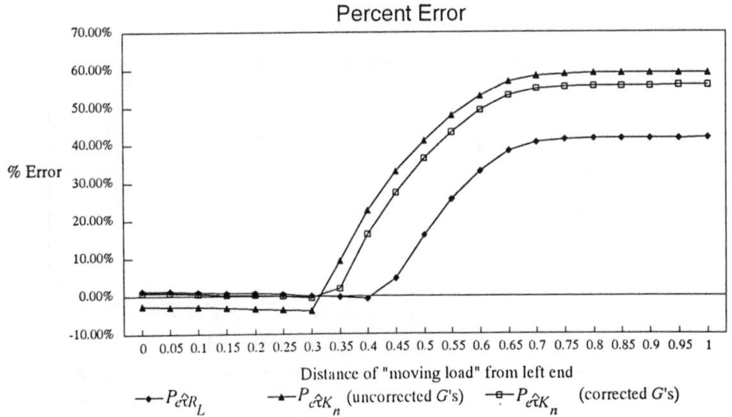

Figure 2.19 Buckling Load of a Story with Columns of Unequal Length

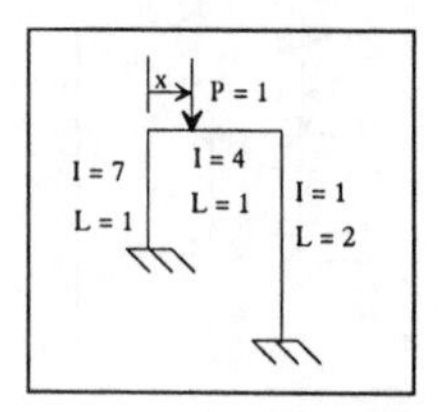

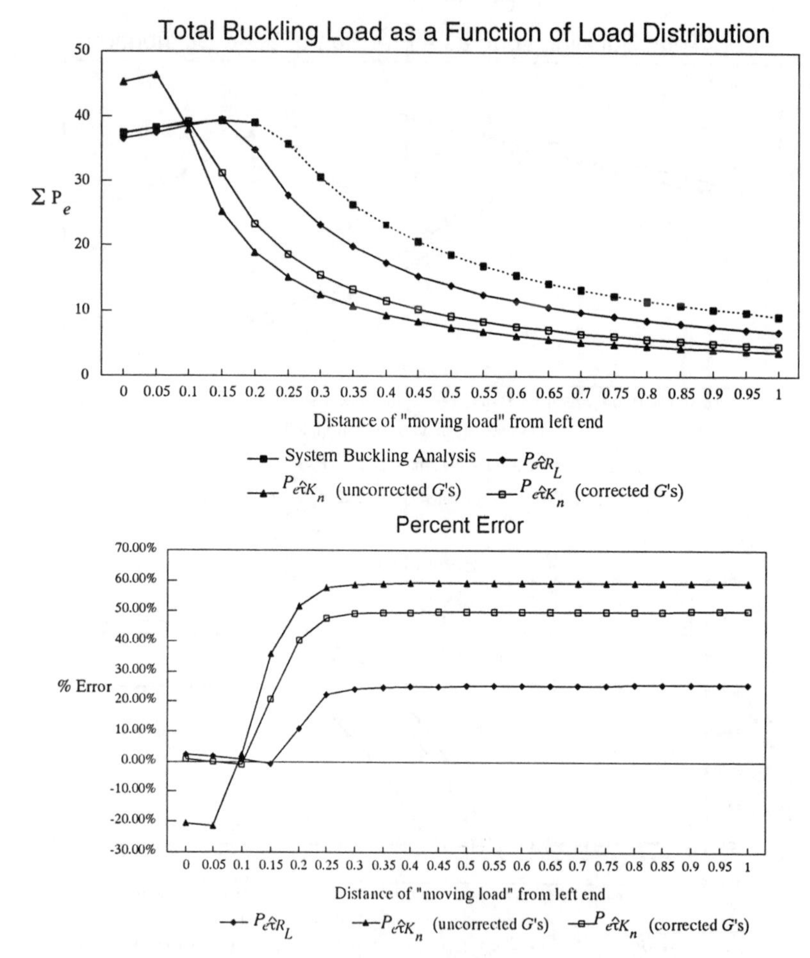

Figure 2.20 Buckling Load of a Story with Columns of Unequal Length and a Weak Long Column

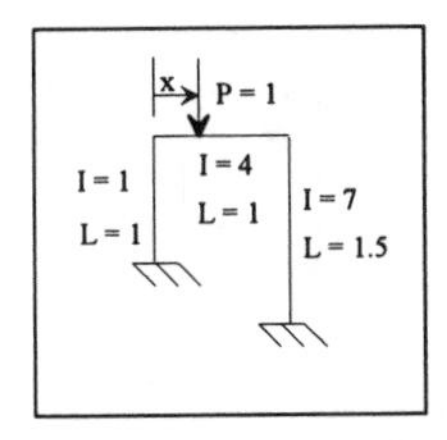

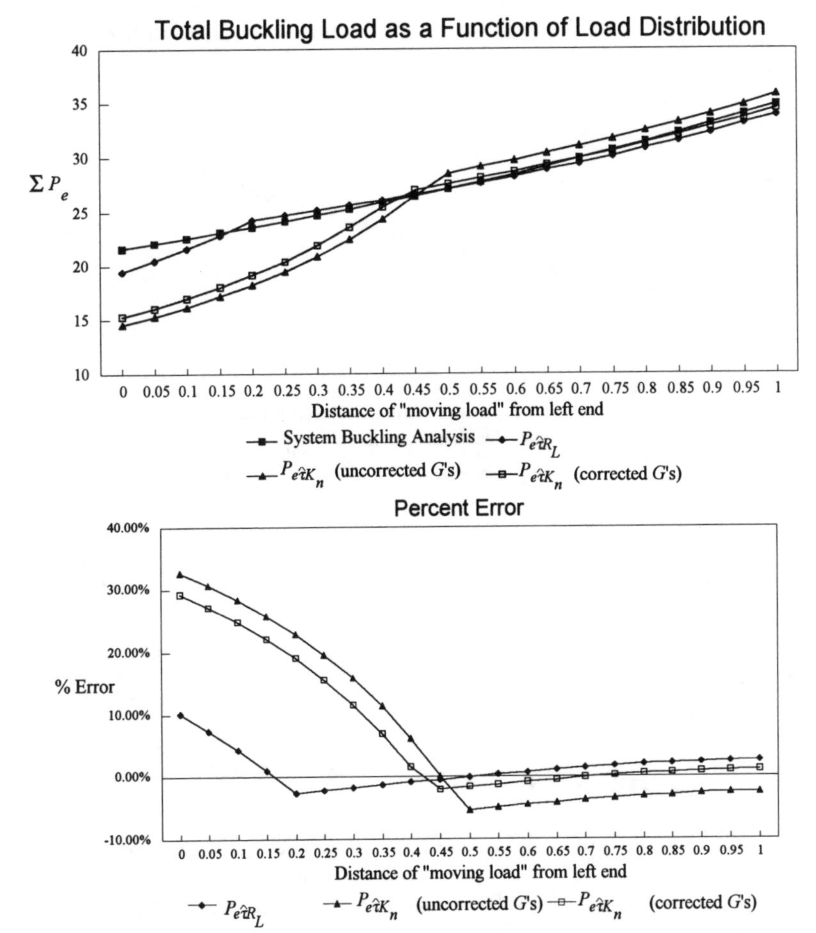

Figure 2.21 Buckling Load of a Story with Columns of Unequal Length and a Strong Long Column

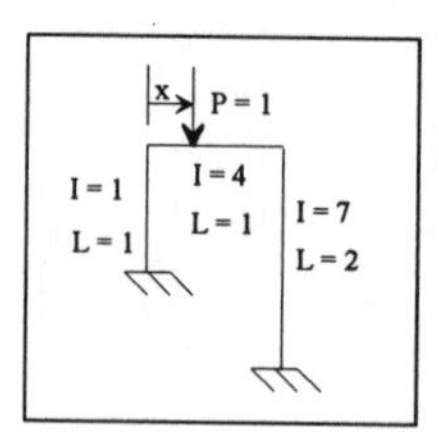

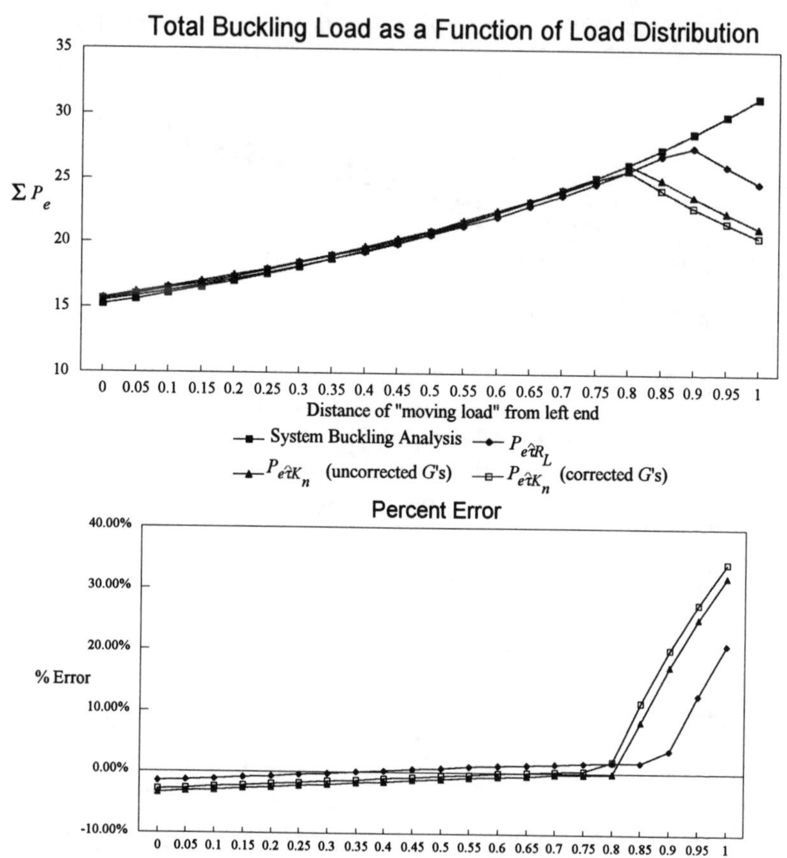

Figure 2.22 Buckling Load of a Story with Columns of Unequal Length and a Weak Long Column

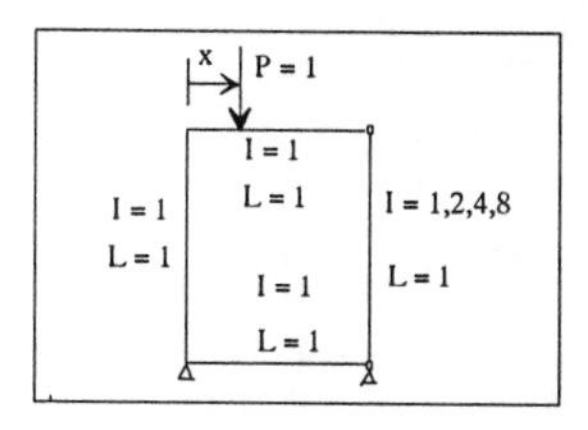

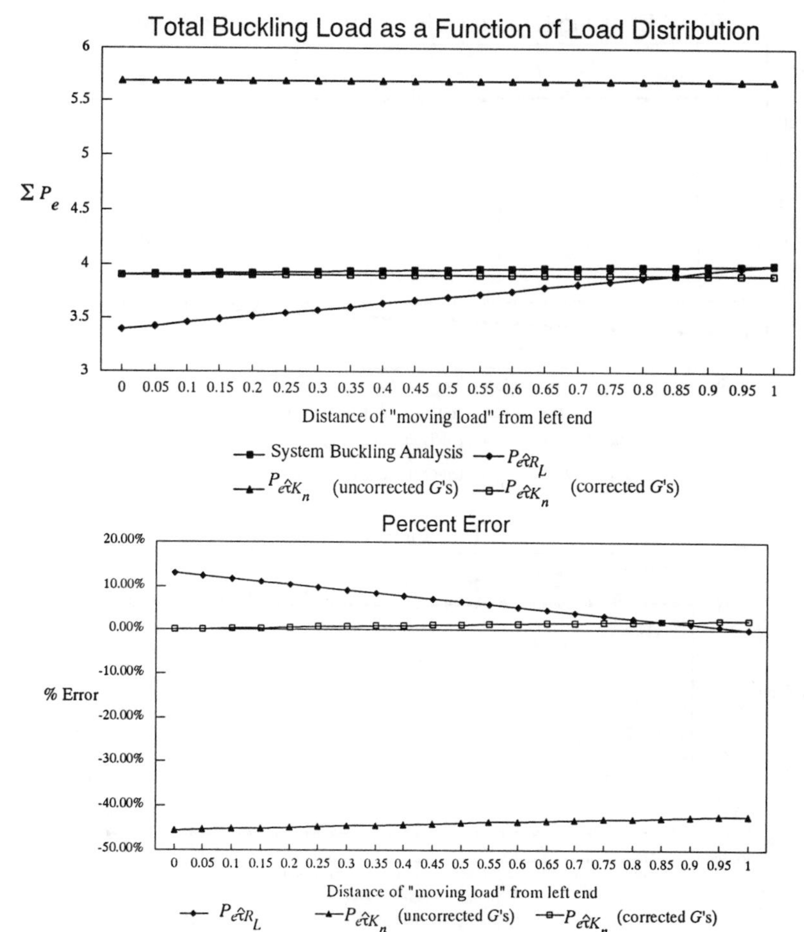

Figure 2.23 Buckling Load of a Frame Containing a Leaner Column

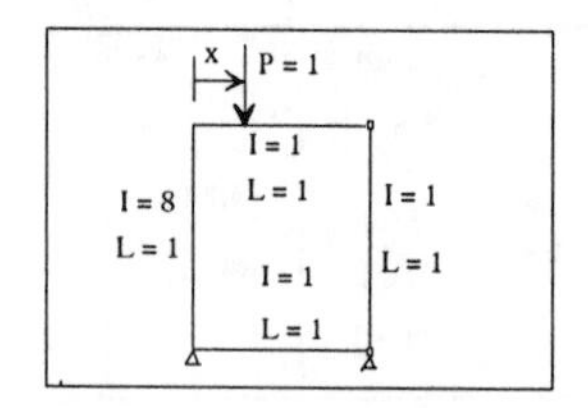

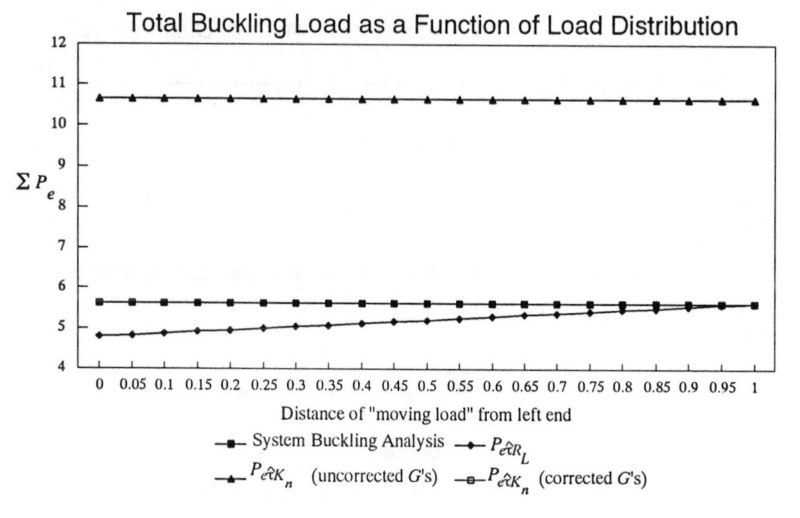

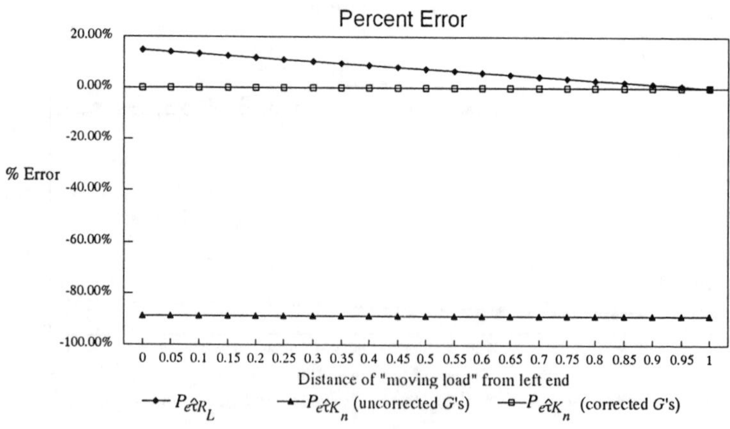

Figure 2.24 Buckling Load of a Frame Having a Strong Rigidly-Connected Column and a Leaner Column

The capacity predicted using $P_{e\hat{\tau}(K_n)}$ in these problems assumes that the total frame capacity equals the capacity of only the non-leaning column. The solution is accurate only when corrected G factors are used because the inflection point is not at the middle of the restraining girder when one end is pinned, and the uncorrected G factors do not account for this. The accuracy of $P_{e\hat{\tau}(K_n)}$ using corrected G factors is evident as the non-leaning column's moment of inertia is increased, as shown in Fig. 2.24. As the left column's moment of inertia is increased, the change in the frame's P-δ effect as the load is redistributed to the leaning column is small because the initial P-δ effect of the left column is small. Thus the change in total frame capacity as the load is redistributed becomes small. The capacity predicted by $\sum_{non-leaner} P_{e\hat{\tau}(K_n)}$ using corrected G factors is nearly exact under these circumstances, and slightly conservative when the restraining girders' moments of inertia and the left column's moment of inertia are small. However, $\sum_{non-leaner} P_{e\hat{\tau}(K_n)}$ is assumed to be constant, which is in contrast to $\sum_{non-leaner} P_{e\hat{\tau}(R_L)}$, which is slightly conservative under all circumstances for Frames 10 through 12 of Fig. 2.18.

Figs. 2.13 through 2.24, the results from LeMessurier [1993, 199?], and of Frames 1 through 15 of Fig. 2.18 from Squarzini and Hajjar [1993] confirm that $P_{e\hat{\tau}(R_L)}$ and $P_{e\hat{\tau}(K_n)}$ conservatively approximate the phenomena of an individual column buckling in a braced mode. $P_{e\hat{\tau}(R_L)}$ consistently proves to be more accurate than $P_{e\hat{\tau}(K_n)}$, and is rarely more than 5% inaccurate. Nevertheless, if corrected G factors are used, and if unequal length columns in a story are accounted for properly (see Appendix B), then $P_{e\hat{\tau}(K_n)}$ also provides reasonably accurate and conservative results for a wide variety of frames [LeMessurier, 1993; LeMessurier, 199?].

References

AISC [1986], *Load and Resistance Factor Design Specification for Structural Steel Buildings*, First Edition, American Institute of Steel Construction, Inc., Chicago, Illinois.

AISC [1989], *Specification for Structural Steel Buildings: Allowable Stress Design and Plastic Design*, Ninth Edition, American Institute of Steel Construction, Inc., Chicago, Illinois.

AISC [1993], *Load and Resistance Factor Design Specification for Structural Steel Buildings*, Second Edition, American Institute of Steel Construction, Inc., Chicago, Illinois.

AISC [1994], *Manual of Steel Construction - Load and Resistance Factor Design*, Second Edition, American Institute of Steel Construction, Inc., Chicago, Illinois.

Abdelrazaq, A., Baker, W., Hajjar, J. F., and Sinn, W. [1993], "Column Buckling Considerations in High-rise Buildings with Mega-Bracing," *Is Your Structure Suitably Braced?, Proceedings of the 1993 Annual Technical Session*, Milwaukee, Wisconsin, April 6-7, 1993, SSRC, Bethlehem, Pennsylvania, 155-169.

Allen, H. G. and Bulson, P. S. [1980], *Background to Buckling*, McGraw-Hill, New York, 582 pp.

Aristizabal-Ochoa, J. D. [1994], "K-Factor for Columns in Any Type of Construction: Nonparadoxical Approach," *Journal of Structural Engineering*, ASCE, 120(4), 1272-1290.

Baker, W. F. [1987], "Design of Steel Buildings for Second Order Effects," *Materials and Member Behavior*, Proceedings of the ASCE Structures Congress '87, Ellifritt, D. S. (ed.), Orlando, Florida, August 17-20, 1987, ASCE, New York, 534-554.

Bathe, K. J. [1982], *Finite Element Procedures for Engineering Analysis*, Prentice Hall, Englewood Cliffs, NJ, 735 pp.

Bridge, R. Q. and Trahair, N. S. [1977], "Effects of Translational Restraint on Frame Buckling," *Civil Engineering Transactions*, Institute of Engineers, Australia, 176-183.

Bridge, R. Q. and Fraser, D. J. [1987], "Improved G-Factor Method for Evalulating Effective Lengths of Columns," *Journal of Structural Engineering*, ASCE, 113(6), 1341-1356.

Bridge, R. Q. [1989], "In-Plane Buckling and Its Influence on Frame Design," *Proceedings of the 1989 Annual Technical Session*, New York, New York, April 17-19, 1989, SSRC, Bethlehem, Pennsylvania, 185-196.

Chen, W. F. and Lui, E. M. [1987], *Structural Stability: Theory and Implementation*, Elsevier, New York, 1987, 490 pp.

Cheong-Siat-Moy, F. [1986], "K-Factor Paradox," *Journal of Structural Engineering*, ASCE, 112(8), 1747-1760.

Council on Tall Buildings and Urban Habitat [1979], *Structural Design of Tall Steel Buildings*, Vol. SB, Gaylord, C. N. and Watabe (eds), Council on Tall Buildings and Urban Habitat and American Society of Civil Engineers, New York.

Disque, R. Q. [1973], "Inelastic K-Factor for Column Design," *Engineering Journal*, AISC, 10(2), 33-35.

ECCS [1976], *Manual for Stability of Steel Structures*, Second Edition, ECCS Committee 8 -- Stability, European Convention for Constructional Steelwork, 328 pp.

Geschwindner, L. F. [1994], "A Practical Approach to the 'Leaning' Column," *Engineering Journal*, AISC, 31(4), 141-149.

Hajjar, J. F. and White, D. W. [1994], "The Accuracy of Column Stability Calculations in Unbraced Frames and the Influence of Columns with Effective Length Factors Less Than One," *Engineering Journal*, AISC, 31(3), 81-97.

Hancock, G. J. [1984], "Structural Buckling and Vibration Analyses on Microcomputers," *Civil Engineering Transactions*, Institute of Engineers, Australia, CE26(4), 327-331.

Hancock, G. J. [1991], "Second Order Elastic Analysis Solution Techniques and Verification," *Structural Analysis to AS4100, A Two Day Post-Graduate Course*, November 6 and 7, 1991, 3-1 to 3-29.

Higgins, T. R. [1965], "Column Stability Under Elastic Support," *Engineering Journal*, AISC, 2(2), 12.

Horne, M. R. [1975], "An Approximate Method for Calculating Elastic Critical Loads of Multi-Storey Frames," *The Structural Engineer*, 53(6), 242-248.

Kavanagh, T. C. [1962], "Effective Length of Framed Columns," *Transactions of the American Society of Civil Engineers*, 127, 81-101.

LeMessurier, W. J. [1976], "A Practical Method of Second Order Analysis / Part 1 - Pin Jointed Systems," *Engineering Journal*, AISC, 13(4), pp. 89-96.

LeMessurier, W. J. [1977], "A Practical Method for Second-Order Analysis. Part 2: Rigid Frames," *Engineering Journal*, AISC, 14(2), pp. 49-67.

LeMessurier, W. J. [1991], Personal communication.

LeMessurier, W. J. [1993], "Discussion of the Proposed LRFD Commentary to Chapter C of the Second Edition of the AISC Specification," Presentation made to the ASCE Technical Committee on Load and Resistance Factor Design, Irvine, California, April 18, 1993.

LeMessurier, W. J. [199?], "A Practical Method for Second-Order Analysis. Part 3," in preparation.

Liew, J. Y. R., White, D. W., and Chen, W.-F. [1991], "Beam-Column Design in Steel Frameworks - Insights on Current Methods and Trends," *Journal of Constructional Steel Research*, 18(4), 269-308.

Livesley, R. K. and Chandler, D. B. [1956], *Stability Functions for Structural Frameworks*, Manchester University Press.

Lui, E. M. [1992], "A Novel Approach for K Factor Determination", *Engineering Journal*, AISC, 29(4), 150-159.

Salem, A. H. [1969], "Discussion of Buckling Analysis of One-Story Frames," by Zweig, A., *Journal of the Structural Division*, ASCE, 95(ST5).

Salmon, C. G. and Johnson, J. E. [1996], *Steel Structures: Design and Behavior*, Fourth Edition, Harper & Row, New York.

Springfield, J. [1991], "Limits on Second Order Elastic Analysis," *Proceedings of the 1991 Annual Technical Session*, Structural Stability Research Council, Bethlehem, Pennsylvania, 89-100.

Squarzini, M. J. and Hajjar, J. F. [1993], "An Evaluation Of Proposed Techniques For Predicting Column Capacity," Structural Engineering Report No. ST-93-3, Department of Civil and Mineral Engineering, University of Minnesota, Minneapolis, Minnesota, 195 pp.

Structural Stability Research Council [1976], *Guide to Stability Design Criteria for Metal Structures*, Johnston, B. G. (ed.), Third Edition, John Wiley & Sons, New York, 616 pp.

Tide, R. H. R. [1985], "Reasonable Column Design Equations," *Proceedings of the 1985 Annual Technical Session*, Structural Stability Research Council, Bethlehem, Pennsylvania, 47-56.

White, D. W. and Hajjar, J. F. [1991], "Application of Second-Order Elastic Analysis in LRFD: Research to Practice," *Engineering Journal*, AISC, 28(4), 133-148.

Wittrick, W. H. and Williams, F. W. [1971], "A General Algorithm for Computing Natural Frequencies of Elastic Structures," *Quarterly Journal of Mechanics and Applied Mathematics*, XXIV(3), 264-285.

Wu, H.-B. [1985], "Determination of Effective Length of Unbraced Framed Columns," *Proceedings of the 1985 Annual Technical Session,* Structural Stability Research Council, Bethlehem Pennsylvania, 105-116.

Yura, J. A. [1971], "The Effective Length of Columns in Unbraced Frames," *Engineering Journal*, AISC, 37-42.

Yura, J. A. [1972a], "Discussion of The Effective Length of Columns in Unbraced Frames," *Engineering Journal*, AISC, 40-48.

Yura, J. A. [1972b], "Discussion of The Effective Length of Columns in Unbraced Frames," *Engineering Journal*, AISC, 167-168.

Ziemian, R. D. [1990], "Advanced Methods of Inelastic Analysis in the Limit States Design of Steel Structures," Ph.D. Dissertation, School of Civil and Environmental Engineering, Cornell University, Ithaca, New York.

Ziemian, R. D., McGuire, W., and Deierlein, G. G. [1992], "Inelastic Limit States Design: Part I - Planar Frame Studies," *Journal of Structural Engineering*, ASCE, 118(9), 2532-2549.

Zweig, A. [1968], "Buckling Analysis of One-Story Frames," *Journal of the Structural Division*, ASCE, 94.

CHAPTER 3

STABILITY ANALYSIS OF PARTIALLY RESTRAINED FRAMES

3.1 Introduction

The AISC LRFD Specification [AISC 1993] designates two types of construction in its provisions: Type FR (fully restrained) construction and Type PR (partially restrained) construction. In the design of fully restrained (or rigid) frames, a designer assumes that full continuity exists between the connected members. The beam-to-column connections used for this type of construction are moment (or rigid) connections. These connections must possess sufficient rigidity to hold the original angles between the connected members virtually unchanged.

In the design of partially restrained (or semi-rigid) frames, the restraint characteristics of the connections must be considered in the design of the connected members. The moment-rotation characteristics of the connections must therefore be established in advance by analytical or empirical means. A special case of partially restrained framing (referred to as simple framing [ASD 1989]) is to ignore the rotational restraint effect of the connections in the design of beams. The beams are designed as if their ends were pinned in so far as gravity loads are concerned. However, the connections and connected members must have adequate capacity to resist wind moments. Furthermore, the connections must possess sufficient inelastic rotation capacity to avoid overstress of fasteners or welds under the combined effect of factored gravity and lateral loads.

While Type FR construction is unconditionally permitted under the AISC LRFD Specification, the use of Type PR construction depends upon a predictable proportion of full end restraint. In other words, the restraint characteristics of the connections used in the design of partially restrained frames must be known. Over the past six decades, a number of tests have been performed to study the restraint characteristics of semi-rigid connections. Empirical formulas relating the moment transmitted and the rotation experienced by various connection types have also been proposed. This voluminous set of data has been gathered and summarized in three connection data bases. They are described briefly as follows:

Goverdhan [1983] collected an extensive set of experimental data gathered since 1950. This data base contains not only experimental data, but empirical equations describing the moment-rotation behavior of the following types of connections: double web angle, single web angle and single plate, header plate, end plate, and top and seat angles with or without web angles.

Nethercot [1985a, 1985b] reviewed data of over 70 experimental studies on over 700 steel beam-to-column connection tests. Empirical equations describing the moment-rotation behavior were presented. Comparative studies on the role different joint parameters play in influencing the moment-rotation characteristics of a number of semi-rigid joints were conducted. The Nethercot data base contains information on the following ten types of connections: single web angle, single web plate, double web angle, flange angle, header plate, flush/extended end plate, combined web and flange angles, t-stubs, top and seat angles, and T-stubs with web angles.

Kishi and Chen [1986] extended Goverdhan's database and conducted a comprehensive literature review on tests and experimental data collected on riveted, bolted and welded connections from 1936 to 1986. The data were compared with two empirical moment-rotation equations: the Frye-Morris [Frye and Morris 1976] equation and the Kishi-Chen [Kishi and Chen 1986; Chen and Kishi 1989] equation. The Frye-Morris equation is a polynomial equation expressed in terms of a standardization parameter that is a function of various connection parameters such as connector size/spacing, angle/plate thickness, etc. The Kishi-Chen equation is an extension of Lui's equation [Lui 1985], which is a curve-fitting equation consisting of exponential and linear terms.

These data bases provide designers with a rather comprehensive set of information on the restraint characteristics of a variety of connections commonly used in steel frame construction. Although no design recommendations are given in their reports, the data bases generated by Goverdhan, Nethercot, and Kishi-Chen serve as valuable references for obtaining essential connection restraint characteristics necessary for the design of semi-rigid frames. In addition to these published data bases, two computer programs [PRCONN 1993, Chen and Toma 1994] are available from which information on a number of partially restrained connection types can be retrieved. The PRCONN [1993] database is an interactive PC program that allows the user to predict the behavior of a variety of commonly used PR steel connections for composite and noncomposite beams. The Chen and Toma [1994] database is an updated version of the Kishi-Chen [1986] data base.

When one utilizes the moment-rotation (M-θ) equations from the above-mentioned databases for design, extreme care must be exercised not to extrapolate too far from the test results used for the development of the equations. The Frye and Morris [1976] equations were developed from a relatively small data base. They were intended primarily as a resource for a parametric analytical study on frame behavior. They were not originally intended as a design tool. Even the more extensive Ang-Morris [1984] and Kishi-Chen [1986] equations were developed mostly from tests that utilized small to medium size beams (W5x16 to W24x76) and small columns (W5x16 to W12x96). Extreme care must be exercized if extrapolation of results beyond the available experimental data, particularly for cases with deep beams (W27 or above), is attempted.

During the past two decades, extensive studies on the effect of connection flexibility on partially restrained frame response have been published. Most of these works have been summarized in book and/or monograph form; see, for example, Chen [1987]; Chen [1988]; Chen and Lui [1991]; and Lorenz, Kato and Chen [1993]. More recent works on partially restrained frame analyses include those of Barakat and Chen [1990, 1991a], Dhillon and Abdel-Majid [1990], Geschwindner [1991], Kishi et al. [1993a, 1993b], Goto et al. [1993], and King and Chen [1993]. Over the past ten years, various methodologies for designing flexural and compressive members in partially restrained steel frames have been proposed; see, for example, Disque [1975]; Dewkett [1984]; Ackroyd [1985, 1987]; Lui [1985]; Lindsey, Ioannides and Goverdhan [1985]; Cronembold [1986]; Nethercot, Davison and Kirby [1988]; Gerstle and Ackroyd [1990]; Zandonini and Zanon [1991]. These are summarized in Chen and Lui [1991]. Analytical and experimental studies on semi-rigid action of connections in composite floor systems have also been reported [Leon and Ammerman 1990; Ammerman and Leon 1990].

In the following section, a succinct discussion of the restraint characteristics of connections in terms of their moment-rotation behavior is provided. This is followed by an extensive discussion of the issues involved with and procedures for calculating effective length factors for columns in PR frames. The chapter concludes with a brief summary of various simplified analysis procedures for the design of PR frames. The focus of the chapter is on the stability behavior of PR frames and the appropriate procedures to account for stability (i.e., calculation of effective length factors) within the context of any of the procedures for calculating design forces.

3.2 Connection Behavior

Although connections can experience axial, flexural, shear and torsional deformation under loads, flexural deformation is perhaps the most important in the context of plane frame design. Experiments on connections conducted in the past six decades have focused primarily on their flexural response. The flexural behavior of connections is best described by moment-rotation (M-θ) curves. A connection moment-rotation curve relates the beam moment transmitted by the connection to the in-plane rotation experienced by the connection. A schematic set of moment-rotation curves for some typical connections used in steel frame construction is shown in Fig. 3.1. Empirical equations describing the moment-rotation behavior of various connection types have been proposed by a number of researchers using parametric or curve-fitting techniques. A summary of these equations can been found in Chen and Lui [1991].

Connection behavior needs to be addressed in terms of stiffness, strength, and rotational ductility. Each of these attributes will be briefly examined in the following sections.

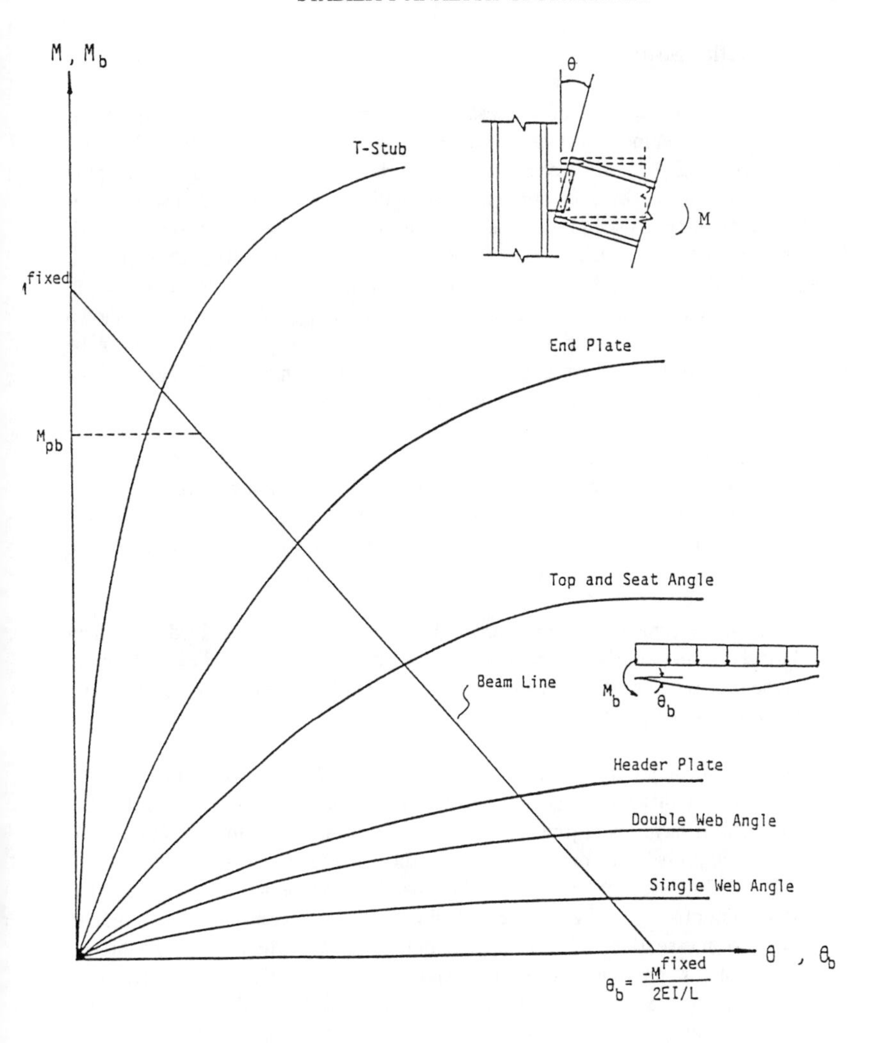

Figure 3.1 Schematic Moment-Rotation Curves and Beam Line

3.2.1 Connection Stiffness

Despite the broad range of stiffness exhibited by a variety of connection types, almost all connection moment-rotation curves are nonlinear practically from the onset of loading (see Fig. 3.1). However, for design purpose it may be desirable to employ linear approximations. One such approximation is illustrated in Fig. 3.2 in which a tri-linear representation is used to describe the connection response. In applying this tri-linear model for the design PR frames using the present AISC-LRFD Specification [AISC, 1993], the designer must first define the initial stiffness of the connection (R_{ki} in Fig. 3.2), and the other connection stiffness values such as R_{kser}, R_{kf}, and R_{ku} as shown in the figure. The determination of these stiffness values is at the discretion of the designers. Nevertheless, three important points need to be made with regard to the determination of the initial (or elastic) connection stiffness :

(1) For design calculations, it is not advisable to obtain the value of the initial (elastic) stiffness R_{ki} as the derivative of the M-θ equation at a zero rotation. This is because most of the proposed equations were derived using curve fitting techniques that forced the equation to pass through the origin. This results, in some cases, in fictitiously large stiffness at the origin.

(2) It should be remembered that many of the equations were derived from data digitized from published sources, introducing both systematic and random errors in the process. These errors are larger at small deformations as a result of data cluttering in the region.

(3) In PR construction under AISC-LRFD (and Type 2 construction under ASD), it is implicit that the connections have "shaken down". This means that the connections have undergone many cycles of loading and unloading (Fig. 3.3). There will probably be many cycles in the service range (shown as a single cycle in Fig. 3.3 for clarity) and a few in the range between the service (M_{ser}) and the design limit (M_{des}). When subjected to this loading, the connections will unload and load in essentially a linear fashion up to 90-95% of the load level achieved in any of the previous cycles (the magnitude of this effect is exaggerated in Fig. 3.3 for clarity). This linear behavior implies that the connection has "shaken down" and further response will be linear until 90% to 95% of the maximum previous connection moment is exceeded.

The elastic stiffness to which the connections shake down at low levels of rotation (less than 0.002 radian) has not been adequately studied, although some data is available. For semi-rigid composite connections, for example, peak to peak secant stiffness under fully reversed cyclic loads at the design level (M_{des}) are approximately 70% of the initial connection stiffness [Ammerman 1986]. More recently, tests on half-scale non-composite top and seat angle connections showed decreases of as much as 50% from the initial stiffness in a monotonic

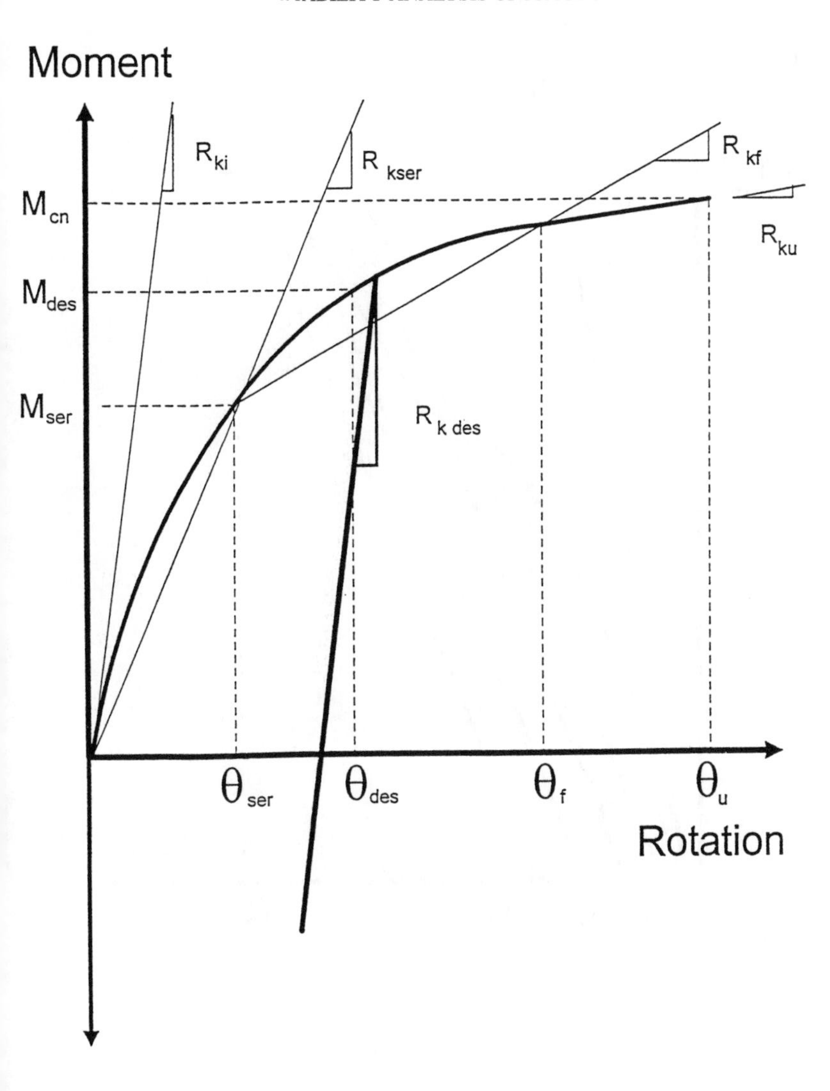

Figure 3.2 - Stiffness characteristics of a typical PR connection.

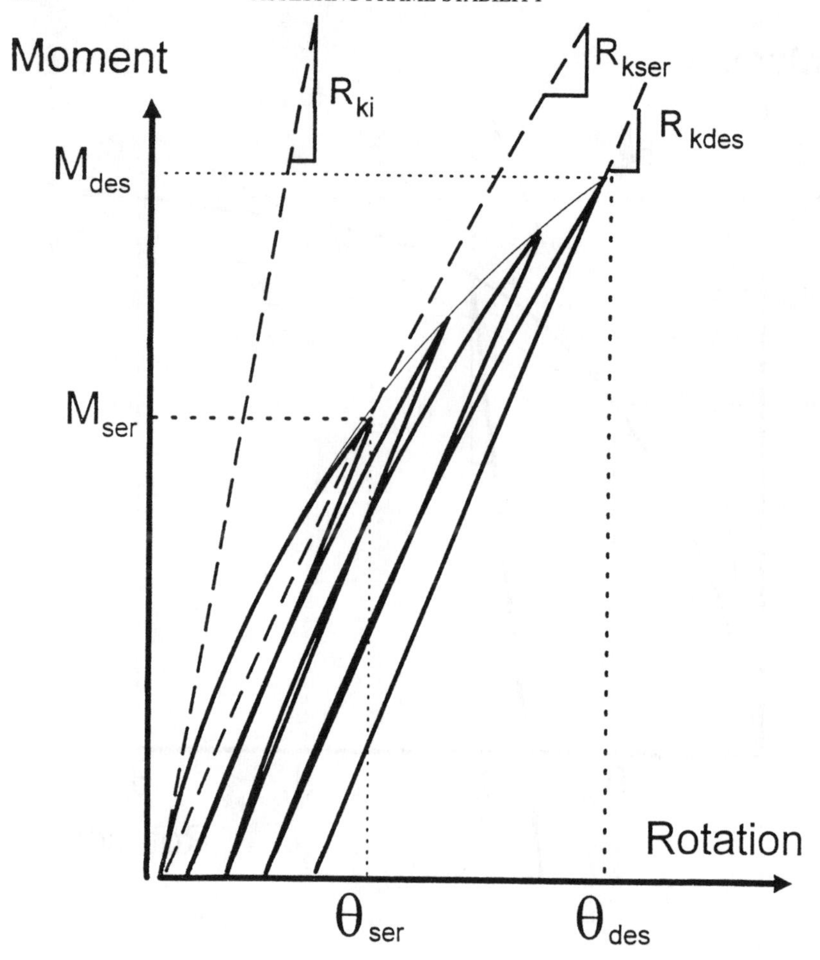

Figure 3.3 - Shakedown behavior of PR connections.

test to the corresponding elastic stiffness for a connection subjected to fully reversed cyclic loading to the design level [Leon and Shin 1995].

The amount of elastic stiffness reduction is strongly tied to the type of connection and the load history. Thus general rules regarding the reduction of this stiffness for connections loaded to the design level cannot be provided. In the case of the semi-rigid composite connections this reduction is dependent on the level at which cracking of the slab and slip of the bolts begins. In the case of all-steel connections, it is related primarily to the slip of the bolts and fit-up clearances. Since all of the tests used in the development of M-θ equations refer to monotonic tests, it is unconservative to assume for design that the connections shake down to the associated initial stiffness R_{ki}. As Fig. 3.3 shows, the elastic stiffness for a connection that has been loaded to the design level (R_{kdes}), falls somewhere between R_{ki} and R_{kser}, where R_{kser} is the secant stiffness associated with an appropriately selected M_{ser} and θ_{ser}. The unloading (or elastic) stiffness at the design level is probably closer to R_{kser} than R_{ki} Based on the above discussion, it can be concluded that the connection initial stiffness and/or the unloading stiffness can be conservatively taken as R_{kser}.

As shown in Fig. 3.2, in addition to R_{kser} two additional stiffnesses R_{kf} and R_{ku} are required to define the tri-linear connection model. At the present time, these stiffnesses and their associated moment and rotation values are defined only for one type of connection [AISC 1996], and even in this case the definitions are somewhat arbitrary. The implementation of the tri-linear model thus requires careful judgement on the part of the engineer to set reasonable values for these parameters.

3.2.2 Connection Strength

A connection is said to have reached its nominal moment capacity (M_{cn}) if the moment in the connection is at or near the flat portion of the M-θ curve (i.e., if the rotation is near or greater than θ_u in Fig. 3.2). Connection strength, however, is only relevant when compared to the strength of the framing beam. Connection strength can be defined as either full or partial strength depending on whether the connection strength exceeds or does not exceed the plastic flexural strength of the connecting beam (M_{pb}). A connection is full strength if $M_{cn}/M_{pb} \geq 1.0$ and partial strength otherwise. An important warning for designers with regard to the strength predicted by many of the M-θ equations in the data bases is that those equations are generally tied to only one mode of failure (typically flexural yielding of the connecting elements). For example, the equations proposed in most data bases for top and seat angle connections do not contain the bolt size as a variable. Consequently, the equations cannot predict when bolt shear or bolt tension, as opposed to yielding of the angle may begin to govern the strength of the connection (i.e., there is not a stated cap on the flexural strength as governed by the bolt strengths).

The distinction between partial and full strength connections is important with regard to the required ductility. By and large, most PR connections are partial strength, resulting in important consequences in design [Leon 1994].

3.2.3 Connection Rotational Ductility

Most partial strength PR connections, if properly designed, exhibit very ductile behavior. The inelastic deformation capacity, $\mu = \theta_u/\theta_{ser}$ (Fig. 3.2), is often greater than 10 in monotonic tests. In a full-strength PR connection, however, the moment introduced in the joint will be capped by M_{pb}, resulting in a shift of most of the deformations to a plastic hinge in the beam adjacent to the connection. In this case the ductility of the connection is probably less of a concern than if the PR connection is partial strength. In this latter case, the large ductility demands result in both the need for careful detailing of the connections and the need for a simplified check to insure that θ_u is not exceeded. It should be remembered that ductility demand is highly dependent on load history. The remarks in this chapter are meant explicitly for monotonic loading cases and not for stability calculations under large seismic loads.

3.2.4 Beam Line

The inclined line plotted in Fig. 3.1 is a beam line [Batho and Rowan 1934]. A beam line relates the end moment to the end rotation of an elastic beam when the beam is subjected to a symmetrical in-span loading. It has the form

$$M_b = \frac{2EI}{L}\theta_b + M_b^{fixed} \tag{3.1}$$

(where the subscript b denotes the beam and M^{fixed} is the fixed-end beam moment. E, I and L are the modulus of elasticity, moment of inertia and length of the beam, respectively. The horizontal dashed line in the figure represents a limit point for the beam line when the plastic moment capacity of the beam cross-section (M_{pb}) falls below the fixed-end moment M^{fixed}. The intersection of the connection moment-rotation curve and the beam line gives the moment exerted on and the rotation experienced by the connection[1].

---

[1] Beam lines can also be drawn for first yield condition of the beam and used as a limit envelope to determine whether failure of the connection will precede yielding of the beam. If a connection moment-rotation curve does not cross this limit envelope, the connection fails before the attainment of first yield at some point on the beam. See Kennedy [1969] for a thorough discussion of the subject.

According to Salmon and Johnson [1990], a connection is considered to be fully restrained (or rigid) if it can carry a moment in excess of 90% of M^{fixed}. This definition predicates that connection stiffness and strength are correlated. Although strength and stiffness for most connections are indeed correlated, it is more appropriate to quantify PR connections directly in terms of their stiffness. Ackroyd and Gerstle [1982] have recommended that a connection be considered as fully restrained if $EI/R_kL<0.05$, and partially restrained if $0.05<EI/R_kL<2.0$. In the preceding inequalities, R_k is the rotational stiffness of the connection and EI/L is the flexural rigidity of the girder to which the connection is attached. In addition, Eurocode 3 for steel structures [Eurocode 1992] and Bjorhovde et al. [1990] have established bounds in terms of the connection to beam plastic moment ratio (M/M_{pb}) and nondimensional connection rotation ($\theta/(M_{bp}L/EI)$ in Eurocode 3, and $\theta/(M_{bp}*5d/EI)$, d=beam depth in Bjorhovde et al.) to demarcate fully restrained from partially restrained joint behavior. In Eurocode 3, the bounds were established so that the reduction in load carrying capacity of the PR frame will not exceed 5% of its FR counterpart.

The amount of restraint that a PR connection can impose on the adjoining structural members depends to a large extent on the rotational stiffness of the connection, R_k. With reference to Fig. 3.2, when a connection is subjected to an increasing moment and undergoing loading, R_k can be approximated by a small value of R_{ku} if the nominal connection moment capacity M_{cn} is approached. This stiffness may equal zero for some connection types. If the direction of moment is reversed and the connection is undergoing unloading, R_k can be approximated by the secant stiffness R_{kser} as shown in Figs. 3.2 and 3.3. This recommendation is based on the assumptions that (1) the connection is symmetrical about the beam axis, (2) the slip resistance of the bolts has not been exceeded, and (3) complete unloading and continued reversed loading (i.e., large negative M values in Fig. 3.2) does not occur.

Because of the loading-unloading phenomenon, similar connections will exhibit dissimilar restraint characteristics. For instance, consider the case in which a beam with two identical connections at opposite ends is subjected to a uniformly distributed gravity load. The two connections will undergo loading simultaneously and so their responses are identical under the gravity load. Suppose a wind load is now applied. The leeward connection will continue to load, but the windward connection will begin to unload. The stiffness, and hence the degree of restraint provided by the two connections, will now be different. To properly design the frame, this loading/unloading characteristic of the connection must be considered. A methodology by which this effect can be accounted for is described in the following section.

3.3 Rotational Stiffness of a Beam-Connection Assembly

In the analysis and design of partially restrained frames, one needs to consider not only the stiffness of the connections, but the combined stiffness of the beams and the contiguous connections. To illustrate this conceptually, consider the partially restrained simple portal frame shown in Fig. 3.4. The frame is subjected to a symmetric gravity load prior to the application of the wind load. The two connections, A and B, are identical connections, so both will be subjected to the same gravity moment, M_g. The moments are equal but opposite; they are denoted by point P in the moment-rotation curves of the connections. When the wind load is applied, the leeward connection will load while the windward connection will unload. The stiffness of the leeward connection will now be less than that of the windward connection. The result is that the leeward column will experience less rotational restraint from the beam-connection assembly than the windward column even though both columns are connected to the same beam using identical connections. In the extreme case when the leeward connection has reached its nominal moment capacity, its stiffness will be near zero and the degree of restraint imposed by the beam-connection assembly on the leeward column will be almost negligible.

The connection loading and unloading behavior described above for the wind loading case is readily applicable to the story buckling behavior of sidesway-uninhibited PR frames. Connections are often loaded by gravity moments in the beams prior to instability. At incipient sidesway instability, the connections on one side of the girders will continue to load, while the connections on the other side unload. The rotational restraints and frame behavior assumed in the effective length factor computations are typically based on some approximation of the tangent (or instantaneous) loading or unloading stiffness of the connections at the design load levels (i.e., factored gravity or combined factored gravity and lateral loads). Several methods for approximating these tangent stiffness are discussed in this and several of the subsequent sections of this chapter.

It is important to note the implications of the above philosophy for the stability design of PR frames versus the stability design of FR frames based on effective length methods. For either type of frame, the effective lengths are calculated for the purpose of obtaining the column axial strength P_n in the absence of any bending moment. This column strength, and the flexural strength M_n of the member in the absence of any axial load, are used as "anchor points" for checking member capacity in the AISC-LRFD beam-column interaction equation. As discussed in Chapter 2, for Type FR frames, P_n is based most rigorously on the tangent state of the structure at incipient inelastic buckling. It is commonly assumed that any yielding in the beams is negligible, and thus the full elastic stiffness of the beams is available to restrain the inelastic columns when the structure buckles. The effect of any yielding due to bending actions in the beams or in the columns is neglected in calculating the column axial strength P_n , with the exception of capturing the initial imperfection effects by use of the AISC-LRFD column curve expressions. When based on an iterative inelastic buckling procedure, the inelasticity calculated in the columns at incipient inelastic buckling may far exceed the actual level of inelasticity to which the actual beam-column member is subjected.

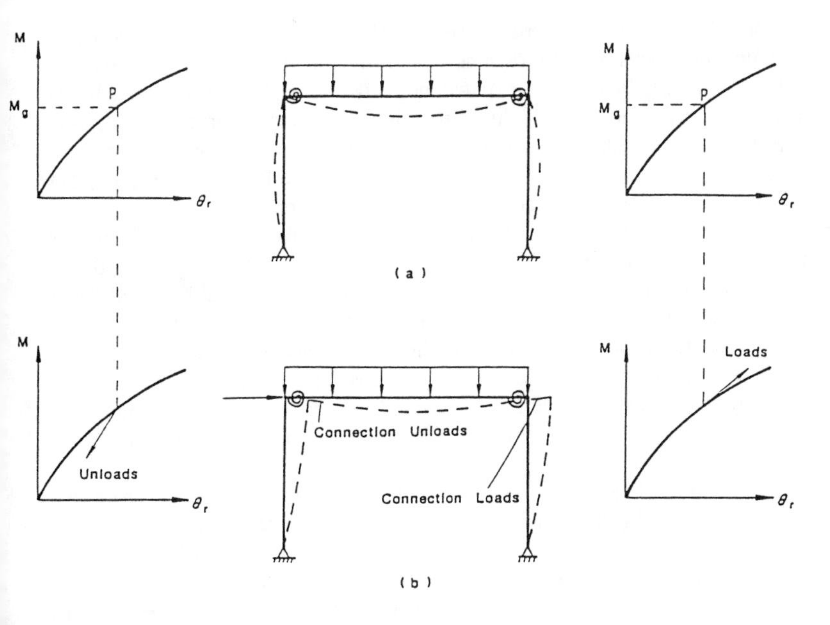

Figure 3.4 Loading and Unloading Behavior of Connections

For PR frames, it is assumed that significant nonlinearity will occur in the beam-to-column connections prior to reaching the factored design load levels. Therefore, the effect of the bending actions in reducing the rotational restraint provided by the connections must be considered. Implicit within the consideration of these bending effects is the philosophy, or assumption, that the connection tangent loading or unloading stiffness corresponding to the factored design load levels are the appropriate values for the stability assessment (i.e., for the calculation of P_n). Therefore, if a procedure parallel to that used for the most rigorous calculation of P_n in FR framing is followed, the inelasticity in the columns should be based on a load level that in general can be significantly higher than the axial force in the actual beam-column at the design load levels; however, the inelastic state of the connections of a PR frame is held at these design levels. Fortunately, the connection tangent stiffness properties recommended in this chapter for design calculations are not likely to be significantly affected by the above subtleties involving at what level of loading the connections should be when the "buckling" of the structure is evaluated for determination of P_n.

In Section 3.3.1, a general rotational stiffness equation for prismatic members, capable of incorporating the combined restraint effect of the beam-connection assembly on the adjoining column in the analysis of partially restrained frames, will be discussed [Lui 1993, 1995]. The rotational stiffness equation developed in this section will be shown in Section 3.3.2 to be a general equation encompassing a number of other rotational stiffness equations proposed in the past by other researchers.

3.3.1 Derivation of the Rotational Stiffness Equation

Consider a prismatic beam shown in Fig. 3.5. The beam has a flexural rigidity of EI and length L. Two connections, A and B, with tangent rotational stiffness R_{kA} and R_{kB} are present at opposite ends of the beam. Denoting dM_A and dM_B as the corresponding incremental connection moments, and $d\theta_A$ and $d\theta_B$ as the incremental column rotations at ends A and B, respectively, the slope-deflection equation for the beam can be written as

$$dM_A = \frac{EI}{L}\left[s_{ii}(d\theta_A - \frac{dM_A}{R_{kA}}) + s_{ij}(d\theta_B - \frac{dM_B}{R_{kB}})\right] \tag{3.2a}$$

$$dM_B = \frac{EI}{L}\left[s_{ij}(d\theta_A - \frac{dM_A}{R_{kA}}) + s_{ii}(d\theta_B - \frac{dM_B}{R_{kB}})\right] \tag{3.2b}$$

where s_{ii} and s_{ij} are stability functions. For sway-uninhibited frames, the axial force in the beam is rather small. As a result, s_{ii} and s_{ij} can be taken as 4 and 2, respectively, giving

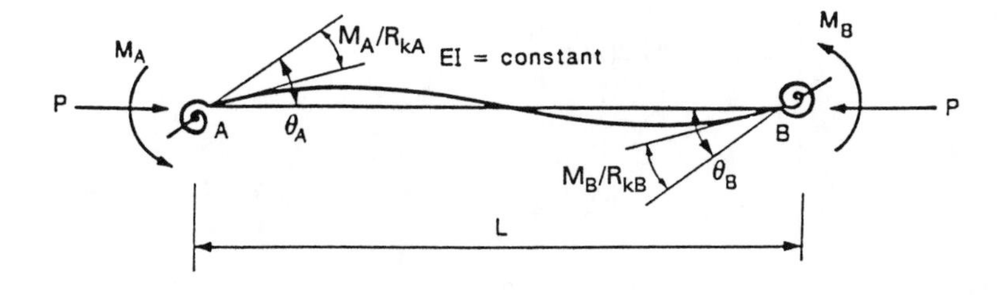

Figure 3.5 A Beam-Connection Assembly

$$dM_A = \frac{EI}{L}\left[4(d\theta_A - \frac{dM_A}{R_{kA}}) + 2(d\theta_B - \frac{dM_B}{R_{kB}})\right] \tag{3.3a}$$

$$dM_B = \frac{EI}{L}\left[2(d\theta_A - \frac{dM_A}{R_{kA}}) + 4(d\theta_B - \frac{dM_B}{R_{kB}})\right] \tag{3.3b}$$

Solving the above equations for $d\theta_A$ and $d\theta_B$, we obtain

$$d\theta_A = \left[\frac{L}{3EI}(1 + \frac{3}{\bar{R}_{kA}}) - \frac{(dM_B/dM_A)L}{6EI}\right] dM_A \tag{3.4a}$$

$$d\theta_B = \left[-\frac{(dM_A/dM_B)L}{6EI} + \frac{L}{3EI}(1 + \frac{3}{\bar{R}_{kB}})\right] dM_B \tag{3.4b}$$

where

$$\bar{R}_{kA} = \frac{R_{kA}L}{EI} \quad \text{and} \quad \bar{R}_{kB} = \frac{R_{kB}L}{EI} \tag{3.5}$$

The rotational restraint that the beam-connection assembly imposes on the columns at the A and B ends of the beam can be written respectively as

$$\left(\frac{dM}{d\theta}\right)_A = \frac{6EI}{L}\left[\frac{1}{2(1 + \frac{3}{\bar{R}_{kA}}) - (\frac{dM_B}{dM_A})}\right] \tag{3.6a}$$

$$\left(\frac{dM}{d\theta}\right)_B = \frac{6EI}{L}\left[\frac{1}{2(1 + \frac{3}{\bar{R}_{kB}}) - (\frac{dM_A}{dM_B})}\right] \tag{3.6b}$$

The above equations can be expressed in a compact form as

$$\left(\frac{dM}{d\theta}\right)_N = \frac{6EI}{L}\left[\frac{1}{2(1 + \frac{3}{\bar{R}_{kN}}) - \left(\frac{dM_F}{dM_N}\right)}\right] \tag{3.7}$$

where the subscripts N and F denote the near end and far end of the beam, respectively.

Since the moment-rotation curves of typical connections are nonlinear, the rotational restraint expressed in the above equation, for the purposes of stability assessment, is most logically calculated based on the tangent (or instantaneous) stiffness of the connections at a specified load value (e.g., factored gravity or combined factored gravity and lateral loads). Furthermore, the term dM_F/dM_N is most logically the ratio of the incremental moments associated with this connection tangent loading or unloading behavior under the buckling displacements of the structure.

Consider the case in which the far end connection has reached its nominal moment capacity M_{cn}. Assuming that the incremental far end moment dM_F is essentially zero at this load level and substituting zero for dM_F/dM_N in Eq. (3.7), we have

$$\left(\frac{dM}{d\theta}\right)_N = \frac{6EI}{L}\left[\frac{1}{2(1+\frac{3}{\overline{R}_{kN}})}\right] \tag{3.8}$$

An appropriate (e.g., conservative) estimate of the connection unloading stiffness should be used for the near-end connection (see the discussion in Section 3.2.1).

On the other hand, if the near end connection has reached its nominal moment capacity, and if we approximate the tangent stiffness of the connection at this state by $\overline{R}_{kN} = 0$, the incremental near end moment dM_N will be zero, giving

$$\left(\frac{dM}{d\theta}\right)_N = 0 \tag{3.9}$$

In physical terms, $(dM/d\theta)_N = 0$ indicates that the near-end connection is not imposing any rotational restraint on the adjoining column.

Equations (3.2) through (3.7) represent a general formulation of the slope-deflection equations for a prismatic member with PR connections. Many similar expressions have been proposed. For example [Rathbun 1936; Johnston and Mount 1942; Baker 1954; Pippard and Baker 1957; Lothers 1960; DeFalco and Marino 1966; Disque 1975; Driscoll 1976; Wang 1983; and Bjorhovde 1984] can be shown to be special cases or subsets of these equations. This will be demonstrated in what follows for the Driscoll [1976], and DeFalco and Marino [1966] equations.

3.3.2 Specialization of the Rotation Stiffness Equation

A set of three rotational stiffness equations for a beam-connection assembly similar to the one shown in Fig. 3.5 was derived by Driscoll [1976]. Each equation represents a special kinematic condition for the beam-connection assembly. The equations, together with their associated kinematic conditions, are given below.

For $d\theta_B=0$,

$$\left(\frac{dM}{d\theta}\right)_A = \frac{4EI}{L}\left[\frac{3(1+\frac{3}{\overline{R}_{kB}})}{4(1+\frac{3}{\overline{R}_{kA}})(1+\frac{3}{\overline{R}_{kB}})-1}\right] \tag{3.10}$$

For $d\theta_B = d\theta_A$,

$$\left(\frac{dM}{d\theta}\right)_A = \frac{6EI}{L}\left[\frac{2(1+\frac{3}{\overline{R}_{kB}})+1}{4(1+\frac{3}{\overline{R}_{kA}})(1+\frac{3}{\overline{R}_{kB}})-1}\right] \tag{3.11}$$

For $d\theta_B=-d\theta_A$,

$$\left(\frac{dM}{d\theta}\right)_A = \frac{6EI}{L}\left[\frac{2(1+\frac{3}{\overline{R}_{kB}})-1}{4(1+\frac{3}{\overline{R}_{kA}})(1+\frac{3}{\overline{R}_{kB}})-1}\right] \tag{3.12}$$

A special form of Eq. (3.10) was used by DeFalco and Marino [1966] based on the work reported by Tall [Tall 1964] for the case when $\overline{R}_{kA}=\overline{R}_{kB}=\overline{R}_k$. The DeFalco equation is thus a special case of the Driscoll equations.

In what follows, it will be shown that all the above equations can be derived from Eq. (3.7). Denoting the near end of the beam as end A and the far end of the beam as end B, the subscripts *N* and *F* in Eq. (3.7) will become *A* and *B*, respectively.

For $d\theta_B=0$, we have, from Eq. (3.4b),

$$\left(\frac{dM_B}{dM_A}\right) = \frac{1}{2(1+\frac{3}{\overline{R}_{kB}})} \tag{3.13}$$

Substituting the above equation for (dM_F/dM_N) in Eq. (3.7), we obtain Eq. (3.10).

For $d\theta_B=d\theta_A$, we have, by equating Eq. (3.4a) to Eq. (3.4b) and solving for (dM_B/dM_A),

$$\left(\frac{dM_B}{dM_A}\right) = \frac{2\left(1+\frac{3}{R_{kA}}\right)+1}{2\left(1+\frac{3}{R_{kB}}\right)+1} \tag{3.14}$$

Substituting the above equation for (dM_F/dM_N) in Eq. (3.7), we obtain Eq. (3.11).

For $d\theta_B$=$-d\theta_A$, we have, by setting the sum of Eq. (3.4a) and Eq. (3.4b) equal to zero and solving for (dM_B/dM_A),

$$\left(\frac{dM_B}{dM_A}\right) = \frac{-2\left(1+\frac{3}{R_{kA}}\right)+1}{2\left(1+\frac{3}{R_{kB}}\right)-1} \tag{3.15}$$

Substituting the above equation for (dM_B/dM_A) in Eq. (3.7), we obtain Eq. (3.12). Equation (3.7) is therefore a more general expression for the rotational stiffness of the beam-connection assembly. In the following section, a procedure which makes use of Eq. (3.7) for calculating effective length factors of columns in partially restrained frames will be presented. The mechanics and validity of the proposed approach will be demonstrated by examples.

3.4 *K* Factor Computation for Columns in Partially Restrained Frames

In using the alignment charts of Fig. 2.2b and Fig. 2.3b to determine column effective length factors, one must first compute the column end restraint parameters G_A and G_B according to Eqs. (2.10a) and (2.10b), or Eqs. (2.10c) and (2.10d) if column inelasticity is to be accounted for. These *G* factors assume that the beams are rigidly connected to the column in question. As a result, only the stiffness of the beam and column are accounted for in calculating *K*. These *G* factors must be modified if the effect of connection flexibility is to be taken into consideration. The required modifications are discussed below for columns in sidesway-inhibited and sidesway-uninhibited PR frames.

3.4.1 Sidesway Uninhibited Frames

Recall that an assumption used in developing the alignment chart for the sidesway uninhibited frame subassemblage shown in Fig. 2.2a is that the restraining beams bend in reverse curvature with the near-end and far-end rotations equal. As a result, the amount of rotational restraint the near end of a beam can impose on the adjoining column is

$$\left(\frac{dM}{d\theta}\right)_N = \frac{6EI}{L} \tag{3.16}$$

where EI is the flexural rigidity of the beam.

Upon comparison of Eq. (3.16) with Eq. (3.7), it can be seen that the effect of connection flexibility can be incorporated into the G factor computation by replacing the beam moment of inertia I in the denominator of Eqs. (2.10a-d) by a modified value I' [Lui 1993, 1995], where

$$I' = I\left[\frac{1}{2(1+\frac{3}{\overline{R}_{kN}}) - (\frac{dM_F}{dM_N})}\right] \tag{3.17}$$

This expression for I' is convenient for use in computing the stiffness factors (G) for effective length calculations in PR or FR (i.e., rigidly connected) frames. For FR frames, use of this expression is equivalent to the modification used to obtain L'_g or $(EI/L')_g$ in Eqs. (2.11). However, when Eq. (3.17) is used, the modification of the girder stiffness parameter is based simply on an adjustment to I rather than an adjustment to the girder length, L. It is important to note that I' can be positive as well as negative. A positive I' means the beam is imposing restraint on the adjoining column and a negative I' means the beam is destabilizing the adjoining column at incipient instability. If I' is negative, the corresponding G factor will also be negative. Negative G factor is theoretically possible but is beyond the scope of the alignment chart given in the current specifications [AISC 1989, AISC 1994]. Therefore, if negative G factors are encountered, K must be obtained directly by solving the transcendental equations [AISC 1994].

A plot of the variation of I'/I with $\overline{R}_{kN}$ ranging from 2 to 20, for several specific values of dM_F/dM_N, is shown in Figure 3.6. Note the dependency of I'/I on both $\overline{R}_{kN}$ and dM_F/dM_N. Also, it is important to note that for sidesway-uninhibited PR frames, the values of dM_F/dM_N will rarely equal one (i.e., the value assumed in the development of the effective length alignment charts). This is because in most cases, a connection at one end of the beam will tend to load while another connection at the other end of the beam will tend to unload at incipient sidesway instability of the structure.

In general, accurate values for $\overline{R}_{kN}$ and dM_F/dM_N needed for Eq. (3.17) are not readily obtainable. Nevertheless, three possible approaches for determining $\overline{R}_{kN}$ and dM_F/dM_N will be discussed in the following. In decreasing order of sophistication, they are:

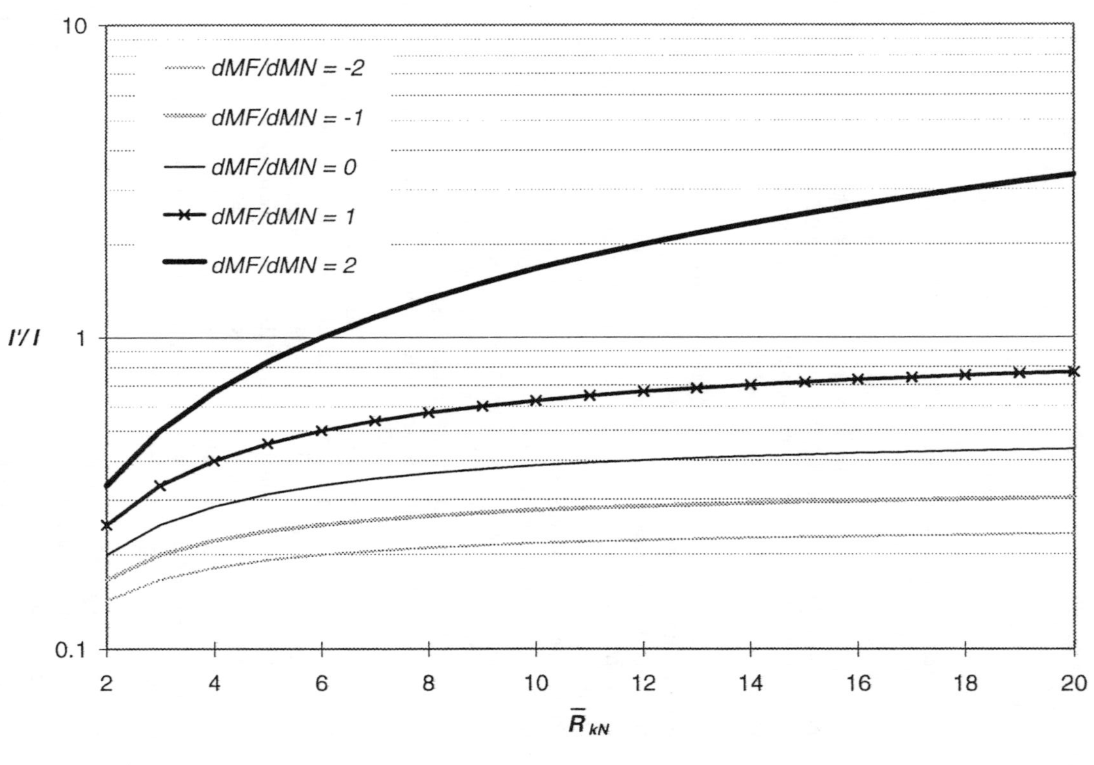

Figure 3.6 Values of I'/I from Eq. (3.17) for specific values of dM_F / dM_N and $\overline{R}_{kN}$ ranging from 2 to 20.

(1) Perform an advanced second-order inelastic analysis and obtain values of $\overline{R}_{kN}$ and dM_F/dM_N directly. This analysis must include all the significant connection and member non-linearities. Of course, if one performs such an analysis, the analysis in itself would be verification of the structure's stability, and the application of Eq. (3.17) would be unnecessary. Therefore, this procedure is only useful for assessing the ability of Eq. (3.17) to predict a frame's stability behavior, given "exact" values for the relevant parameters. This report does not address the possible direct use of explicit nonlinear frame analysis as part of a design procedure. Rather, the focus is solely on presentation and discussion of proper procedures for assessing frame stability within the context of an elastic analysis of the frame -- with appropriate representation of the effects of the connection nonlinearities in the case of PR type construction.

(2) Use tabulated values for R_{kN} based on simplified bi-linear or tri-linear M-θ curves. In this approach one may assume that the loading connection is in the R_{kf} range and the unloading one is in the R_{kser} range (see Fig. 3.2). This approach has been developed for frames with semi-rigid composite connections, and both continuous and tri-linear M-θ curves have been proposed [AISC 1996]. Strictly speaking it is necessary to incorporate in this procedure a check to insure that all the connections have not exceeded the rotation θ_f . These additional calculations can be made based on a simple second-order plastic analysis [Leon and Hoffman 1995, AISC 1996], or waived altogether if further parametric studies in PR frame analysis and design research indicate that the maximum rotations at factored loads are always well below θ_f . Alternately, in drift-controlled frames, it is possible that one could perform a direct analysis using a bi-linear or tri-linear curve to show that all the connections have not exceeded M_{ser} at factored loads. In this case it is possible to use R_{kser}, rather than R_{kf}, for the loading connections resulting in a substantial reduction in the K factors. Unfortunately, this procedure has been studied thoroughly for only one type of structure. For it to gain widespread acceptance, similar tri-linear M-θ relationships will need to be developed for other types of PR connections, and research studies will need to be conducted to ascertain the implications of the M-θ assumptions on the frame design and behavior.

(3) Employ a simplified behavioral model in which the connection stiffness and incremental moment are assumed to be zero for all loading connections, and R_{ki} are used for all unloading connections. Designers should note that unless the connections have reached their nominal moment capacities under beam gravity loads, this approach often leads to an overly conservative design. This over-conservatism is due primarily to the fact that the connection stiffness for many loading connections is quite appreciable at the design load level (although the amount of conservatism is somewhat lowered by the use of R_{ki} rather than R_{kser} for the unloading connections). The assumption of zero stiffness for these connections tends to undermine the stiffness contribution these connections have on the overall

stiffness of the frame. Nevertheless, at the present time (1995), this approach remains the most logical for everyday design since the first approach suggested above is not practical for routine use, and the second approach has not been thoroughly evaluated for other than composite PR frames. In addition, an advantage of the third approach over the first two approaches is that one does not need to have knowledge of the complete moment-rotation behavior of the connection. In applying the third approach, only two connection parameters: R_{ki} and M_{cn} need be known, and analytical expressions for these two parameters are available in the literature (see, for example, Kishi and Chen [1990], Chen and Lui [1991], Kishi et al. [1993b, 1994]).

Because of the ease with which the third approach can be applied to PR frame analysis, the rest of this chapter pre-supposes its use. The equations that follow, therefore, are predicated on the assumptions of the third approach. For this situation, the modified beam moment of inertia I' becomes:

If the near end connection unloads and the far end connection loads

$$I' = I\left[\frac{1}{2(1+\frac{3}{\bar{R}_{kN}})}\right] \tag{3.18}$$

where $\bar{R}_{kN}=R_{kdes}L/EI$, R_{kdes} being an appropriate approximation of the unloading connection stiffness .

If the near end connection loads and the far end connection unloads

$$I' = 0 \tag{3.19}$$

For the special case when the connection stiffness are the same at both ends of the beam (i.e., $R_{kA}=R_{kB}$), dM_F/dM_N most likely would be equal to unity (e.g., the connections are identical on each end of the beam, and the total moments M_F and M_N are the same at each of the ends). Substituting $dM_F/dM_N=1$ into Eq. (3.17), we obtain

$$I' = I\left[\frac{1}{1+\frac{6}{\bar{R}_{kN}}}\right] \tag{3.20}$$

However, it should be noted that the condition $R_{kA}=R_{kB}$ is seldom realized because of the loading/unloading characteristic of the connections. Eq. (3.17) is a more general equation and thus it is more appropriate for use in computing effective length factors for columns in PR frames.

3.4.2 Sidesway Inhibited Frames

For the sidesway-inhibited frame subassemblage shown in Fig. 2.3a, an assumption commonly used is that the restraining beams bend in single curvature with the near end and far end rotations equal but opposite. Thus the amount of rotational restraint the near end of a beam can impose on the adjoining column is

$$\left(\frac{dM}{d\theta}\right)_N = \frac{2EI}{L} \tag{3.21}$$

where EI is the flexural rigidity of the beam.

Comparing the above equation with Eq. (3.7), one can account for the effect of connection flexibility by replacing I in the denominator of Eqs. (2.10a-d) by I' where [Lui 1993]

$$I' = 3I\left[\frac{1}{2(1+\frac{3}{\bar{R}_{kN}}) - (\frac{dM_F}{dM_N})}\right] \tag{3.22}$$

For sidesway-inhibited frames, the connections at opposite ends of the beam normally undergo either loading or unloading simultaneously at incipient instability. As a result, one can circumvent the difficulty of having to determine the tangent stiffness and the incremental end moments of the loading connections simply by assuming that these connections have zero stiffness. Therefore, only the stiffness of the unloading connections needs to be considered. With this assumption, one can conservatively use:

If both the near and far end connections unload

$$I' = I\left[\frac{1}{(1+\frac{2}{\bar{R}_{kN}})}\right] \tag{3.23}$$

where $\bar{R}_{kN}=R_{kdes}L/EI$, R_{kdes} being an appropriate estimate of the unloading connection stiffness (e.g. one might use R_{kser} as discussed in Section 3.2.1).

If both the near and far end connections load

$$I' = 0 \tag{3.24}$$

If we substitute $I'= I'=0$ into Eqs. (2.10a-d), we obtain $G_A=G_B=\infty$, giving K=1. This condition represents the most severe case for the column in a braced frame. For simplicity and conservativeness, an effective length factor of unity is often used for the design of sidesway-inhibited columns[2].

3.4.3 Procedure for Evaluating *K* using the Modified Moment of Inertia Approach

The steps for employing the modified moment of inertia approach for determining effective length factors for columns in partially restrained frames are outlined below.

1. Calculate the modified moment of inertia I' using Equation (3.17) (or, if a simplified PR frame behavioral model is used as per Procedure 3 in Section 3.4.1, calculate I' using the appropriate equation in Section 3.4.1 or 3.4.2.)

2. Replace I by I' and calculate G factors for the column using Eqs. (2.10a) and (2.10b), or Eqs. (2.10c) and (2.10d) if column inelasticity is to be accounted for.

3. Use the appropriate alignment chart in Fig. 2.2b or 2.3b to obtain a nomograph effective length factor, K_n, for the column.

If the columns in a story have different values of $L\sqrt{(P_u/EI)}$, or if the values of end restraints are significantly different, or if leaning columns are present (In general, one or more of these conditions will often be the case for a PR frame), proceed to step 4.

4. Based on the K_n values obtained in steps 1-3, use Eq. (2.25b) to calculate the effective length factor for the column in question. For clarity, this equation is reproduced below:

$$K_{K_n} = \sqrt{\frac{\sum_{all} P_u}{P_u} \frac{\hat{\tau} I}{\sum_{non\text{-}leaner} \frac{\hat{\tau} I}{K_n^2}}} \qquad (2.25b)$$

If leaning columns are present in the story, the K_n values must be calculated only for the non-leaning columns. Non-leaning columns are columns which can provide lateral stability to the frame. All columns which have a predictable amount of rotational end restraint are classified as non-leaning columns.

---

[2] If $K<1$ is used for the design of the column, the adjoining beam must be designed with due consideration given to the second-order effect on the beam. See Section 3.11 in Chen and Lui [1987].

3.5 Illustrative Examples

Example 1 As an illustrative example, consider the partially restrained portal frame shown in Fig. 3.7a. The connections at joints B and C are assumed to be linear (i.e. they exhibit linear moment-rotation behavior) with $\bar{R}_k = R_k L/EI = 10$. The elastic effective length factor (i.e., $\hat{\tau}=1$)[3] for column AB is to be determined.

Step 1: Determine I'
Using Eq. (3.17) with $dM_F/dM_N=1$ (because the connections are linear) and $R_{kN}=\bar{R}_k=10$, we obtain

$$I' = \frac{I}{1.6}$$

Step 2: Calculate G_A and G_B
At end A:

$$G_A = \infty$$

(The theoretical G value rather the recommended G value of 10 is used in this example so the result can be compared with the theoretical solution directly.)
At end B:

$$G_B = \frac{(EI/L)}{(EI'/L)} = 1.6$$

Step 3: Determine K
From the alignment chart of Fig. 2.2b, we obtain K=2.5. Since the two columns of the frame are identical (they have the same EI, subjected to the same loading and have the same rotational restraints), the solution is complete. This calculated value of K compares favorably with the theoretical K value of 2.52 [Lui and Chen 1988].

Before proceeding any further, it is of interest to see how K varies with $\bar{R}_k$ for a variety of G values calculated at the beam-to-column joints. The variation of K is shown in Figure 3.8 as $\bar{R}_k$ changes from a near simple to a near rigid connection for several values of column to beam stiffness ratio $G=(I/L)_{col}/(I/L)_{beam}$. The exact solution (designated as E in the legend) is shown for the case of G=1, while approximate solutions (designated as Z in the legend) are shown for G=0.1, 0.25, 1, 2, and 10. The approximate

---

[3] The procedure described in this chapter applies equally well for calculating inelastic K factor when $\hat{\tau} \neq 1$.

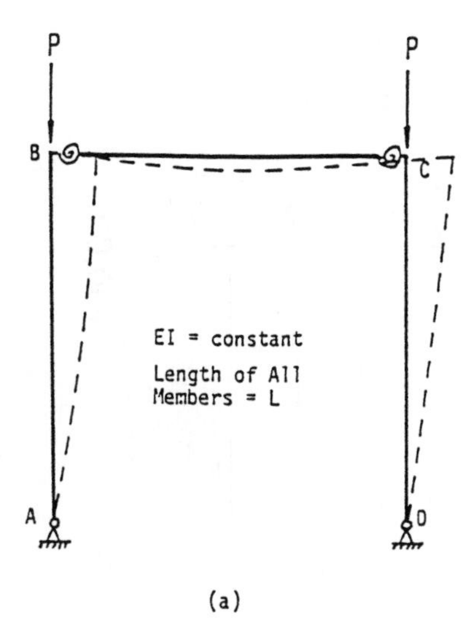

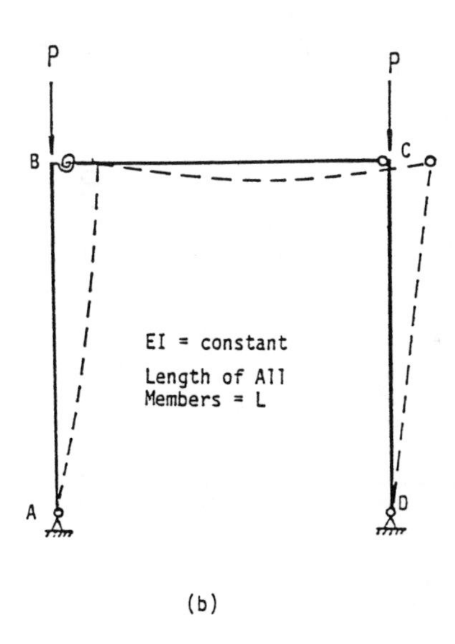

Figure 3.7 Buckling Analyses of Partially Restrained Frames

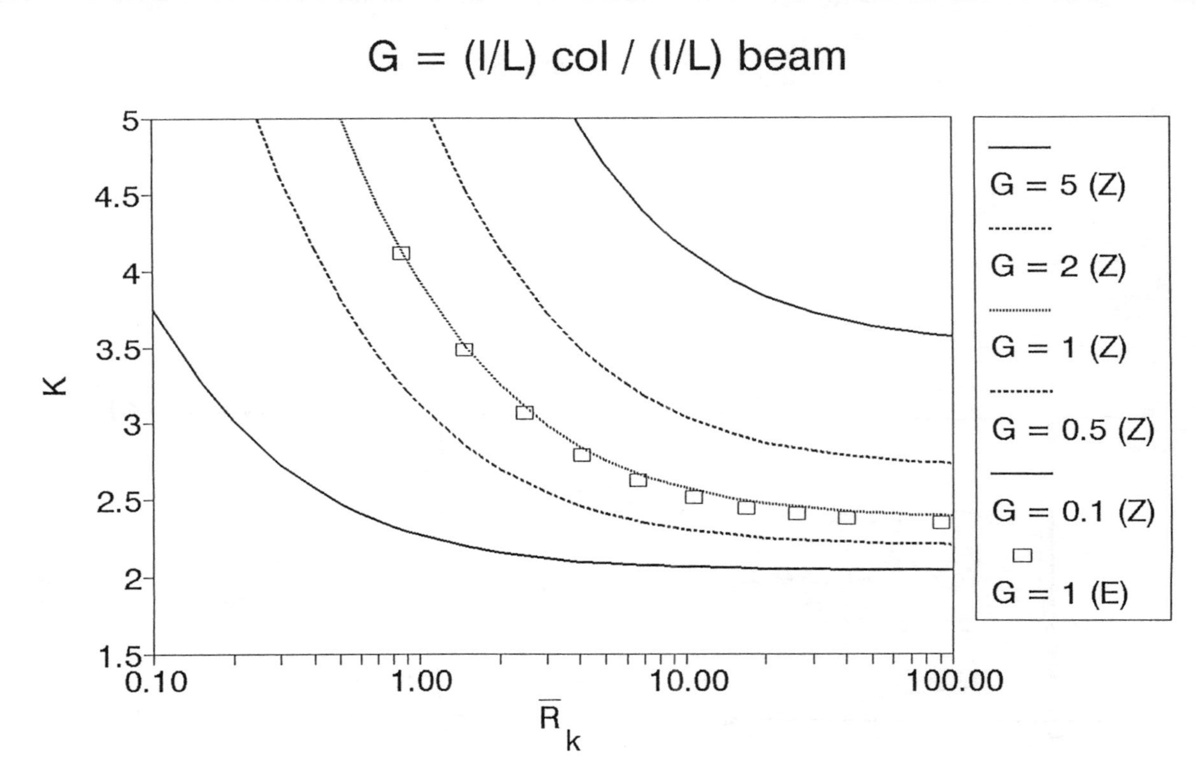

Figure 3.8 Effect of $\overline{R}_k$ and G on K for the pinned-base portal frame of Example 1

solutions were calculated using the equation

$$K = \sqrt{\frac{1}{R}} \tag{3.25}$$

where R is defined as the ratio of the axial force in the columns at incipient instability to $P_e = \pi^2 EI/L^2$. For this simple case, R is given by [Zoetmeijer 1989]

$$R = \frac{1}{4}\left[\frac{1}{1 + \dfrac{\pi^2 G(\bar{R}_k + 6)}{24\bar{R}_k}}\right] \tag{3.26}$$

From Fig. 3.8 it can be seen that very little change in K occurs when $\bar{R}_k$ is increased beyond 20 since the frame is rapidly approaching the rigid condition. The figure also demonstrates that the uncertainty arises in determining connection stiffness (which experimental studies have shown can vary as much as ±20% for nominally identical connections) does not affect K significantly unless $\bar{R}_k$ is small or G is large.

Figure 3.9 shows similar results as Fig. 3.8 but for the cases of either fixed or pinned bases (as originally assumed in this example), with or without the consideration of inelastic buckling characteristics. Note that the data are plotted on a log-log scale. From the figure it can seen that while the effect of inelasticity on K is small for pinned-base frames, the effect is quite noticeable for fixed-base frames. In reality, ideally pinned and fully fixed support conditions are rarely encountered. However, it suffices to say that the significance of inelasticity will lie somewhere in between the two extreme cases.

Example 2 Now, consider the frame in Fig. 3.7b. The connection at joint C is a pinned connection, making column CD a leaning column. Again, the elastic effective length factor of column AB is to be determined.

Step 1: Determine I'

Using Eq. (3.17) with dM_F/dM_N=0 and $\bar{R}_{kN}$=$\bar{R}_k$=10, we obtain

$$I' = \frac{I}{2.6}$$

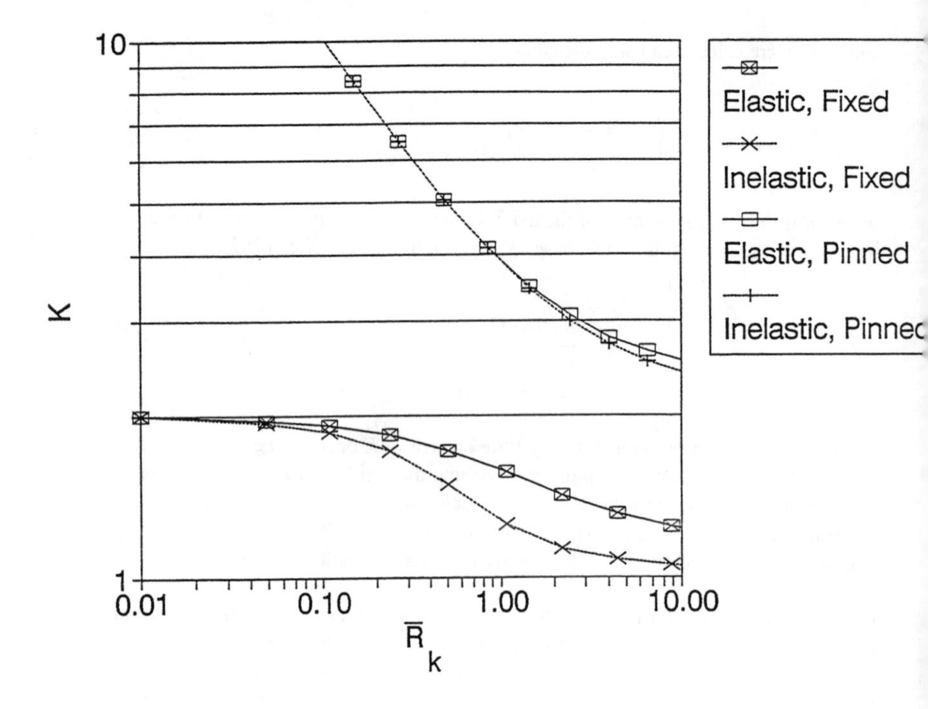

Figure 3.9 Effect of base fixity and inelasticity on K

Step 2: Calculate G_A and G_B
At end A:

$$G_A = \infty$$

At end B:

$$G_B = \frac{(EI/L)}{(EI'/L)} = 2.6$$

Step 3: Determine K
From the alignment chart of Fig. 2.2b, we obtain K=2.8. Because of the presence of the leaning column CD, further modification to K is necessary.

Step 4:
Not applicable for this problem since the only non-leaning column is column AB.

Step 5:
Using Eq. (2.25b) with ℓ=1 and recognizing that the only non-leaning column is column AB, we have

$$(K_{K_n})_{AB} = \sqrt{\frac{2P}{P}\frac{I}{(\frac{I}{2.8^2})}} = 3.96$$

Note that the summation in the denominator consists of just one term because column AB is the only non-leaning column in the frame. The theoretical K value for column AB obtained from a system buckling analysis (Section 2.3.3) using stability functions (Appendix A) is 3.93. Thus, good agreement is obtained.

Using Eq. (3.25) with R for this case given by [Zoetmeijer 1989]

$$R = \frac{1}{8}\left[\frac{1}{1 + \frac{\pi^2 G(\bar{R}_k + 3)}{12\bar{R}_k}}\right] \tag{3.27}$$

where $G=(I/L)_{column}/(I/L)_{beam}$=1 and $\bar{R}_k$=10, we obtain K=4.07, which is quite close to the theoretical solution.

Consider again the frame in Fig. 3.7a. If the connections are nonlinear, the "windward" connection (connection B) will unload and leeward connection (connection C) will load as the frame buckles in the indicated sway mode. The incremental moment ratio dM_F/dM_N will fall between 0 and 1, giving a K value for column AB in the range of 2.52

to 3.93. K=3.93 is thus a conservative estimate of the effective length factor for column AB.

Example 3 As a third example, consider the pinned-based simple portal frame shown in Fig. 3.10. Using β=0.5, γ=1, a W24x76 section (A_b=22.4 in^2, I_b= 2100 in^4) for the beam and a W14x74 section (A_c=21.8 in^2, I_c=796 in^4) for the columns, the elastic effective length factors of Columns AB and CD are to be calculated for nine cases. Each case corresponds to a specific value of rotational stiffness for the connection at C. Case 1 and case 9 correspond to the extreme conditions when the connection at C is fully rigid and ideally pinned, respectively. The connection at B is assumed to be rigid for all cases.

Step 1-3: The calculations are presented in tabular form as follows for Column AB:

Case	$\bar{R}_{kN}$ $=R_{kB}L_b/EI_b$	dM_F/dM_N $=dM_C/dM_B$	I_b' (from Eq. (3.17)) (in^4)	G_A	G_B	K_{AB} (from nomograph)
1	∞	1	2100	∞	0.975	2.32
2	∞	0.998	2096	∞	0.977	2.32
3	∞	0.987	2073	∞	0.987	2.32
4	∞	0.893	1897	∞	1.079	2.35
5	∞	0.385	1300	∞	1.575	2.51
6	∞	7.77x10^{-2}	1092	∞	1.874	2.60
7	∞	8.36x10^{-3}	1054	∞	1.942	2.62
8	∞	8.42x10^{-4}	1050	∞	1.949	2.62
9	∞	0	1050	∞	1.949	2.62

Note that for Column AB the near end of the beam is at B and the far end is at C. Thus, the moment ratio dM_F/dM_N is equal to dM_C/dM_B. This moment ratio was obtained from a first-order analysis in which the frame was subjected to a fictitious lateral load *only*, the magnitude of which is proportional to the applied story gravity loads. The rationale behind subjecting the frame to this fictitious lateral load is to create a deflected geometry for the frame which closely resembles its sidesway buckled shape [Lui 1992]. Infinity rather than 10 (as recommended in the AISC Manual) was used for G_A for all cases so that a direct comparison with the theoretical K factors could be made later.

Since the axial loads and the rotational restraints are different for the two columns, modification to the effective length factors calculated above must be made. This

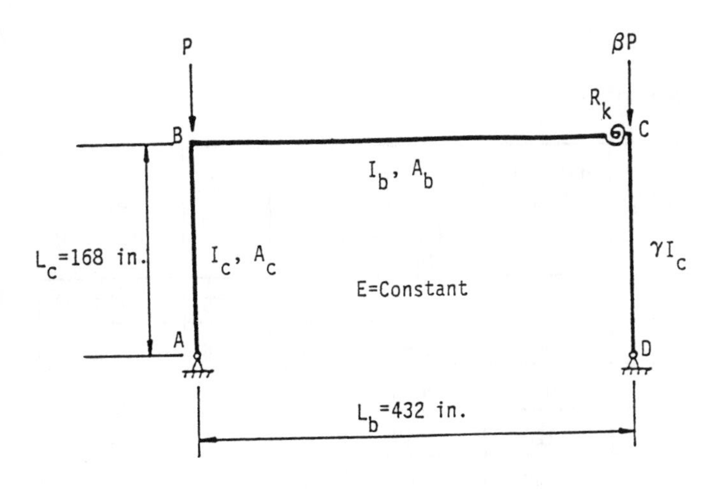

Case	$\overline{R}_k = R_k L_b / EI_b$
1	∞ (Rigid Connection)
2	1000
3	100
4	10
5	1
6	0.1
7	0.01
8	0.001
9	0 (Pinned Connection)

Figure 3.10 Buckling Analysis of a Partially Restrained Frame with Rigid Connection at B and Flexible Connection at C

modification is carried out in Steps 4 and 5 as shown below.

Step 4: Repeat Steps 1-3 for Column CD.

Case	$\bar{R}_{kN}$ $=R_{kC}L_b/EI_b$	dM_F/dM_N $=dM_B/dM_C$	I_b' (from Eq. (3.17)) (in^4)	G_D	G_C	K_{CD} (from nomograph)
1	∞	1	2100	∞	0.975	2.32
2	1000	1.002	2092	∞	0.978	2.32
3	100	1.013	2006	∞	1.020	2.33
4	10	1.120	1419	∞	1.442	2.47
5	1	2.597	389	∞	5.262	3.48
6	0.1	12.87	42.74	∞	47.89	9.06
7	0.01	119.6	4.353	∞	470.2	27.9
8	0.001	1188	0.4362	∞	4693	87.9
9	0	∞	0	∞	∞	∞

For Column CD the near end of the beam is at C and the far end is at B, the moment ratio dM_F/dM_N is therefore dM_B/dM_C, which is the reciprocal of the moment ratio for Column AB. Infinity rather than 10 was used for G_D to allow for a direct comparison with the theoretical values of K.

Step 5: Use Eq. (2.25b) to calculate K_{K_n} for Columns AB and CD.

Case	$(K_{K_n})_{AB} = \sqrt{\dfrac{1.5P}{P}\dfrac{I_c}{I_c(\frac{1}{K_{AB}^2}+\frac{1}{K_{CD}^2})}}$	$(K_{K_n})_{CD} = \sqrt{\dfrac{1.5P}{0.5P}\dfrac{I_c}{I_c(\frac{1}{K_{AB}^2}+\frac{1}{K_{CD}^2})}}$
1	2.01	2.84
2	2.01	2.84
3	2.02	2.85
4	2.09	2.95
5	2.49	3.52
6	3.06	4.33
7	3.19	4.51
8	3.21	4.53
9	3.21	4.54*
*Use K=1 for design, see explanation in Section 3.6		

In the following table, the K values computed using the proposed procedure are compared with those calculated from a system buckling analysis (Section 2.3.3) using stability functions (Appendix A). Excellent correlations are observed.

Case	$\bar{R}_{kC}$ $=R_{kC}L_b/EI_b$	K_{AB}		K_{CD}	
		Proposed	Theory	Proposed	Theory
1	∞	2.01	2.01	2.84	2.84
2	1000	2.01	2.01	2.84	2.84
3	100	2.02	2.02	2.85	2.86
4	10	2.09	2.09	2.95	2.96
5	1	2.49	2.50	3.52	3.53
6	0.1	3.06	3.04	4.33	4.29
7	0.01	3.19	3.16	4.51	4.47
8	0.001	3.21	3.18	4.53	4.49
9	0	3.21	3.18	4.54*	4.49*

*Use K=1 for design, see explanation in Section 3.6

The variation of K for Column CD with the nondimensional connection stiffness at C, $\bar{R}_{kC}$, is plotted in Fig. 3.11. It can be seen that the proposed method gives rather accurate results over a large range of connection stiffness, ranging from the near pinned condition to the near rigid condition.

The above example shows that the effective length factors for both columns increase as the stiffness of Connection C, and hence the stiffness of the frame, decreases. A larger K value means that the design will result in a heavier column. However, it will be demonstrated in Section 3.6 that a *threshold value* of K exists for Column CD. Designing the column for an effective length factor greater than this threshold value *will not increase* the sidesway uninhibited critical load of the frame.

For the nine cases shown, Case 9 is of particular interest. When Connection C is a pinned connection, Column CD becomes a leaning column. The K factor for this column was calculated to be 4.54 when the frame buckled in the sidesway uninhibited mode shown in Fig. 3.12a. However, K becomes 1 when the frame buckles in a sidesway inhibited mode as shown in Fig. 3.12b. Regardless of the mode of buckling, designing Column CD for K>1 is meaningless because a leaning column can not contribute any lateral stiffness to a frame. The proper procedure is to design the column using K=1, even when the sidesway uninhibited mode controls. It will be shown in the following section that the threshold K value for a leaning column is one. Therefore, designing the column for K>1 is unwarranted because no increase in P_{cr} for the frame will result.

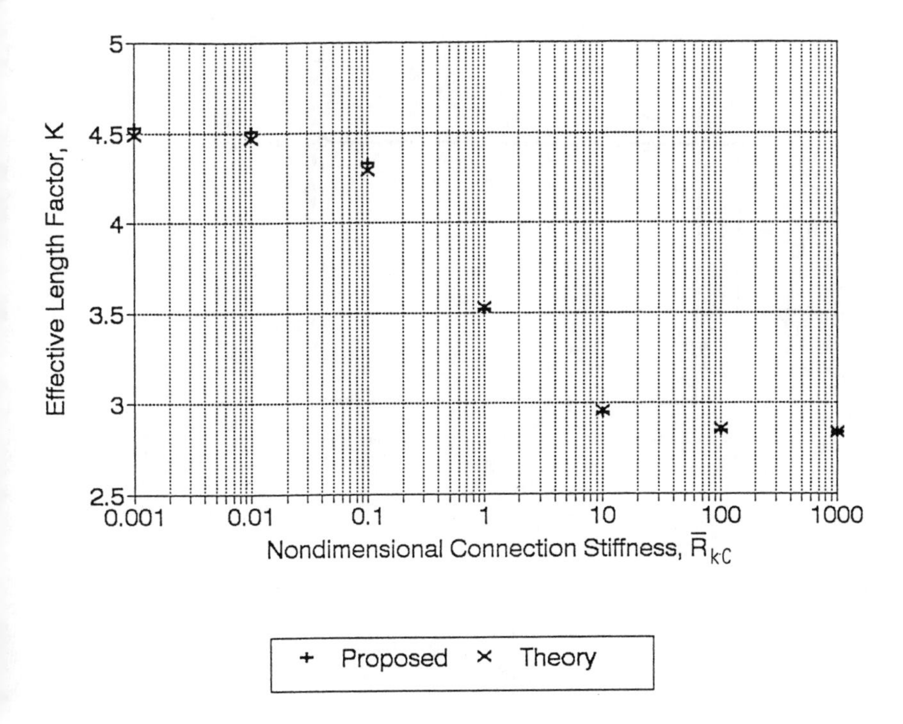

Figure 3.11 Variation of *K* factor with Connection Stiffness

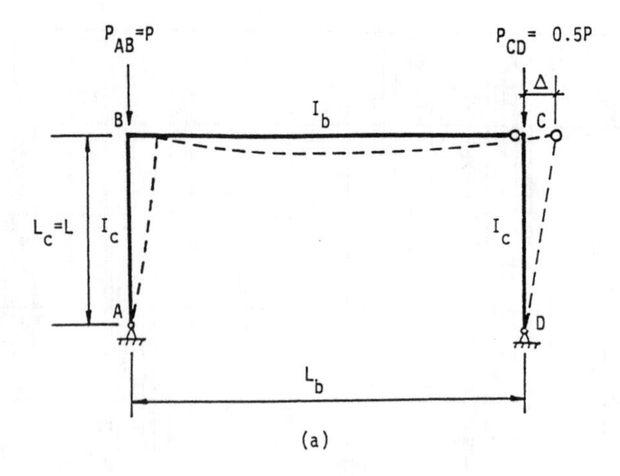

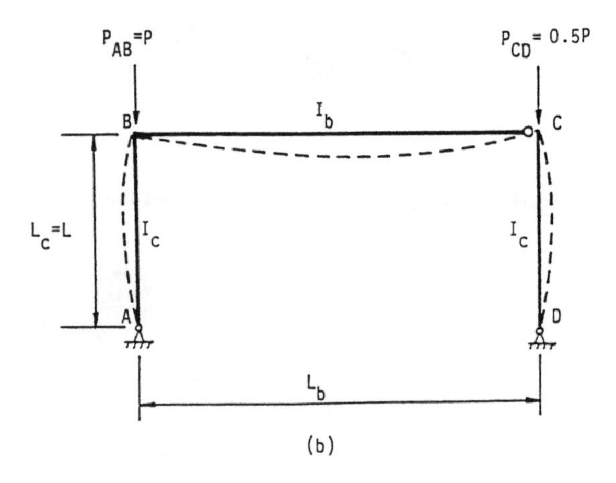

Figure 3.12 Sidesway Uninhibited and Sidesway Inhibited Modes of Instability

3.6 The *K* Factor Paradox

The "*K* Factor Paradox" [Cheong-Siat-Moi 1986] arises when one uses the effective length approach to design a column that is flexibly connected at one end and either pinned or flexibly connected at the other end. Column CD of Fig. 3.10 is an example of such a column. As can been seen in Example 3 of the preceding section, if the connection at C is rigid, the *K* factor to be used for the design of the column is 2.84. When the stiffness of the connection and hence the lateral stiffness of the frame decreases, a larger *K* factor is required for the design of the column. In the extreme case when the stiffness of the connection vanishes, Column CD becomes a leaning column and a (design) *K* factor of 1 will be used. This sudden change of *K* factor when one designs Column CD as the connection stiffness decreases gradually from a finite value to zero has generated much controversy regarding the validity of the effective length approach for frame design. In an attempt to quell the controversy, Aristizabal-Ochoa [1994] proposed the use of K>1 (i.e., the theoretical *K* value) for the design of such columns. However, the use of K>1 is unwarranted because the resulting heavier column will not enhance the sidesway uninhibited critical load of the frame. It will further be demonstrated that a threshold value of *K* exists [Lui 1993, 1995] beyond which an increase in column size will not result in an increase of the sidesway uninhibited critical load of the frame.

Consider the frame shown in Fig. 3.10. If a system buckling analysis (see Section 2.3.3 and Appendix A) is performed on the frame using β=0.5, I_b/I_c=2.638, and $\overline{R}_{kC}$=∞, 10, 0.1, 0, one can obtain the sidesway uninhibited critical load P_{cr} of the frame for various values of γ. The results of the analysis, together with the corresponding effective length factors for Columns AB and CD are presented in the following table.

$\bar{R}_{kC}$	γ	K_{AB}	K_{CD}	$P_{cr}L_c^2/EI_c$
∞ (Rigid)	0.1	2.96	1.32	1.69
	0.5	2.29	2.29	2.84
	1.0	2.01	2.84	3.66
	5.0	1.66	5.26	5.35
	10.0	1.61	7.20	5.72
	20.0	1.58	9.99	5.93
10	0.1	2.96	1.32	1.69
	0.5	2.33	2.33	2.73
	1.0	2.09	2.95	3.39
	5.0	1.80	5.69	4.57
	10.0	1.76	7.85	4.81
	20.0	1.73	10.95	4.94
0.1	0.1	3.08	1.38	1.56
	0.5	3.04	3.04	1.60
	1.0	3.03	4.29	1.61
	5.0	3.03	9.58	1.61
	10.0	3.03	13.55	1.61
	20.0	3.03	19.16	1.61
0 (Pinned)	0.1	3.18	1.42	1.47
	0.5	3.18	3.18	1.47
	1.0	3.18	4.49	1.47
	5.0	3.18	10.05	1.47
	10.0	3.18	14.21	1.47
	20.0	3.18	20.09	1.47

For the case when the connection at C is rigid, an increase in moment of inertia for Column CD increases the sidesway uninhibited critical load P_{cr} of the frame. This phenomenon is also observed for the case when $\bar{R}_{kC}$=10, although the rate of increase for P_{cr} is somewhat slower. For the case when $\bar{R}_{kC}$=0.1, a large increase in the moment of inertia for Column CD only results in a slight increase in P_{cr}, and when P_{cr} reaches a value of $1.61EI_c/L_c^2$, no further increase in P_{cr} occurs. For the case when $\bar{R}_{kC}$=0, P_{cr} remains constant regardless of the value of the moment of inertia for Column CD.

If Connection C is rigid, any increase in moment of inertia for Column CD will enhance the lateral stiffness of the frame. The result is a rise in the sidesway uninhibited critical load for the frame. Of course, P_{cr} can not increase beyond the sidesway inhibited critical load. If the stiffness of Connection C is finite, an increase in moment of inertia for Column CD will raise P_{cr} only if the column buckles in the mode shown in Fig. 3.13a, in which the column bends as well as sways at incipient instability. When the moment of inertia is increased to such an extent that the column becomes unstable in the mode shown in Fig. 3.13b, no further increase in P_{cr} will occur. Note that the column in Fig. 3.13b experiences very little or no bending deformation at incipient instability. As a result, the flexural rigidity *EI*, which governs the bending resistance of the column, will have negligible effect on the sidesway uninhibited critical load of the frame. The exact magnitude of the moment of inertia that will cause the column to experience instability of the form shown in Fig. 3.13b is not readily obtainable because it depends on a number of factors such as frame geometry, load distribution as well as connection, beam and column stiffness. Nevertheless, it can be shown [Lui 1993, 1995] that P_{cr} will assume a stationary value for an *end-restrained* (i.e., a non-leaning) column when

$$(P_{cr})_{sidesway\ uninhibited} = S_{k,system}L + \frac{R_{k,system}}{L} \ngtr (P_{cr})_{sidesway\ inhibited} \tag{3.28}$$

where $S_{k,system}$ and $R_{k,system}$ are the system translational and rotational restraints on the column, and L is the length of the column[4].

The effective length factor that corresponds to this stationary value of P_{cr} is referred to as the threshold effective length factor, $K_{threshold}$, where

$$K_{threshold} = \sqrt{\frac{\pi^2 EI}{S_{k,system}L^3 + R_{k,system}L}} \tag{3.29}$$

Although the theoretical K factor may exceed $K_{threshold}$, designing the end-restrained column for $K>K_{threshold}$ is unjustifiable because no increase in the sidesway uninhibited critical load of the frame will result. This is because the column simply undergoes an essentially rigid body rotation at incipient instability as shown in Fig. 3.13b.

For the special case of a leaning column in which there is no rotational restraint exerted at either end of the member, the column will experience instability in the form

[4] A finite stationary value will exist for P_{cr} only if both $S_{k,system}$ and $R_{k,system}$ are finite. For example, if a column is fixed at one end, $R_{k,system}$ will be infinite and P_{cr} will not have a finite stationary value.

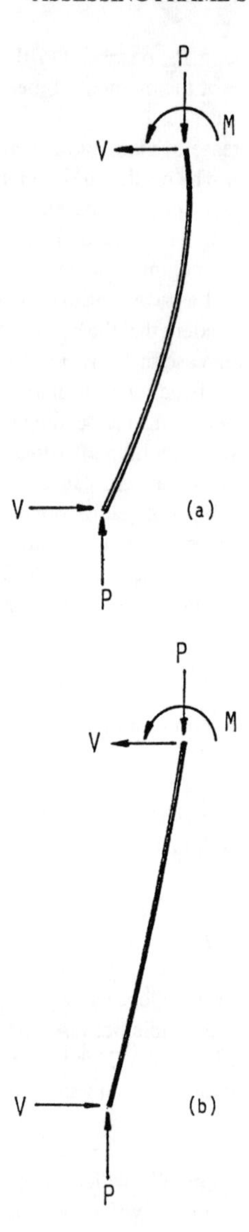

Figure 3.13 Sidesway Instability Modes for Column CD at P_{cr}

shown in Fig. 3.13b if the frame buckles in the sidesway uninhibited mode. The sidesway uninhibited critical load of the frame will be *independent* of the flexural rigidity of the column. The threshold effective length factor of the leaning column is therefore equal to 1. A leaning column only needs to be designed for stability against the sidesway inhibited mode of buckling using K=1. Designing the column for K>1 only increases the size of the column but it will have no effect on the sidesway uninhibited critical load of the frame.

Generally speaking, it is rather onerous to calculate this stationary value of P_{cr} and the corresponding threshold K value because $S_{k,system}$ and $R_{k,system}$ are not easy to calculate. However, it should be noted that proportioning a column for $K>K_{threshold}$ will only result in a *conservative* design.

In what follows, it will be demonstrated using the frame in Fig. 3.10 that one can design Column CD (i.e., the partially restrained column) using K=1 regardless of the stiffness of the connection at C, provided that proper adjustment is made to the K factor of Column AB [Lui 1993]. The procedure is demonstrated as follows:

For P_{cr} to remain unchanged for the frame, the set of K factors used for design and the theoretical K values obtained from a system buckling analysis must satisfy the condition

$$(P_{cr,story})_{design} = (P_{cr,story})_{theory} \tag{3.30a}$$

or

$$\sum_{all}\left(\frac{\pi^2 E\bar{I}}{(KL)^2}\right)_{design} = \sum_{all}\left(\frac{\pi^2 EI}{(KL)^2}\right)_{theory} \tag{3.30b}$$

where the bar above I denotes new values for the moment of inertia of the columns.

If the load distribution across the story is to remain unchanged (i.e., for a constant β), we must have

$$\frac{\pi^2 E\bar{I}_{AB}}{(K_{AB}^2)_{design}L^2} = \frac{\pi^2 EI_{AB}}{(K_{AB}^2)_{theory}L^2} \tag{3.31a}$$

and

$$\frac{\pi^2 E\bar{I}_{CD}}{(K_{CD}^2)_{design}L^2} = \frac{\pi^2 EI_{CD}}{(K_{CD}^2)_{theory}L^2} \tag{3.31b}$$

If one chooses to design Column CD using $(K_{CD})_{design}$=1, $(K_{AB})_{design}$ can not be arbitrarily chosen. This is because $(K_{AB})_{design}$ and $(K_{CD})_{design}$ must satisfy Eq. (2.25b).

As an illustrative example, consider the case in which $\bar{R}_{kC}$=0.1, $\gamma = I_{CD}/I_{AB}$=0.5, we have, from the above table, $(K_{AB})_{theory}=(K_{CD})_{theory}$=3.04. If we design the frame using these theoretical K values, the resulting columns will have moment of inertia of I_c and $0.5I_c$, respectively, for the left and right columns. Now, suppose we are to design Column CD using K=1, we have from t Eq. (3.31b),

$$\bar{I}_{CD} = \frac{I_{CD}}{3.04^2} = \frac{0.5I_c}{3.04^2} = 0.0541I_c$$

and, from Eq. (3.31a), we have

$$\frac{\bar{I}_{AB}}{(K_{AB}^2)_{design}} = \frac{I_{AB}}{3.04^2} = \frac{I_c}{3.04^2} = 0.108I_c$$

The two unknowns, $\bar{I}_{AB}$ and K_{AB}, can be solved by using the above equation and Eq. (2.25b) giving $\bar{I}_{AB}$=1.18I_c and $(K_{AB})_{design}$=3.31. If P_{cr} is to remain unchanged and one designs Column CD using K=1, one should design Column AB using K=3.31. The two options for designing the frame are summarized in the following table.

Option	Moment of Inertia for Column AB, I_{AB}	Moment of Inertia for Column CD, I_{CD}	Sidesway Uninhibited Critical Load, P_{cr}
Option 1: Use K_{AB}=3.04 and K_{CD}=3.04	I_c	$0.5I_c$	$1.60EI_c/L_c^2$
Option 2: Use K_{AB}=3.31 and K_{CD}=1	$1.18I_c$	$0.0541I_c$	$1.60EI_c/L_c^2$

Note that P_{cr} *is* the same for both idealizations. In either case, the sidesway uninhibited critical load for each column can not exceed their respective sidesway inhibited value. This condition is satisfied as long as the imposed limits discussed in Section 2.4.4 are satisfied.

The above example demonstrates that the story based procedure inherently accounts for the stability interaction effect among columns in a story when the frame buckles in a sidesway uninhibited mode. The decrease in lateral resistance of a column as a result of a reduction in its flexural rigidity can be compensated by an increase in flexural rigidity of another column in the story.

It should be pointed out that the two design options shown above are valid only if no primary moments are acting in the columns. If primary moments exist due to the presence of lateral loads, the columns must be designed in accordance with the beam-column interaction equations.

3.7 Design of Partially Restrained Frames

The simplest approach to design Type PR frames is to use the "simple framing" provision of the AISC LRFD Specification. Under this provision, beams are designed as simply-supported members insofar as gravity loads are concerned; no consideration is given to the beneficial effect of any negative moments which may present at the beam ends due to the restraining effect of the connections. The columns, on the other hand, are designed assuming that the connections can provide full rotational restraint for lateral loads. This approach, while historically safe, often results in a design in which the beams are oversized and the columns are undersized. A more rational approach, in light of the limit state philosophy of LRFD, should therefore be used.

If a second-order analysis computer program capable of modelling the restraint characteristic of a connection [Lui 1985; Goto and Chen 1987; Lindsey 1987; Chen and Toma 1993] is available, the program should be used as an analysis tool for the design of the partially restrained frame. In lieu of such programs, simplified analysis methods [Lui, 1988, King and Chen 1993] can be used to determine member forces and moments necessary for the design.

Over the years, various simplified approaches for the design of PR frames have been proposed(see, for example, Disque [1975], Lui [1985, 1996]). In Disque's method of PR frame design [Disque 1975], all loading connections are assumed to have zero stiffness and all unloading connections are assumed to be rigid. Interior columns are designed for moments equal to one-half of the difference between (1) the connections' nominal moment capacities M_{cn} on either side of the column, or (2) the larger of the two connection nominal moment capacities and the fixed end beam moment due to gravity load, whichever is larger. Exterior columns are design for moments of $M_{cn}/2$. The factor of one-half is used as a result of the assumption that the beam moments are distributed equally to the columns above and below. Lui's approach for PR frame design [Lui 1985] is an extension of Disque's approach in which special attention is paid to the influence of the connections' loading and unloading characteristics on frame behavior. Details of these two approaches for PR frame design are described in Chen and Lui [1991] and will not be repeated here.

Additional approaches for partially restrained frame design include those by Dewkett [Dewkett 1984], Ackroyd [Ackroyd 1985] and Xu et al. [1993, 1995], among others. Dewkett's approach for partially restrained frame design is quite similar to Disque's approach. The only difference between the two approaches is in the design of the exterior columns. In Dewkett's approach the exterior columns are designed to develop the full connection moment capacity M_{cn} instead of $M_{cn}/2$. According to the study by Dewkett, exterior columns tend to be under-designed by the Disque's method. By designing these columns for a larger moment, the problem of under-design can be somewhat alleviated.

Ackroyd's partially restrained frame design method aims to redistribute the connection stiffness between the interior and exterior columns. In Ackroyd's approach, girders and columns are designed as in Disque's method. However, the exterior beam-to-column connections are deliberately designed to be more flexible. The intent is to reduce the amount of gravity moment induced in the exterior columns. To compensate for the loss in frame stiffness as a result of these weaker exterior connections, the designer would design stronger interior connections. Studies by Ackroyd have shown that this approach would lead to a higher reliability than Disque's method. In designing the columns, an effective length factor of 1.5 was recommended [Gerstle and Ackroyd 1990]. The approach has been refined [Cronembold 1986; Ackroyd 1987] and implemented in an electronic spreadsheet program to facilitate the design of partially restrained frames [Ackroyd 1990].

Xu et al. [1993, 1995] utilize a computer-automated optimization scheme for the design of PR frames. The nonlinear behavior of the connections is modelled using a four-parameter power model. An iterative procedure based on secant connection stiffness is used to obtain member forces and moments for design.

References

Ackroyd, M.H. [1985], *Design of Flexibly Connected Steel Frames*, Final Research Report to the American Iron and Steel Institute for Project No. 333, Rensselaer Polytechnic Institute, Troy, NY.

Ackroyd, M.H. [1987], "Simplified Frame Design of Type PR Construction," *Engineering Journal*, American Institute of Steel Construction, 24(4), 141-146.

Ackroyd, M.H. [1990], "Electronic Spreadsheet Tools for Semi-Rigid Frames," *Engineering Journal*, American Institute of Steel Construction, 27(2), 69-78.

Ackroyd, M.H., and Gerstle, K.H. [1982], "Strength and Stiffness of Type 2 Steel Frames," *Trans, Journal of the Structural Division*, ASCE, 108(7), 1541-1556.

AISC [1989], *Specifications for Structural Steel Buildings: Allowable Stress Design and Plastic Design*, 9th Edition, American Institute of Steel Construction, Chicago, IL.

AISC [1993], *Load and Resistance Factor Design Specification for Structural Steel Buildings*, 2nd edition, American Institute of Steel Construction, Chicago, IL.

AISC [1994], *Manual of Steel Construction, Load and Resistance Factor Design*, 2nd edition, Volume I and II, American Institute of Steel Construction, Chicago, IL.

AISC [1996], *Design of PR Composite Frames*, Steel Design Guide Series, American Institute of Steel Construction, Chicago, 1996 (in press).

Ammerman, D.J [1986], *An Experimental Investigation into the Behavior of Semi-Rigid Composite Connections*, M.S. Thesis, The Graduate School, University of Minnesota, Minneapolis, MN.

Ammerman, D.J. and Leon, R.T. [1990], "Unbraced Frames with Semi-Rigid Composite Connections," *Engineering Journal*, American Institute of Steel Construction, 27(1), 12-21.

Ang, K.M. and Morris, G.A. [1984], "Analysis of Three-Dimensional Frames with Flexible Beam-Column Connections," *Canadian Journal of Civil Engineers*, 11, 245-254.

Aristizabal-Ochoa, J.D. [1994], "K-Factor for Columns in any Type of Construction: Nonparadoxical Approach," *Journal of Structural Engineering*, ASCE, 120(4), 1272-1290.

Baker, J.F. [1954], *The Steel Skeleton, Vol 1: Elastic Behavior and Design*, Cambridge University Press, London, England.

Barakat, M. and Chen, W.F. [1990], "Practical Analysis of Semi-Rigid Frames," *Engineering Journal*, American Institute of Steel Construction, 27(2), 54-68.

Barakat, M. and Chen, W.F. [1991], "Design Analysis of Semi-Rigid Frames Evaluation and Implementation," *Engineering Journal*, American Institute of Steel Construction, 28(2), 55-64.

Batho, C. and Rowan, H.C. [1934], "Investigation of Beam and Stanchion Connections," 2nd Report, Steel Structures Research Committee, Department of Scientific and Industrial Research of Great Britain, His Majesty's Stationery Office, London, England.

Bjorhovde, R. [1984], "Effect of End Restraint on Column Strength - Practical Applications," *Engineering Journal*, American Institute of Steel Construction, 21(1), First Quarter, 1-13. (Also see the discussions of the paper by K.H. Gerstle, E.M. Lui and W.F. Chen, and closure by the author in *Engineering Journal*, AISC, 22(1), First Quarter, 1985, 41-45.)

Bjorhovde, R., Brozzetti, J.and Colson, A. [1990], " A Classification System for Beam to Column Connections," *Journal of Structural Engineering*, ASCE, 116(11), 3059-3076.

Chen, W.F. [1987], Editor, *Joint Flexibility in Steel Frames*, Elsevier, New York, NY.

Chen, W.F. [1988], Editor, *Steel Beam-to-Column Building Connections*, Elsevier, New York, NY.

Chen, W.F. and Kishi, N. [1989], "Semi-Rigid Steel Beam-to-Column Connections: Data Base and Modeling," *Journal of Structural Engineering*, ASCE, 115(1), 105-119.

Chen, W.F. and Lui, E.M. [1987], *Structural Stability - Theory and Implementation*, Elsevier, New York, NY.

Chen, W.F. and Lui, E.M. [1991], *Stability Design of Steel Frames*, CRC Press, Boca Raton, FL.

Chen, W.F. and Toma, S. [1994], *Advanced Analysis of Steel Frames-Theory, Software and Applications*, CRC Press, Boca Raton, FL.

Cheong-Siat-Moi, F. [1986], "K-Factor Paradox," *Journal of Structural Engineering*, ASCE, 112(8), 1747-1760.

Cronembold, J.R. [1986], *Evaluation and Design of Type 2 Steel Building Frames*, M.S. Thesis, Department of Civil Engineering, Rensselaer Polytechnic Institute, Troy, NY.

DeFalco, F. and Marino, F.J. [1966], "Column Stability in Type 2 Construction," *Engineering Journal*, American Institute of Steel Construction, 3(2), 67-71.

Dewkett, K.A. [1984], *An Evaluation of Disque's Directional Moment Connection Design Method*, M.S. Thesis, Department of Civil Engineering, Rensselaer Polytechnic Institute, Troy, NY.

Dhillon, B.S. and Abdel-Majid, S. [1990], "Interactive Analysis and Design of Flexibly Connected Frames," *Computers and Structures*, 36(2), 189-202.

Disque, R.O. [1975], "Directional Moment Connections - A Proposed Design Method for Unbraced Steel Frames," *Engineering Journal*, American Institute of Steel Construction, 12(1), 14-18.

Driscoll, G.C. [1976], "Effective Length of Columns with Semi-Rigid Connections," *Engineering Journal*, American Institute of Steel Construction, 13(4), 109-115.

Eurocode 3, *Design of Steel Structures, Part 1.1 (ENV 1993-1-1)*, European Committee for Standardization, Brussels.

Frye, M.J. and Morris, G.A. [1976], "Analysis of Flexibly-Connected Steel Frames," *Canadian Journal of Civil Engineers*, 2(3), 280-291.

Gerstle, K.H. and Ackroyd, M. H. [1990], "Behavior and Design of Flexibly-Connected Building Frames," *Engineering Journal*, American Institute of Steel Construction, 27(1), 22-30.

Geschwindner, L.F. [1991], "A Simplified Look at Partially Restrained Beams," *Engineering Journal*, American Institute of Steel Construction, 28(2), 73-78.

Goto, Y. and Chen, W.F. [1987], "On the Computer-Based Design Analysis for Flexibly-Connected Building Frames," *Journal of Constructional Steel Research, Special Issue on Joint Flexibility in Steel Frames* (W.F. Chen, Ed.), 8, 203-231.

Goto, Y., Satsuki, S. and Chen, W.F. [1993], "Stability Behavior of Semi-Rigid Sway Frames," *Computers and Structures*, 15(3), 209-228.

Goverdhan, A.V. [1983], *A Collection of Experimental Moment-Rotation Curves and Evaluation of Prediction Equations for Semi-Rigid Connections*, Master's Thesis, Vanderbilt University, Nashville, TN.

Johnston, B. and Mount, E.H. [1942], "Analysis of Buildings with Semi-Rigid Connections," *Transactions of ASCE*, No. 2152, 993-1019.

Kennedy, D.J.L. [1969], "Moment-Rotation Characteristics of Shear Connections," *Engineering Journal*, American Institute of Steel Construction, October, 105-115.

King, W.-S. and Chen, W.F. [1993], "LRFD Analysis for Semi-Rigid Frame Design," *Engineering Journal*, American Institute of Steel Construction, 30(4), 130-140.

Kishi, N. and Chen, W.F. [1986], *Data Base of Steel Beam-to-Column Connections*, Structural Engineering Report No. STR-86-26, School of Civil Engineering, Purdue University, West Lafayette, IN.

Kishi, N. and Chen, W.F. [1990], "Moment-Rotation Relations of Semi-Rigid Connections with Angles," *Journal of Structural Engineering*, ASCE, 116(7), 1813-1834.

Kishi, N, Chen, W.F., Goto, Y. and Matsuoka, K.G. [1993a], "Analysis Program for the Design of Flexibly Jointed Frames," *Computers and Structures*, 49(4), 705-714.

Kishi, N., Chen, W.F., Goto, Y. and Matsuoka, K.G. [1993b], "Design Aid of Semi-Rigid Connections for Frame Analysis," *Engineering Journal*, American Institute of Steel Construction, 30(3), 90-107.

Kishi,N. , Hasan, R., Chen, W.F. and Goto, Y. [1994], "Power Model for Semi-Rigid Connections," *Steel Structures*, Journal of Singapore Structural Steel Society, 5(1), 37-48.

Leon, R.T. and Ammerman, D.J. [1990], "Semi-Rigid Composite Connections for Gravity Loads," *Engineering Journal*, American Institute of Steel Construction, 27(1), 1-11.

Leon, R.T. [1994], "Composite Semi-Rigid Construction," *Engineering Journal*, American Institute of Steel Construction, 31(2), 57-67.

Leon, R.T. , and Shin, K.J. [1995], " Performance of Semi-Rigid Frames," *Proceedings of Structures Congress XIII*, ASCE, New York, 1020-1035.

Leon, R.T. and Hoffman, J.J. [1995], "Plastic Design of Semi-Rigid Frames," *Proceedings of the Third Int. Workshop on Connections in Steel Structures,* Trento, Italy, (to be published by Elsevier in 1996).

Lindsey, S.D., Ioannides, S. and Goverdhan, A.V. [1985], "LRFD Analysis and Design of Beams with Partially Restrained Connections," *Engineering Journal*, American Institute of Construction, 22(4), 157-162.

Lindsey, S.D. [1987], "Design of Frames with PR Connections," *Journal of Construction Steel Research*, 8, 251-260.

Lorenz, R.F., Kato, B. and Chen, W.F. [1993], Editors, *Semi-Rigid Connections in Steel Frames*, Council on Tall Buildings and Urban Habitat, McGraw-Hill, New York, NY.

Lothers, J.E. [1960], *Advanced Design in Structural Steel*, Prentice-Hall, Englewood Cliffs, NJ.

Lui, E.M. [1985], *Effects of Connection Flexibility and Panel Zone Deformation on the Behavior of Plane Steel Frames*, Ph.D. Dissertation, School of Civil Engineering, Purdue University, West Lafayette, IN.

Lui, E.M. [1988], "A Practical P-Delta Analysis Method for Type FR and PR Frames," *Engineering Journal*, American Institute of Steel Construction, 25(3), 85-98.

Lui, E.M. [1992], " A Novel Approach for K Factor Determination," *Engineering Journal*, American Institute of Steel Construction, 29(4), 150-159.

Lui, E.M. [1993], "A Simplified Analysis Method for Partially Restrained Steel Frames," Structural Engineering Report, Department of Civil and Environmental Engineering, Syracuse University, Syracuse, NY.

Lui, E.M. [1995], "Column Effective Length Factor for Semi-Rigid Frames," *Steel Structures*, Journal of Singapore Structural Steel Society, 6(1), 3-20.

Lui, E.M. [1996], "Stability Design of Partially Restrained Frames," *Proceedings, Fifth International Colloquium on Stability of Metal Structures*, Chicago, IL, 259-268.

Lui, E.M. and Chen, W.F. [1988], "Behavior of Braced and Unbraced Semi-Rigid Frames," *International Journal of Solids and Structures*, 24(9), 893-913.

Nethercot, D.A. [1985a], *Steel Beam-to-Column Connections - A Review of Test Data and its Applicability to the Evaluation of Joint Behavior in the Performance of Steel Frames*, CIRIA Project Record, RP338.

Nethercot, D.A. [1985b], "Utilization of Experimentally obtained connection data in assessing the performance of steel frames, in *Connection Flexibility and Steel Frames* (W.F. Chen, Ed.), Proceedings of a Session Sponsored by the Structural Division, ASCE, Detroit, MI, 13-37.

Nethercot, D.A., Davison, J.B. and Kirby, P.A. [1988], "Connection Flexibility and Beam Design in Non-sway Frames," *Engineering Journal*, American Institute of Steel Construction, 25(3), 99-108.

Pippard, A.J.S. and Baker, J.F., *The Analysis of Engineering Structures*, 3rd Edition, Edward Arnold Publishers, London, England.

PRCONN [1993], "Moment-Rotation Curves for Partially Restrained Connections," RMR Design Group, Inc., 4421 E. Coronado Drive, Tucson, AZ.

Rathbun, J.C. [1936], "Elastic Properties of Riveted Connections," *Transactions of ASCE*, 101, 524-563.

Salmon, C.G. and Johnson, J.E. [1990], *Steel Structure - Design and Behavior*, 3rd edition, Harper and Row, New York, NY.

Tall, L. [1964], Editor-in-Chief, *Structural Steel Design*, The Ronald Press, New York, NY.

Wang, C.K. [1993], *Intermediate Structural Analysis*, International Editions, McGraw-Hill, NY, Chapter 20.

Xu, L. and Grierson, D.E. [1993], "Computer-Automated Design of Semirigid Steel Frameworks," *Journal of Structural Engineering*, American Society of Civil Engineers, 119(6), 1740-1760.

Xu. L., Sherbourne, A.N. and Grierson, D.E. [1995], "Optimal Cost Design of Semi-Rigid, Low-Rise Industrial Frames," *Engineering Journal*, American Institute of Steel Construction, 32(3), 87-97.

Zandonini, R. and Zanon, P. [1991], "Beam Design in PR Braced Steel Frames," *Engineering Journal*, American Institute of Steel Construction, 28(3), 85-97.

Zoetemeijer, P. [1989], "Influence of Joint Characteristics on Structural Response of Frames," in *Structural Connections-Stability and Strength* (R. Narayanan, ed.), Elsevier Applied Science, NY, 121-152.

CHAPTER 4

NOTIONAL LOAD APPROACH FOR THE ASSESSMENT OF FRAME STABILITY

4.1 Introduction

The previous two chapters investigated the use of effective lengths for the determination of the strength of columns in frames. An alternative to computing effective length factors to take into account differing end-restraints in sway columns is to use the *actual* column length (i.e., "$K = 1.0$") in conjunction with a notional lateral load and a beam-column interaction equation. For sway columns, this *notional load* approach, also termed the *equivalent imperfection* approach, accounts for many of the imperfection effects that influence column strength.

Imperfections in steel building frames can arise from a number of sources. The residual stresses in the members may be regarded as material imperfections, and the geometric imperfections relevant to the in-plane behavior of frames composed of compact cross-sections include member out-of-straightness, story out-of-plumbness, and overall non-verticality of the frame. In addition, when the cross-sections are slender, local geometric imperfections of the component plates may influence the local instability of the section. For frames which are not fully braced laterally, out-of-plane geometric imperfections may affect the behavior of the frame with respect to flexural-torsional instability.

At least three different approaches are currently in use for the design of columns in unbraced frames. The AISC LRFD Specification [AISC, 1993b] uses an effective length concept to determine the axial resistance term of the beam-column interaction equations. Second-order elastic analysis is performed on the structure assuming perfectly straight and perfectly plumb members. Out-of-straightness imperfections, out-of-plumbness imperfections and residual stresses are accounted for solely through the column curve for an equivalent pin-ended member. The Australian Standard AS4100-1990 [SA, 1990], the Canadian Standard CSA-S16.1-M94 [CSA, 1994] and the British Standard BS5950:Part 1:1990 [BSI, 1990] use a notional lateral load with second-order elastic analysis of the geometrically perfect structure to account for story out-of-plumbness under gravity loads (i.e., sway) and then use the actual member length in the beam-column interaction equations to account for member out-of-straightness and residual stresses. When second-order elastic analysis is used to determine the amplified moments, the Australian, British and Canadian Standards therefore enable beam-column design, at least as far as in-plane strength is concerned, to be performed without the need to calculate effective length factors. The European specification Eurocode 3 [CEN, 1992] is similar to AS4100-1990, CSA-S16.1-M94 and BS5950:Part 1:1990 in that a notional lateral load is used with second-order elastic analysis of the geometrically perfect frame except that the effective length for the *non-sway* (braced) mode may be used for the axial term in the beam-column interaction equations, rather than the *actual* column length.

The aim of this chapter is to investigate the calibration and use of a notional lateral load (which can be thought of as an equivalent out-of-plumbness imperfection) to account for frame imperfections in the design of columns for the strength limit state using elastic second-order analysis. To facilitate the calibration, the concept of "advanced analysis" of structural members and frames is employed. In the present context, the term advanced analysis refers to the nonlinear analysis of steel frames composed of members of compact section which have full lateral restraint [SA, 1990]. The advanced analysis should model all relevant behavioral phenomena including geometric and material nonlinearities, and imperfections and residual stresses, to such an extent that the "true" behavior and strength of the structure is captured and separate specification checks of member strength are rendered superfluous. The theoretical basis of the advanced analysis referred to throughout this chapter is described in Clarke [1994].

In the calibration process, the important distinction between an *equivalent pin-ended* column (the basis of the effective length approach) and a *physically imperfect* column (employed in advanced analysis and consequently in the calibration of the notional load approach) is emphasized. An overview of the notional load approach as it may be interpreted in the context of the AISC LRFD Specification is given in Section 4.2. The influence of imperfections on column strength, and a survey of the imperfection related provisions of several steel design specifications, is introduced in Section 4.3. In Section 4.4, the issue of the calibration and appropriate definition of the various imperfection parameters for use in advanced analysis, is discussed. Such emphasis on imperfections is important since it provides the motivation for the notional load approach as a rational design procedure, and lays the foundations for its appropriate calibration. For reasons of simplicity and comparison with the "exact" column strengths obtained from an advanced inelastic analysis of steel frames (permitted in AS4100-1990 for the design of steel structures), attention in this chapter is primarily restricted to in-plane member behavior and compact cross-section behavior. Although not addressed specifically in this report, there would appear to be no reason in principle, however, why the notional load approach could not be applied to assess the in-plane strength of members of non-compact cross-section.

In the calibration process and the examples presented in this chapter, major and minor axis bending are considered separately. While material and geometric imperfections are considered, the influences of loading imperfections, and semi-rigid connections in partially restrained frames, are not. Individual restrained members, single-story frames and multistory frames are investigated. Both rigidly connected columns and leaning columns are considered. Although not explicitly studied in this chapter, there is no fundamental reason why the principles of the notional load approach expounded within are not directly applicable to partially restrained frames. Comparisons are made between the strength predictions based on advanced analysis of the physically imperfect member or frame, the effective length approach, and the notional load (equivalent imperfection) approach. Although all the frame examples presented in this chapter have equal-length columns within a story, the notional load approach is also straightforwardly applicable to frames comprising stories with unequal column lengths. Also, as in the rest of the report, the contribution of the slab to the story lateral stiffness and strength is not considered in this chapter.

The practical application of the notional load approach usually involves the application of a notional lateral load N at each and every story level of a frame. In this case, the magnitude of the notional load N acting at each story level is expressed in the form $N = \xi Q$, in which Q represents the sum of all gravity (vertical) loads acting on the story, and ξ is a constant termed the notional load parameter or notional load coefficient. A "tiered" notional load calibration entailing three levels of sophistication—a "simple" calibration, a "modified" calibration, and a "refined" calibration—is undertaken in Section 4.6 of this chapter. The objective of the simple calibration is to determine a unique value for ξ which can be universally adopted with the assurance that the deduced story or member strength is either accurate or conservative when compared with advanced analysis results. In the calibration process, the influence of slenderness, yield stress, end-moment ratio and end-restraint stiffness is considered. The intent of the modified calibration is to modify the simple notional load parameter in an elementary manner for the effect of steel yield stress.

The "refined" notional load calibration was undertaken primarily to improve the accuracy of the simple and modified approaches for beam-columns of low to intermediate slenderness which are bent in double-curvature. This scenario is inevitable for the columns in multistory frames, particularly those in the lower stories. Full details of the refined notional load calibration are given in Appendix F. The essence of the calibration, however, is that the notional load parameter ξ is expressed as a function of the yield stress of the columns in the story, the slenderness of the story, the lateral stiffness of the story, and the number of columns per plane. In this way, different values of ξ can be used for each story of a frame to reflect more accurately the true story or frame strength. The application of the refined notional load approach to multistory frames, together with other issues concerning the advanced and notional load analysis of multistory frames, are covered in Sections 4.10 to 4.12. In particular, a more rigorous "story based" notional load procedure, which can be used to assess the design strength of a single story in a multistory frame, is discussed. Through this "story based" notional load philosophy, the strength and stability of an entire frame can be assessed merely by applying suitable notional loads to the one or few stories which are known to be "stability critical" and hence which "govern the design" of the frame. A comprehensive five-story one-bay frame example is presented in Section 4.13.

As presented in this chapter, the notional load approach is developed within the context of sidesway uninhibited rectangular frames. The effective length factors for the columns in such frames are usually greater than unity and thus the additional moments induced by the notional loads are important in offsetting the increase in the beam-column interaction equation axial resistance term from one based on the effective column length to one based on the actual column length. If the frame is partially or fully sidesway inhibited (braced), the notional load approach as described in this chapter is still directly applicable, but possibly with increased design conservatism. For the extreme case of a fully braced frame, notional loads acting at each story level will cause a small change in the axial force distribution throughout the structure but will have negligible effect on the moment distribution. Of course, if the effective length factors of the columns in a story are genuinely less than unity, then any simplified design procedure involving the use of notional loads in conjunction with $K = 1$ will of necessity be conservative.

Examples are interspersed throughout this chapter for demonstration and verification purposes. Since it leads to the most ready comparison between various methods of assessing frame stability and strength, the results for most of these examples are often plotted as interaction diagrams which define the frame "strength" in terms of the ratio of gravity to lateral loads. Several design examples illustrating the practical application of the notional load approach (together with the effective length based approaches) are given in Chapter 6. The simplicity and brevity of the design calculations associated with the notional load approach is clearly evident from these design examples.

While the scope of the current work is primarily restricted to in-plane strength and behavior, towards the end of the chapter (Section 4.14), the three-dimensional strength and behavior of beam-columns which are unbraced in-plane and subjected to in-plane flexure only, but which are braced out-of-plane (at the ends of the member), is addressed briefly. The role of the notional load approach in assessing the three-dimensional strength of such members is outlined, taking into consideration that failure can potentially occur either in an "in-plane or an "out-of-plane" mode. Detailed discussion of the application of the notional load approach to three-dimensional frames comprising members which are sidesway uninhibited in both the in-plane and out-of-plane directions (spatial beam-column behavior) is outside the scope of this chapter. However, some introductory comments on this topic are provided in Section 4.15, and a frame in this category constitutes one of the design examples in Chapter 6. The approach adopted to assess the strength of members in this design example involving biaxial bending was to apply notional loads simultaneously in both orthogonal directions, a philosophy thought to be not inappropriate in view of the current lack of any rigorous theoretical investigation and calibration studies using an appropriate three-dimensional advanced analysis.

4.2 Overview of Notional Load Approach

An overview of the notional load approach is presented in this section. Sections 4.3 and 4.4 provide a detailed exposition on the different types and roles of imperfections in determining the strength of columns and frames, thus laying the foundations and motivation for the use of the notional load approach as a rational engineering design procedure. The theoretical development behind the calibration of the notional load approach is given in Sections 4.5 and 4.6.

An alternative to computing effective length factors to take into account differing end restraints in columns in unbraced frames is to use the *actual* column length (i.e., "$K = 1$") in conjunction with "notional" lateral loads acting at each story level; second-order elastic analysis is then conducted on the geometrically perfect structure. This procedure for the assessment of frame stability is termed the *notional load approach* and is permitted in the Australian Standard AS4100-1990, the Canadian Standard S16.1-M94, the British Standard BS5950:Part 1:1990 and Eurocode 3. The notional lateral load acting on a story contrives to take into account the *story out-of-plumbness* imperfection under gravity loads. The notional load N acting at each story level is thus assumed to be some proportion ξ of the total gravity load Q acting on the story, so that $N = \xi Q$. The appropriate determination of the value of ξ is the objective of the calibrations undertaken in Section 4.6, the simplest of which indicates that $\xi = 0.005$ is an appropriate value which can be applied "universally" with the assurance that the deduced strengths of members and frames are either accurate or slightly conservative compared to the corresponding "advanced analysis" results.

Restricting the present discussion to in-plane strength only[1], if the notional load approach were permitted within the realms of the AISC LRFD Specification, the relevant beam-column interaction equation required to be satisfied by every member is

$$\begin{aligned} &\frac{P_u}{\phi_c P_{n(L)}} + \frac{8}{9}\frac{M_u}{\phi_b M_n} \le 1.0 \qquad \text{for} \quad \frac{P_u}{\phi_c P_{n(L)}} \ge 0.2 \\ &\frac{P_u}{2\phi_c P_{n(L)}} + \frac{M_u}{\phi_b M_n} \le 1.0 \qquad \text{for} \quad \frac{P_u}{\phi_c P_{n(L)}} < 0.2 \end{aligned} \tag{4.1}$$

in which the axial resistance term $P_{n(L)}$ is computed from the column curve using the *actual* member length L; this essentially accounts for the residual stresses and member out-of-straightness imperfections. In the elastic frame analysis, the notional lateral load acting in conjunction with real lateral and gravity loads invariably results in higher design bending moments M_u in the members of the structure compared to those due to real loads alone. From a design perspective, this additional contribution to the design bending moment is counterbalanced by the increase in the axial resistance term from P_n (based on the effective length KL) to $P_{n(L)}$ (based on the actual length L).

Typical load cases for which steel frames need to be designed involve gravity (vertical) load only (e.g., dead + live load) and combined gravity and lateral load (e.g., dead + live + wind). Notional horizontal loads should be employed in all load cases, and, in principle, they need to be considered acting from all directions (from the "left" and from the "right", say, for a two-dimensional frame), but only from one direction at a time. Where real lateral load is present, the real and notional lateral loads are logically assumed to act in the same direction. In the absence of real lateral load, however (i.e., the gravity load cases), and when the frame and/or loading is non-symmetric, the notional loads should be considered acting from both directions independently. This effectively doubles the number of gravity load cases which need to be considered in the structural design relative to what would be employed in a design procedure based on the calculation of effective length factors. The greater number of load cases inherent in the notional load procedure is not believed to be a significant concern in the present era of automated computerized structural analysis. In any event, any tedium associated with the greater number of load cases in the notional load approach is offset by the simpler nature of the interaction equation checks, which themselves often constitute a more accurate (less conservative) assessment of the design strength of the member or frame compared to current AISC LRFD practice based on the use of elastic effective length factors.

---

[1] The assessment of out-of-plane and spatial beam-column strength within the context of the notional load approach is considered in Sections 4.14 and 4.15.

A consequence of the notional load approach is that, even for the gravity load cases, every restrained column in a sidesway uninhibited frame is subjected to some design bending moment M_u, and it is only leaning columns which are designed for pure axial load. This means that the concept of an "anchor point" on the P_n/P_y axis of an interaction diagram, corresponding to the design strength of the member under pure axial load, does not exist in the notional load approach, and there is some portion of the interaction diagram adjacent to the P_n/P_y axis (wherever the moment M_u is less than that generated under the action of notional loads alone) which is "unusable". These particular aspects associated with the notional load approach are not shortcomings of the method, but simply a reflection of the reality that out-of-plumbness imperfections, and thus bending moments induced by vertical loads, are invariably present in frames constructed in the field, and thus should be considered in any design procedure. The notional load approach also has the subtle advantage that it furnishes directly (even in the absence of beam loading) the bending moments acting at the joints of a frame under gravity loads, and for which the connections and beams must be designed to transfer.

4.3 Influence of Imperfections on Column Strength

4.3.1 General

In general, only two types of imperfections have been considered in the determination of column strength as reflected in the column curves of design specifications: member-out-of straightness and residual stresses. The influence of each and both these imperfections is illustrated in Fig. 4.1 from Trahair and Bradford [1991]. For a perfectly straight member with residual stresses, there is no lateral deformation until buckling at the tangent modulus buckling load P_t, and the subsequent deformations follow curve B. For a member with initial out-of-straightness but no residual stresses, lateral deformations occur immediately upon loading and then follow the elastic second-order bending curve until first yield takes place at a load P_l; subsequent deformations follow curve A. If both imperfections are present, the deformations follow curve C and reach a maximum load of P_m, with the subsequent deformations following a path that approaches curve B.

The multiple column curves in AS4100-1990 and the British Standard BS5950:Part 1:1990 are based on the Perry-Robertson equation (see Trahair and Bradford [1991]) which includes an out-of-straightness term. In AS4100-1990, the out-of-straightness term is artificially increased in magnitude to account for residual stresses, thus reducing the calculated strength from P_l to P_m (Fig. 4.1), and is calibrated to the central SSRC Curve 2 [SSRC, 1976]. The basis of the column curves in BS5950:Part 1:1990 is the calibration of the Perry-Robertson equation to a slightly modified version of the European column curves, which are themselves based on a combination of ultimate strength analyses and full-scale testing. Conversely, the column curve in the AISC Allowable Stress Design (ASD) Specification [AISC, 1978] is based on the response for residual stresses alone (tangent modulus buckling, curve B) where the maximum compressive residual stress is assumed to be $0.5F_y$. This value represents an artificial magnification of the typical maximum compressive residual stress encountered in practice ($0.3F_y$) and partially accounts for

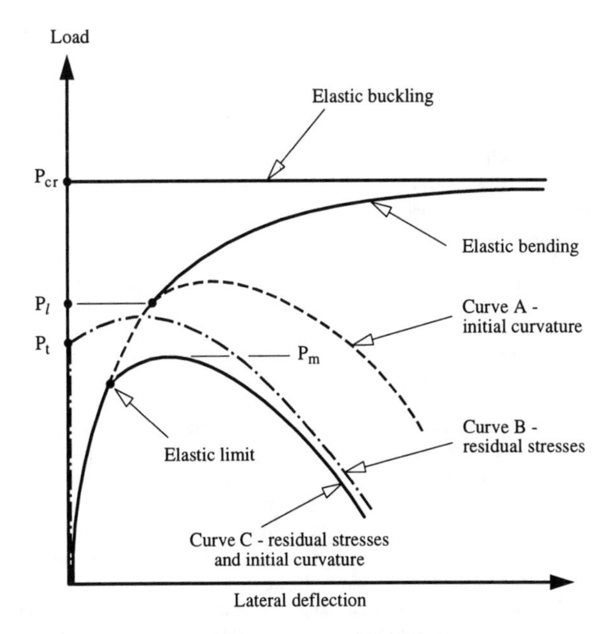

Fig. 4.1 Influence of imperfections on column behavior [Trahair and Bradford, 1991]

the effects of initial curvature. However, this procedure did not adequately account for initial out-of-straightness and so a variable factor of safety was also used in order to reduce the value of P_t to a value corresponding to P_m determined from test results. The column curve in the AISC LRFD Specification represents a fit to the approximate ASD curve which includes residual stress and imperfection effects.

For members in sway frames, the column curves described above do not directly take account of story out-of-plumbness or overall frame non-verticality. All the imperfection effects can be included, however, using advanced inelastic analysis, the results of which can be regarded as "exact" column strengths. The strengths determined using advanced analysis can then be compared with the results of design methods in standards and specifications for calibration purposes.

Besides those imperfections considered above, there are others that could occur in practice; deviations in the plate elements forming the cross-section, inaccuracies in load placement (inadvertent eccentricity of loading), accidental transverse loading, and movement or settlement of restraints are but some examples. None of these effects are considered in this chapter.

4.3.2 Types of Imperfections

4.3.2.1 Residual stresses

Hot-rolled structural members contain residual stresses as a consequence of different parts of the section experiencing different cooling rates during the manufacturing process. For I-section members, the residual stresses are usually compressive at the flange tips. Compared to sections free of residual stresses, the magnitude and distribution of these compressive residual stresses influences the strength of compression members because of the earlier initiation of yielding and subsequent lowering of flexural strength that occurs. The effects of residual stresses are more severe for I-section columns bent about the minor axis than the major axis [Bild and Trahair, 1989].

Residual stresses in hot-rolled I-sections have been measured by numerous researchers and analytical models have been proposed. The residual stress pattern employed by Galambos and Ketter [1959] and Ketter [1961] in their studies of the strength of steel beam-columns is shown in Fig. 4.2(a). This distribution is consistent with measured residual stresses in American wide-flange column-type sections resulting from cooling of the section during and after rolling [Ketter et al., 1955]. Researchers investigating the inelastic lateral buckling of beams and beam-columns [Trahair, 1983; Bradford and Trahair, 1985] represented the residual stresses induced in the section flanges after hot-rolling by parabolic distributions varying from $0.35F_y$ compression at the flange tips to $0.5F_y$ tension at the flange-web junction. The residual stress distribution in the web was represented by a quartic variation. However, based on achieving a better match to experimental distributions, the bi-linear flange distribution and tri-linear web distribution shown in Fig. 4.2(b) was adopted by Bild and Trahair [1989] for studies of the in-plane strength of steel columns and beam-columns.

The residual stress distribution shown in Fig. 4.2(a) has been employed widely in parametric studies of beam-column and frame strength in the context of the calibration and development of the AISC LRFD interaction equations [Kanchanalai, 1977]. For consistency with Kanchanalai's work, this distribution of residual stresses has also been adopted for the advanced analysis studies reported in this chapter, except where otherwise noted.

4.3.2.2 Geometric imperfections

Design specifications may include explicit or implicit fabrication tolerances for member out-of-straightness (δ_0), and frame erection tolerances for local story out-of-plumbness (Δ_0) and global non-verticality (e_0). The two types of imperfections, which may be termed member and frame imperfections, are illustrated in Fig. 4.3. Global non-verticality has been illustrated using the tolerance envelopes specified in AS4100-1990 and Eurocode 3.

Only member out-of-straightness has been closely examined and its effects included in column strength curves. The inclusion of geometric imperfections in the design procedure for frames is much more complex. Eurocode 3 recommends that frame imperfections should be included in the elastic global analysis of the frame. Although the influence of the number of columns in a plane and the number of stories is considered, only limited guidance is given with respect to the shape and distribution of imperfections. AS4100-1990 and the CSA-S16.1-M94 include the effect of frame imperfections through the use of an equivalent notional lateral load, a procedure also allowed in Eurocode 3.

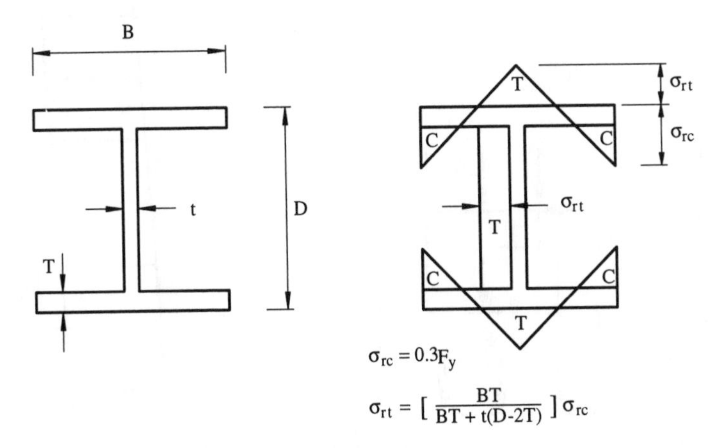

(a) Residual stress distribution of Galambos and Ketter [1959]

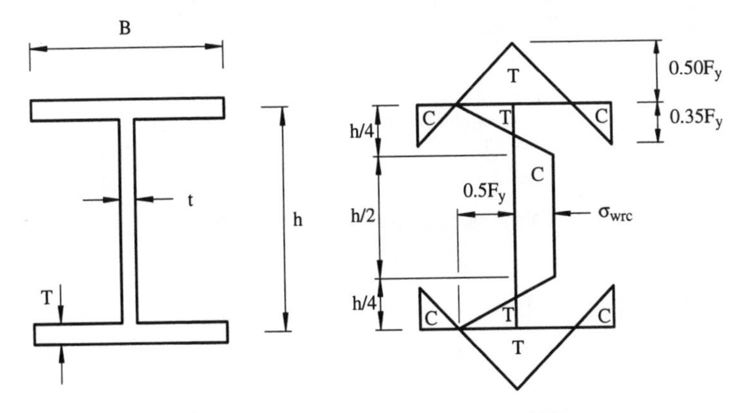

(b) Residual stress distribution of Bild and Trahair [1989]

Fig. 4.2 Examples of residual stress distributions

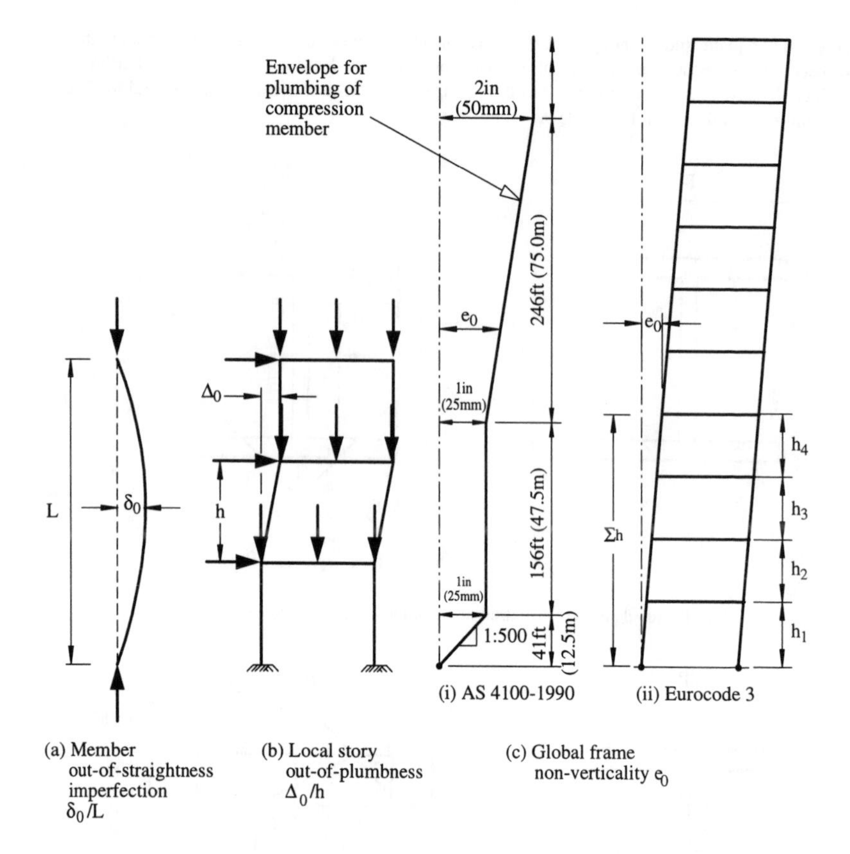

Fig. 4.3 Member and frame geometric imperfections

The modeling of geometric imperfections is not straightforward. It is not just the magnitude, but also the shape, distribution and orientation of the imperfections that is significant. For example, the ramifications of the equivalent pin-ended column (effective length) concept on the implicitly assumed imperfect shapes of beam-columns with end-fixity conditions other than pinned are illustrated in Fig. 4.4. In the absence of geometric imperfections, the beam-columns shown in Fig. 4.4(a) to (d) exhibit identical behavior. Similarly, when geometric imperfections are included according to the equivalent pin-ended column concept with an out-of-straightness δ_0 as shown in Fig. 4.4, the behavior of each of the four beam-columns with the loading shown is also identical. The out-of-plumbness of the sway beam-columns (Fig. 4.4(b) and (c)) is implicitly defined by the member out-of-straightness δ_0 and the column effective length K, and is independent of any erection tolerance. This last point may be illustrated by observing that for the cantilever sway column (Fig. 4.4(b)), for which the effective length factor $K = 2$, the implicitly assumed out-of-plumbness Δ_0 is equal to $\delta_0 = 2\delta_0/K$ while for the sway column fixed at both ends (Fig. 4.4(c)), for which $K = 1$, the implicitly assumed out-of-plumbness is equal to $2\delta_0 = 2\delta_0/K$. The implicitly assumed out-of-plumbness corresponding to the braced columns of Fig. 4.4(a) and (d)) is of little relevance to the present work since imperfections of this type have negligible effect on the strength of braced members and frames [Clarke et al., 1992].

Alternately, from arguments based on independently specified fabrication and erection tolerances (δ_0/L and Δ_0/L respectively), the geometric imperfections existing in the "physical column" representation of the four members of Fig. 4.4 are shown in Fig. 4.5 in which δ_0/L and Δ_0/L have been assumed the same irrespective of the column boundary conditions. The behavior of the four physically imperfect beam-columns of different actual length L in Fig. 4.5 is *not identical*, although, in relation to the particular sway members of Fig. 4.5(b) and (c), if the imperfection magnitudes are chosen such that the erection imperfection Δ_0 is twice the fabrication imperfection δ_0 ($\Delta_0 = 2\delta_0$) then the differences in the ultimate strengths computed by advanced analyses are small.

The modeling of geometric imperfections for a frame, in particular the determination of appropriate orientations, is much more complex than for a single member. Clarke et al. [1992], for example, have shown that a judiciously selected pattern of imperfections can actually enhance the strength of a pin-based portal frame above that of the corresponding geometrically perfect frame. Researchers investigating the local buckling of thin-walled members have used the elastic buckling mode of the member as a basis for considering plate imperfections [Davids and Hancock, 1987] on the assumption that this will have the worst effect on strength.

As an alternative to prescribing member out-of-straightness and frame out-of-plumbness geometric imperfections on an independent "physical" basis for the analysis of sway frames, the concept of an "equivalent" frame imperfection may be employed [De Luca and De Stefano, 1994]. An equivalent imperfection is defined as an enlarged frame imperfection of uniform angular deviation from the vertical which accounts for the combined effects of member and frame imperfections prescribed in design specifications.

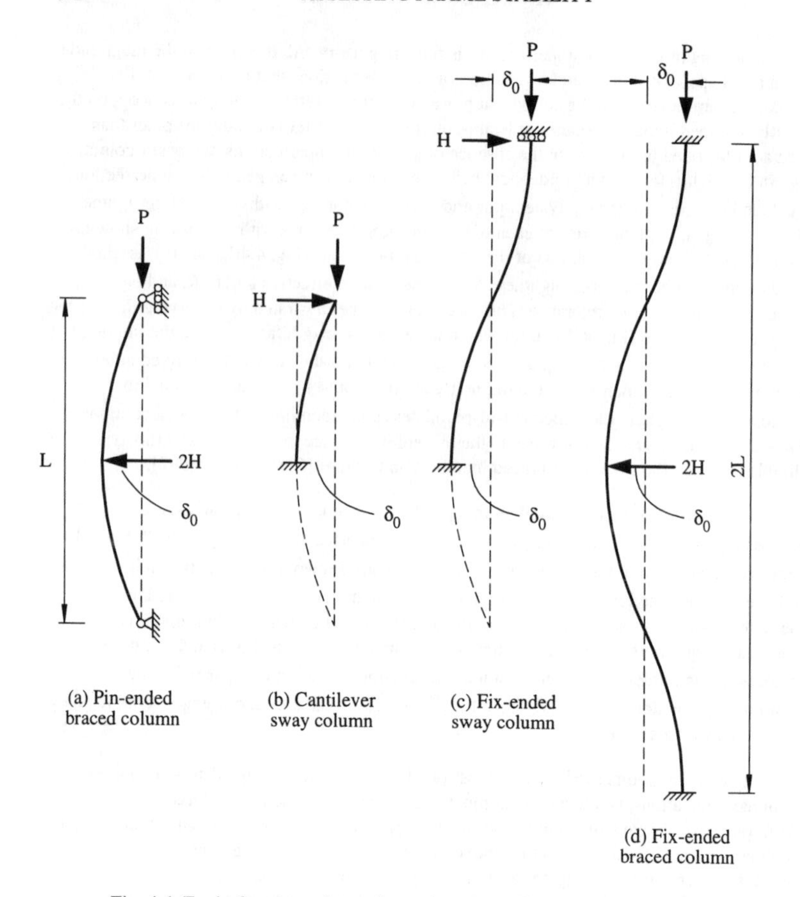

Fig. 4.4 Equivalent imperfect columns based on effective length concept

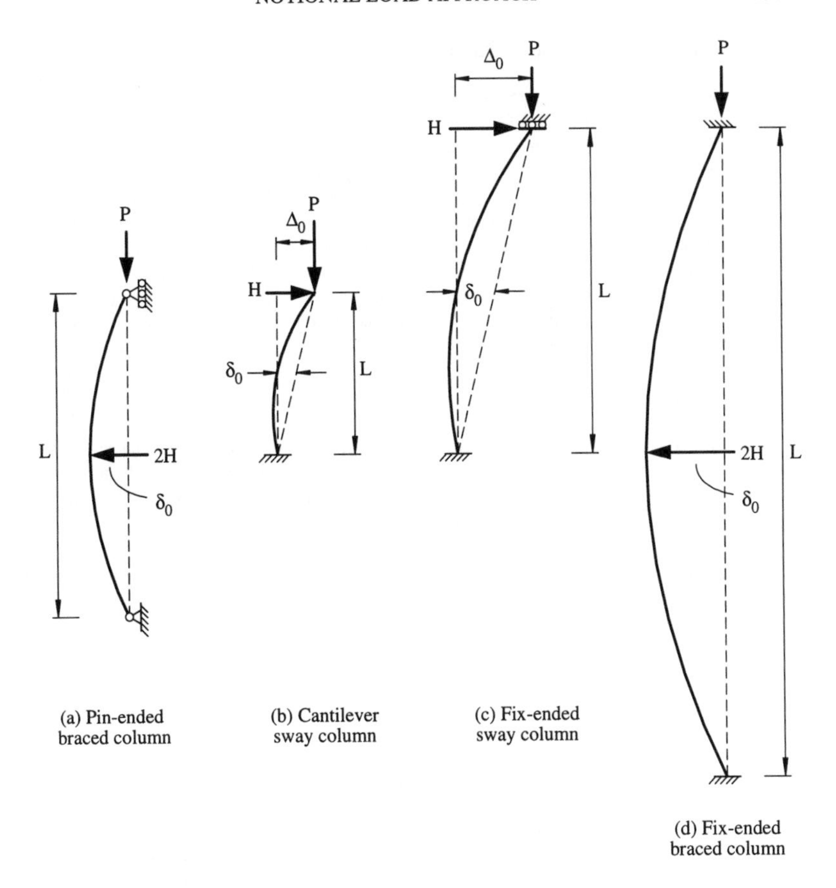

Fig. 4.5 Imperfect columns based on fabrication and erection tolerances

4.3.3 Survey of Imperfections Included in Steel Design Specifications

A review of material and geometric imperfections included in design specifications around the world can be found in the SSRC World View document [SSRC, 1991]. Some design specifications include imperfections directly in the design procedure. Alternately, they may be included in the analysis of the imperfect structure to determine the load action effects and the member resistance. The present emphasis on imperfections, and how they are included in design procedures premised on the use of second-order elastic analysis, is central to understanding the motivation for the notional load approach. The fundamental aim of the latter approach is to include—within the context of a design procedure based on second-order elastic analysis—the effects of residual stresses, member out-of-straightness and story out-of-plumbness imperfections in a manner which is as consistent as possible with the "physically imperfect frame" philosophy employed in advanced analysis.

4.3.3.1 Member out-of-straightness

Design specifications usually neglect the member out-of-straightness in the elastic global analysis of the frame for the determination of load action effects, but include its effect on member design through the use of a column strength curve which also includes the effects of residual stresses. Eurocode 3 requires that a specified member out-of-straightness should be included in the global analysis when the member slenderness exceeds a defined limit related to the ratio of the axial force to the yield load. The member out-of-straightness is accounted for in several design specifications as follows:

AISC LRFD (1993): Included in column strength curve [AISC, 1993b].

CSA-S16.1-M94: Included in column strength curve.

AS4100-1990: Included in column strength curve.

Eurocode 3 (1992): Included in column strength curve; or alternately using second-order analysis of the column with a design value of equivalent initial bow imperfection which has been calibrated to both the method of analysis and the method used to verify the cross-section resistance so as to match the appropriate column strength curve. An extensive table of bow imperfections is given in Eurocode 3 to cover the range of methods.

4.3.3.2 Story out-of-plumbness

Initial frame imperfections can be accounted for in *elastic* frame analysis by means of an *equivalent* geometric imperfection in the form of an initial sway imperfection ψ as shown in Fig. 4.6. It is important to realize that the equivalent out-of-plumbness imperfection ψ is often larger than the corresponding "physical" story out-of-plumbness specified on the basis of erection tolerances. For convenience in the structural analysis, the initial sway imperfection ψ may be replaced by equivalent notional lateral loads $N = \psi P$ (see Section 4.5), where P represents the

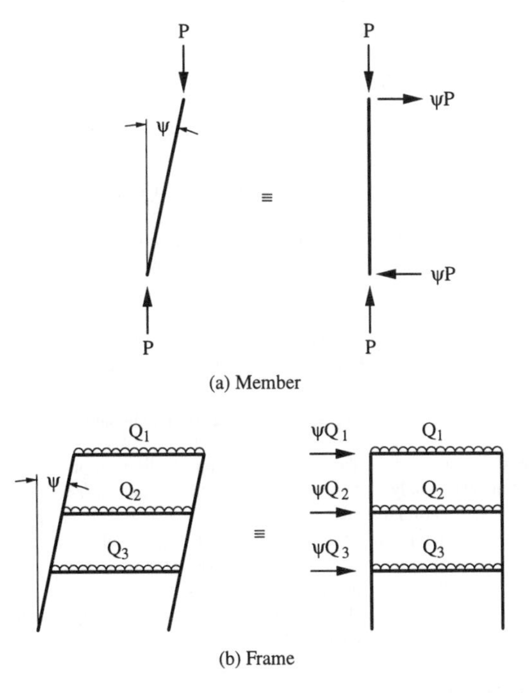

Fig. 4.6 Sway imperfections and equivalent notional lateral loads

gravity load acting on the story. Equivalent notional lateral loads are shown in Fig. 4.6(a) for a single member and Fig. 4.6(b) for a frame. Values of ψ relating to elastic frame analysis which are included in some design specifications are:

AISC LRFD (1993): None; effective length factors are used for sway columns instead, so that ψ is always an implied quantity [AISC, 1993b].

CSA-S16.1-M94: $\psi = 0.005$ in conjunction with an effective length factor $K = 1$ where notional lateral loads are greater than the real lateral loads, otherwise zero.

AS 4100-1990: $\psi = 0.002$ in conjunction with an effective length factor $K = 1$ where notional lateral loads are greater than the real lateral loads, otherwise zero.

Eurocode 3 (1992): $\psi = k_c k_s \psi_0$

with $\psi_0 = 0.005$

$$k_c = \sqrt{0.5 + 1/c} \quad \text{but } k_c \leq 1.0$$

$$k_s = \sqrt{0.2 + 1/s} \quad \text{but } k_s \leq 1.0$$

where c is the number of columns per plane, and
s is the number of stories[2].

The value of ψ can therefore vary from a maximum of 0.005 to a minimum of 0.0016. Note that ψ is used in conjunction with the effective length factor K for the *non-sway* (braced) mode.

4.3.4 Survey of Current Fabrication and Erection Tolerances

An alternative approach to the modeling of imperfections, as ideally should be incorporated in advanced inelastic analysis, could be based on *physical considerations* of erection tolerances for frames and fabrication tolerances for members (Fig. 4.3). Design specifications usually have specific requirements regarding the fabrication and erection tolerances for compression members. While these tolerances are not necessarily the same as the imperfections included, either explicitly or implicitly, in the member strength rules of the specification (which are usually employed in conjunction with elastic analysis), they should ideally reflect the imperfections permitted to achieve the required strength.

4.3.4.1 Member out-of-straightness

While the actual imperfect shape of the unloaded member could be modeled accurately by a number of terms of a Fourier series, it is generally the first term representing a half-sine wave from end-to-end of the column that has the most significant effect on the strength. Therefore, the member fabrication out-of-straightness is usually specified as the ratio δ_0/L where δ_0 is the midheight imperfection and L is the length of the column as shown in Fig. 4.3(a). The fabrication tolerances for compression members in a number of design specifications are:

AISC LRFD (1993): $\delta_0/L = 0.001$ [AISC, 1993b].

CSA-S16.1-M94: $\delta_0/L = 0.001$.

---

[2] Columns which carry a vertical load P_u of less than the mean of the vertical load per column in the plane considered, or columns which do not extend through all the stories included in s, should not be included in c. Also, those floor levels and roof levels which are not connected to all the columns included in c, should not be included in s [CEN, 1992].

AS4100-1990: The lesser of $\delta_0 = 0.12$ in (3 mm) or $\delta_0/L = 0.001$.

Eurocode 3 (1992): $\delta_0/L = 0.001$ generally, or $\delta_0/L = 0.002$ for hollow cross-sections.

4.3.4.2 Story out-of-plumbness

During the erection of the frame, the columns in each story in turn are plumbed to maintain verticality within a tolerance. A measure of the out-of-plumbness of each story (the erection tolerance) is the ratio Δ_0/h where Δ_0 is the relative story displacement in its unloaded state and h is the height of the story as shown in Fig. 4.3(b). The erection tolerances in a number of design specifications are:

AISC LRFD (1993): $\Delta_0/h = 0.002$ [AISC, 1993a].

CSA-S16.1-M94: $\Delta_0/h = 0.001$.

AS4100-1990: $\Delta_0/h = 0.002$.

Eurocode 3 (1992): $\Delta_0/h = 0.002$.

From an analysis viewpoint, all the above specifications except the AISC LRFD permit the frame out-of-plumbness imperfections to be represented by notional lateral story loads $N = (\Delta_0/h)Q$, in which Q corresponds to the total gravity load acting on the story. It is important to note that the value of the equivalent imperfection ψ discussed in Section 4.3.3.2 can be larger than the value of the erection tolerance Δ_0/h considered in the present section; this is discussed further in Section 4.6.

Although in the above discussion the out-of-plumbness has been viewed on a story-by-story basis, in reality it might be more appropriate to consider the limits as applying between "work points" (i.e., for each shipping piece) since the latter is more closely aligned with the construction procedure.

4.3.4.3 Global non-verticality

During the erection of the frame, the columns in the stories must be plumbed to maintain overall verticality of the frame within a tolerance. Examples of the envelope for the maximum allowable plumbing tolerance e_0 for columns in a frame are shown in Fig. 4.3(c), where e_0 is measured from the vertical plumb line above the correct position of the base of the column. The requirements in a number of design specifications are:

AISC LRFD (1993): For exterior columns: 1 in (25 mm) towards the building line or 2 in (50 mm) away from the building line for the first 20 stories; above the 20th story, the deviation may be increased by 1/16 in (1.6 mm) for each additional story up to a maximum of 2 in (50 mm) toward the building line or 3 in (75 mm) away from the building line. For columns adjacent to elevator shafts: 1 in (25 mm); above the 20th story, the deviation may be increased by 1/32 in (0.8 mm) for each additional story up to a maximum of 2 in (50 mm) [AISC, 1993a].

CSA-S16.1-M94: For exterior columns: 1 in (25 mm) towards the building line or 2 in (50 mm) away from the building line for the first 20 stories plus 0.08 in (2 mm) for each additional story, up to a maximum of 2 in (50 mm) toward or 3 in (75 mm) away from the building line. For columns adjacent to elevator shafts: 1 in (25 mm) for the first 20 stories plus 0.04 in (1 mm) for each additional story, up to a maximum of 2 in (50 mm).

AS 4100-1990: For all columns, 0.002 of the height, or the lesser of the following: 1 in (25 mm) for a point up to 197 ft (60 m) above the base, or 1 in (25 mm) plus 0.04 in (1 mm) for every 9.8 ft (3 m) in excess of 197 ft (60 m) up to a maximum frame out-of-plumbness of 2 in (50 mm) (Fig. 4.3(c)).

Eurocode 3 (1992): $0.0035\Sigma h/\sqrt{n}$ where Σh is the total height and n is the number of stories from the base to the floor level concerned (Fig. 4.3(c)).

At an overall height of 414 ft (126 m), equivalent to 36 stories of 11.5 ft (3.5 m) height, the maximum value of e_0 is 2 in (50 mm) for AS4100-1990 and 2.9 in (73.5 mm) for Eurocode 3.

4.4 Calibration of Imperfections Using Advanced Analysis

A sway column of length L and with zero translational restraint and elastic rotational end-restraints (Fig. 4.7) is used in this and subsequent sections as the vehicle for the theoretical investigation of the notional load approach for the assessment of column strength and stability in unbraced frames. The rotational stiffnesses of the elastic restraints at ends A and B of the column are denoted α_A and α_B respectively. These elastic restraints may be expressed conveniently in non-dimensional form as

$$\begin{aligned}\overline{\alpha}_A &= \alpha_A L/EI \\ \overline{\alpha}_B &= \alpha_B L/EI\end{aligned} \tag{4.2}$$

in which EI is the flexural rigidity of the column section. The commonly used G-factors G_A and G_B for sway columns (which are a measure of flexibility rather than stiffness) can be expressed in terms of the non-dimensional rotational restraint stiffnesses $\overline{\alpha}_A$ and $\overline{\alpha}_B$ as

$$\begin{aligned}G_A &= 6/\overline{\alpha}_A \\ G_B &= 6/\overline{\alpha}_B\end{aligned} \tag{4.3}$$

The elastic buckling equations, inelastic buckling equations, and the elastic nonlinear behavior of the sway column shown in Fig. 4.7, are presented in Appendix D.

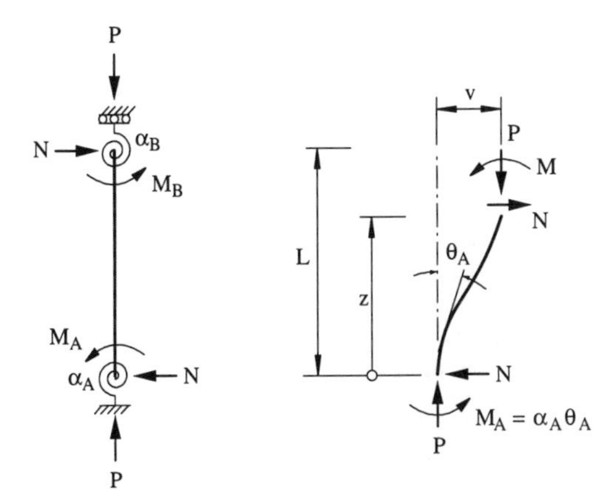

Fig. 4.7 Rotationally restrained sway column with zero translational restraint

Subsequent sections of this chapter refer to the end-moment ratio β when characterizing the behavior of sway columns. This ratio β is defined as the ratio of the smaller end-moment to the larger end-moment, taken positive when the member is bent in double-curvature.

4.4.1 Method of Advanced Analysis

The advanced analysis used for the studies described in this chapter is of the plastic zone type, as distinct from the plastic hinge type [SSRC, 1993], and has been described in detail in Clarke [1994]. Plastic zone advanced analysis is variously referred to in the literature as distributed plasticity analysis, spread-of-plasticity analysis, distributed inelasticity analysis and elastic-plastic analysis. Plastic zone analysis is characterized by the explicit modeling of the spread of yielding within the members of a framework. Other factors that affect frame strength and stability, such as residual stresses and geometric imperfections, may also be modeled explicitly. The advanced analysis employed in the studies presented in this chapter is based on the finite element method using an isoparametric curved beam element, and is suited to the in-plane analysis of arches, beam-columns and plane frame structures [Clarke and Hancock, 1991; Clarke et al., 1992; Clarke, 1994]. The analysis can include the effects of large displacements, residual stresses, material nonlinearity, gradual yielding, elastic unloading, geometric imperfections and non-proportional loading on the strength of members and frames.

4.4.2 Analytical Assessment of the Relative Importance of Imperfections

The effects of geometric imperfections and residual stresses on column strength have been studied using the cantilever column shown in Fig. 4.8. The cantilever column permits the simple inclusion of both out-of-plumbness and out-of-straightness imperfections and hence provides insight into the more complex issue of the inclusion of geometric imperfections in the analysis of framed structures. For most of the analyses, the column slenderness parameter λ_c, defined by

$$\lambda_c = \frac{1}{\pi}\left(\frac{KL}{r}\right)\sqrt{\frac{F_y}{E}} \tag{4.4}$$

in which K is the effective length factor, was chosen as unity since this is the value for which the yield load and elastic buckling load coincide and produce the greatest imperfection sensitivity in columns. The imperfection shapes considered are illustrated in Fig. 4.9, and consist of combinations of an out-of-plumbness imperfection of magnitude Δ_0 and a sinusoidal member out-of-straightness of amplitude δ_0. The cross-sectional shape used in the analyses was a W8×31 bent about the major axis and the material behavior was assumed to be elastic-perfectly plastic with a yield stress $F_y = 36$ ksi (250 MPa). When included in the analysis, the distribution of residual stresses followed the Galambos and Ketter [1959] model with a compressive residual stress at the flange tips equal to $0.3F_y$ (Fig. 4.2(a)). The elastic modulus E was assumed to be 29000 ksi (200000 MPa).

The column strengths P_n obtained from the advanced analyses and the AISC LRFD Specification are summarized in Table 4.1, in which it can be seen that both the geometric imperfections and the residual stress distribution assumed in the analysis affect the computed column strength. The lowest strength estimate of the advanced analysis is $P_n/P_y = 0.650$ (column V with residual stresses) and the strength predicted by the column rules of the AISC LRFD Specification is 0.658. It should be noted that columns IV and V differ only in the orientation of the out-of-straightness imperfection in relation to the out-of-plumbness imperfection. The favorably oriented imperfection in column IV results in a strength of 0.723 compared to 0.650 for the unfavorably oriented imperfection in column V. The imperfection of column VII, which is in the buckling mode, results in an ultimate strength close to but slightly above the strength of column V. This close agreement is only a result of the magnitudes of the chosen frame and member imperfections, for which coincidentally $\Delta_0 = K\delta_0$ for the cantilever where $K = 2.0$. For a column in a frame with different end-restraints and hence effective length factors K, this would not be the case.

By comparing the AISC LRFD strengths with the advanced analysis strengths for $\lambda_c = 0.5$ and 1.0 in Table 4.1, it can be deduced that the implied out-of-straightness of the AISC LRFD column curve is not constant but varies with the length of the column. Therefore, for any particular cross-section, the AISC LRFD column curve is not premised on a particular (unique) set of imperfection parameters.

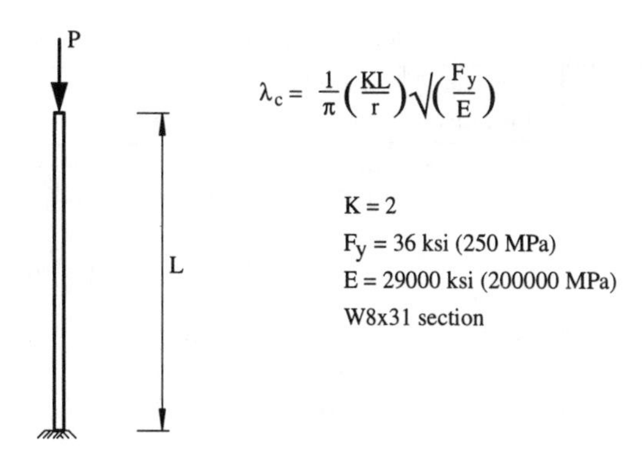

Fig. 4.8 Cantilever column studied with advanced analysis

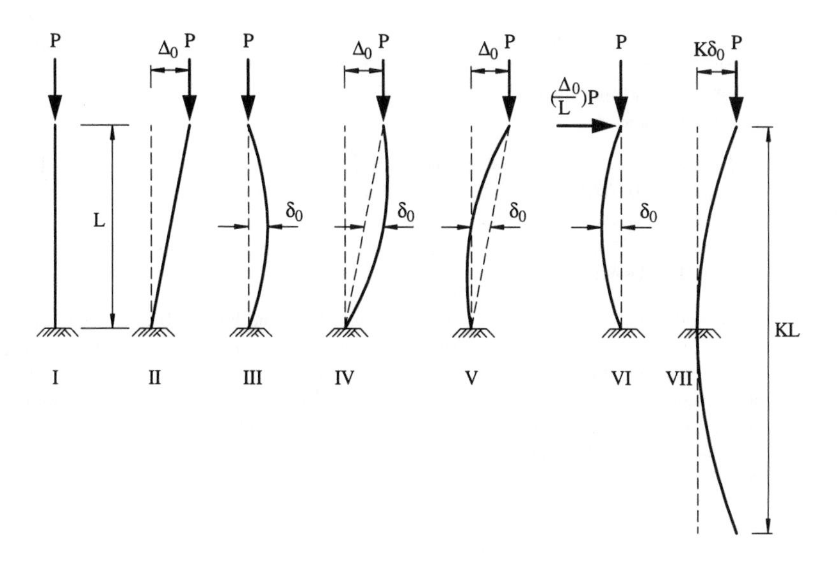

Fig. 4.9 Initially imperfect shapes for cantilever column

Table 4.1 Results for Strength of Cantilever Column

Slenderness λ_c	Column Geometry	Residual Stress[a]	Imperfection δ_0/L	Imperfection Δ_0/L	Strength P_n/P_y
0.5	← AISC LRFD column strength →				0.901
1.0	← AISC LRFD column strength →				0.658
1.0	I	0.0	0	0	1.000[b]
1.0	II	0.0	0	0.002	0.750
1.0	III	0.0	0.001	0	0.823
1.0	IV	0.0	0.001	0.002	0.801
1.0	V	0.0	0.001	0.002	0.712
1.0	VII	0.0	0.001	0	0.732
1.0	II	$0.3F_y$	0	0.002	0.681
1.0	III	$0.3F_y$	0.001	0	0.727
1.0	IV	$0.3F_y$	0.001	0.002	0.723
1.0	V	$0.3F_y$	0.001	0.002	0.650
1.0	VI	$0.3F_y$	0.001	0.002	0.650
1.0	VII	$0.3F_y$	0.001	0	0.666

[a] Compressive residual stress at flange tips
[b] Theoretical value

For the cantilever column of slenderness $\lambda_c = 1.0$, the result of using a notional lateral load of magnitude $N = 0.002P$ (column VI), in which P is the applied axial load, in lieu of the out-of-plumbness imperfection of magnitude $\Delta_0 = 0.002L$ (column V) is also given in Table 4.1. For $\delta_0/L = 0.001$, an identical strength of 0.650 is obtained by both methods (see Section 4.5 for the theoretical explanation of this result).

By comparing the results in Table 4.1 for columns II, III and V (or VII) either with or without residual stresses, it can be clearly seen that the inclusion of the out-of-plumbness Δ_0 has a more significant effect on the reduction in strength of this cantilever sway member than the inclusion of the member out-of-straightness δ_0. Therefore, the influence of out-of-plumbness Δ_0 must be included in any design procedure for sway columns. As expected, residual stresses have a significant effect on the reduction in column strength for any given set of geometric imperfections.

4.4.3 Calibration of Imperfections to AISC LRFD Column Curve

As stated in the Commentary to the AISC LRFD Specification, the AISC LRFD column curve represents a reasonable conversion of research data into a single design curve and is essentially the same as the SSRC column strength curve 2P of the 4th edition of the SSRC Guide [SSRC, 1988]. Using the AISC LRFD curve as the best estimate of the nominal column strength P_n, advanced analyses of rotationally restrained, physically imperfect columns with out-of-straightness δ_0 and out-of-plumbness Δ_0, as in Fig. 4.10(a), and a pattern of residual stresses, as in Fig. 4.2(a), were conducted to determine values of δ_0, Δ_0 and σ_{rc} which give a close fit to the AISC LRFD column curve over the full range of column slenderness λ_c. The AISC LRFD column strength curve is defined by Eqs. (2.1b) and (2.4b) of Chapter 2 (see also Eq. (D.9) of Appendix D), in which P_n/P_y is a function of the modified effective column slenderness λ_c (Eq. (4.4)). For a sway column rotationally restrained at its ends, it will be shown later that for maximum consistency with advanced analysis results, λ_c is best expressed as a function of the *inelastic*, rather than the *elastic*, effective length factor K. Calculation of the inelastic K entails the simultaneous solution of Eqs. (D.8), (D.9) and (D.10) of Appendix D. Comparisons of advanced analysis column strength and the AISC LRFD column strength were therefore done on the basis of the *inelastic* effective length factor K.

As for the cantilever column studied in Section 4.4.2, the cross-sectional shape used for all the analyses was a W8×31 subjected to bending about the major axis; this section has one of the lowest shape factors of all the North American hot-rolled beam and column sections. The material stress-strain curve was assumed to be elastic-perfectly plastic and defined by an elastic modulus of $E = 29000$ ksi (200000 MPa) and a yield stress of $F_y = 36$ ksi (250 MPa). The beneficial effects of strain hardening were neglected.

4.4.3.1 Braced pin-ended column

Firstly, a pin-ended column ($G_A = G_B = \infty$) braced at its ends was analyzed to calibrate the out-of-straightness δ_0 and the level of residual stress σ_{rc} for the AISC LRFD column curve independently of the effective length factor K (which equals 1.0 in this case) and the out-of-plumbness Δ_0 (no effect in this case).

As shown in Fig. 4.11, it was found that a combination of an initial out-of-straightness $\delta_0 = 0.001L$ and compressive residual stress $\sigma_{rc} = 0.3F_y$ at the flange tips gives a close fit to the AISC LRFD column curve over the full range of column slenderness λ_c. These values have therefore been adopted for all further advanced analyses of columns in this chapter. It is interesting to note that the value of $\delta_0 = 0.001L$ corresponds to the fabrication tolerance in a number of design specifications (see Section 4.3.4.1).

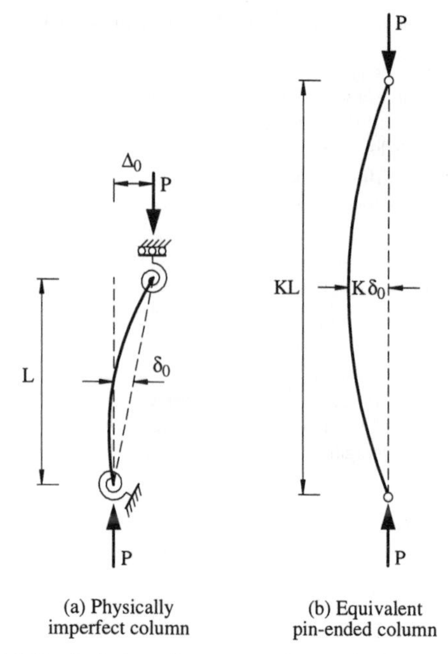

Fig. 4.10 Physically imperfect rotationally restrained sway column and equivalent pin-ended column

Although advanced analyses of columns with an out-of-straightness of $\delta_0 = 0.001L$ combined with $\sigma_{rc} = 0.3F_y$ happened to produce a close match to the AISC LRFD column curve for the W8×31 section buckling about the major axis, it should be borne in mind that the latter column curve is simply a reasonable or representative fit to the actual strengths for many types of columns buckling about either cross-sectional axis. The LRFD column equations are actually based on the strength of an "equivalent" pin-ended column of length KL, with a mean maximum out-of-straightness at its mid-length of approximately $KL/1500$.

To assess the sensitivity of the column strength to the two imperfection variables of out-of-straightness and residual stresses, the analytical results have also been plotted in Fig. 4.11 for the following imperfection parameters: $\delta_0 = 10^{-5}L$ and $\sigma_{rc} = 0.3F_y$; $\delta_0 = 0.002L$ and $\sigma_{rc} = 0.3F_y$; and $\delta_0 = 0.001L$ and $\sigma_{rc} = 0.0$. It can be seen that both variables have a significant effect on column strength and therefore should be included in advanced analyses of imperfect columns.

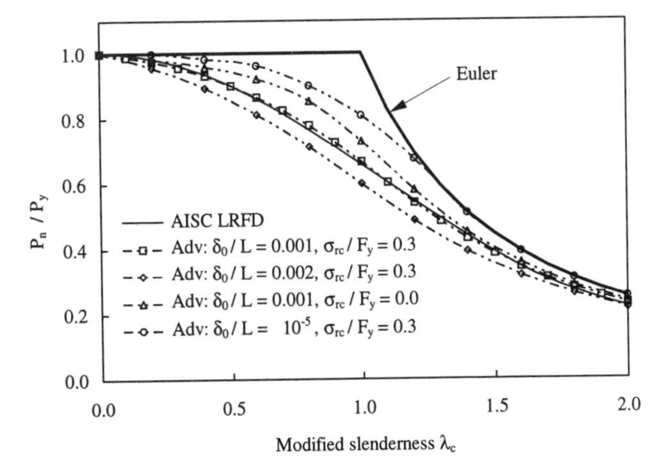

Fig. 4.11 Comparisons of advanced analysis column strengths with AISC LRFD Specification for pin-ended columns

4.4.3.2 Rotationally restrained sway columns

In addition to the out-of-straightness δ_0, physically imperfect columns of length L are characterized by an out-of-plumbness Δ_0 as shown in Fig. 4.10(a). This imperfection arises from erection tolerances and has a significant effect on the strength of the sway column (see Section 4.4.2 above for a cantilever column). While this imperfection also occurs in braced columns, its effect on braced column strength has been shown to be insignificant [Clarke et al., 1992].

Advanced analyses were performed on a number of physically imperfect sway columns of length L with rotational end-restraints and zero translational restraint, as shown in Fig. 4.10(a), in order to determine the "exact" column strength P_n. The cross-sectional shape and material properties employed were the same as those used for the braced pin-ended column investigated above. The previously calibrated values of out-of-straightness $\delta_0 = 0.001L$ and compressive residual stress $\sigma_{rc} = 0.3F_y$ at the flange tips were adopted. The value of out-of-plumbness Δ_0 was then varied until a close fit with the AISC LRFD curve was obtained over the full range of column slenderness λ_c for a practical range of end-restraints. It is again emphasized here that the *inelastic* effective length factor K should be used when plotting the results of the advanced analysis for comparison with the AISC LRFD column curve.

The first sway case considered was a cantilever with a positive rotational restraint at the base (G_A) and zero rotational restraint at the top ($G_B = \infty$) such that the column was bent in single curvature (end-moment ratio $\beta = 0$). Values of G_A between zero and 20 were considered. For an elastic column ($P_n/P_y \leq 0.390$, see Eq. (D.10)), a value of restraint corresponding to $G_A = 20$ results in an elastic effective length factor $K = 6.0$, thereby providing some appreciation of the low degree of restraint in this case. The advanced analysis results shown in Fig. 4.12 for various values of G_A were computed assuming an out-of-plumbness $\Delta_0 = 0.002L$. This value of out-of-plumbness imperfection evidently gives quite a close fit to the AISC LRFD column curve.

The second case considered was a sway column with equal rotational end-restraints (i.e. $G_A = G_B$, end-moment ratio $\beta = 1$) such that the column was bent in double curvature. Values of G_A and G_B in the range from zero to 6 were investigated. The advanced analysis results shown in Fig. 4.13 correspond to an out-of-plumbness $\Delta_0 = 0.002L$. Again, it can be seen that this value of imperfection gives quite a close fit to the AISC LRFD column curve.

Considering the close agreement between the advanced analysis results and the AISC LRFD column curve for the wide range of restraints considered, an out-of-plumbness $\Delta_0 = 0.002L$ has been adopted for all further advanced analyses of columns in this chapter. Again, it is interesting to note that this value of Δ_0 matches closely the erection tolerance from a number of design specifications (see Section 4.3.4.2).

The importance of using the *inelastic* effective length factor K to evaluate the column slenderness parameter λ_c is clearly demonstrated in Fig. 4.14 for a cantilever with $G_A = 20$ and $G_B = \infty$. When the "exact" strengths P_n determined from advanced analyses of the cantilever are plotted using the slenderness parameter λ_c based on the *inelastic* effective length factor K ($2.0 \leq K \leq 6.0$), the advanced results agree closely with the AISC LRFD column curve, the latter relating to the equivalent pin-ended member of length KL. However, if the same "exact" strengths P_n determined from advanced analyses are plotted using λ_c based on the *elastic* ($K = 6.0$) value, the advanced analysis results deviate considerably from the AISC LRFD column curve, particularly in the intermediate slenderness range. Perceiving this issue from the alternate viewpoint of column *design*, it should be noted that if λ_c is based on the elastic K value, the column strength P_n determined from the AISC LRFD column curve is generally less than the corresponding advanced analysis strength, and is therefore conservative. The variation of both the inelastic effective length factor K and the corresponding inelastic stiffness reduction factor τ with the modified inelastic slenderness λ_c is shown in Fig. 4.15. It can be seen in this figure that the value of the inelastic K decreases to 2.0 as the column length tends to zero (column stiffness relative to the restraint being small due to inelasticity in the column) but equals the elastic value of 6.0 for sufficiently slender columns (no inelasticity).

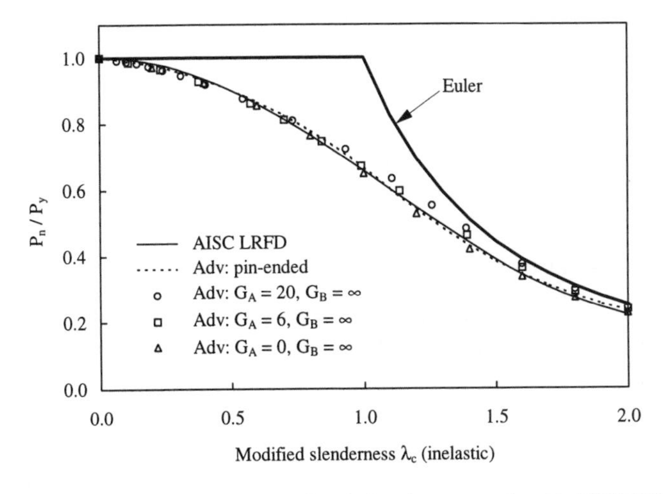

Fig. 4.12 Comparisons of advanced analysis column strengths with AISC LRFD Specification for sway columns with restraint at one end ($\beta = 0$)

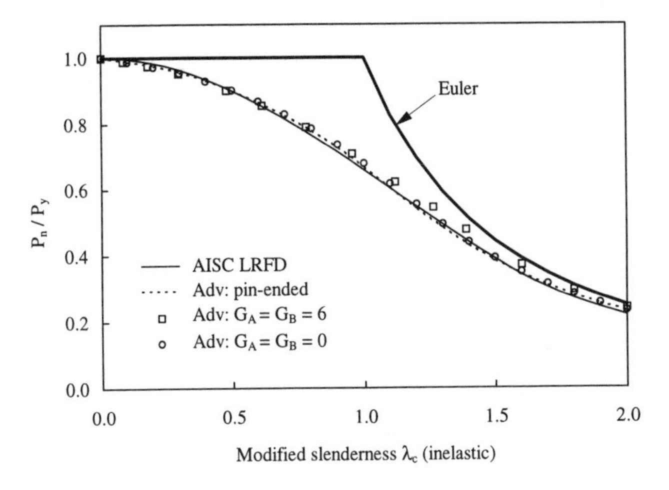

Fig. 4.13 Comparisons of advanced analysis column strengths with AISC LRFD Specification for sway columns with restraints at both ends ($\beta = 1$)

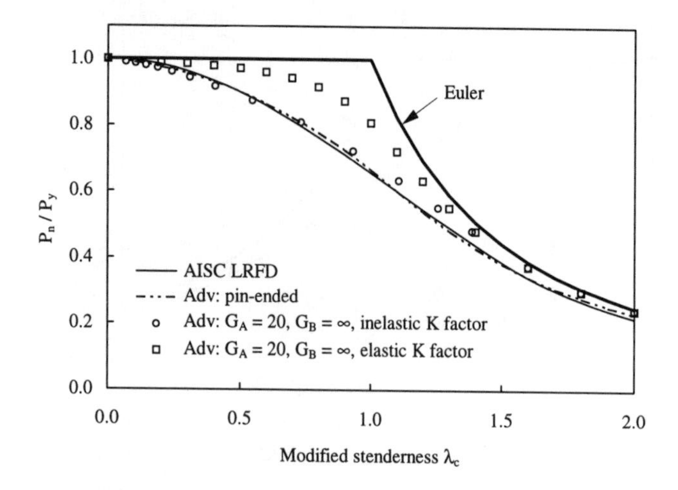

Fig. 4.14 Comparison of column curves for elastic and inelastic effective lengths

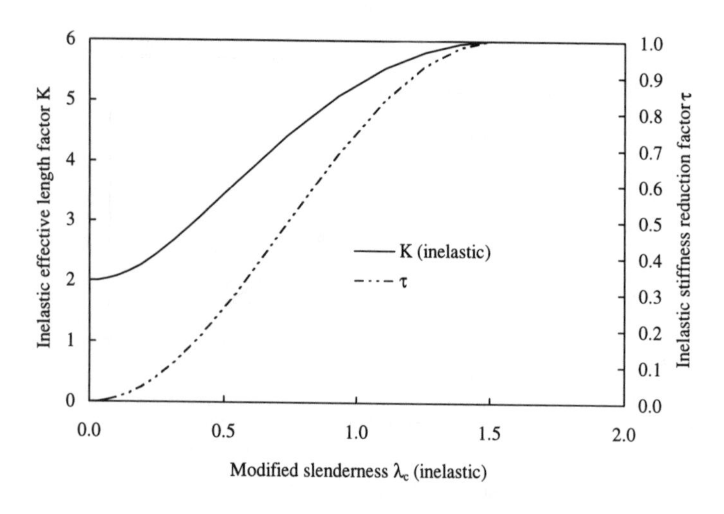

Fig. 4.15 Variation of inelastic K and τ with slenderness for cantilever with $G_A = 20$, $G_B = \infty$

The above discussion demonstrates that the AISC LRFD column curve, in conjunction with the *inelastic* effective length factor K, matches closely the results of accurate advanced analyses of physically imperfect sway members failing about the major axis for a wide range of practical end-restraints. As far as the calibration of the equivalent notional lateral load (see Section 4.6) is concerned, the most "correct" procedure would be to compare the notional load results to the corresponding strengths determined from advanced analyses of "physically imperfect" members. However, to simplify considerably the mechanics of performing the comparisons, and to facilitate algebraic manipulations pertaining to the notional load calibration, the AISC LRFD column curve can instead be used, with little loss of accuracy, as the basis for calibrating the notional load. It should be emphasized that this procedure is only justified by virtue of the fact that the AISC LRFD column curve, in conjunction with the slenderness parameter λ_c based on the correct inelastic effective length, closely reflects the strength curves determined from advanced analysis of physically imperfect members for a wide range of end-restraints.

4.5 Notional Loads for Equivalent Elastic and Inelastic Behavior

The *notional load* approach to the determination of column strength in frames, also termed the *equivalent imperfection* approach, is an alternative to the commonly used effective length approach. It can be argued that, compared to the effective length approach, the notional load approach accounts for the physical nature of the column and frame imperfections in a more direct way, the effects of member and story imperfections being considered independently. Member out-of-straightness and residual stresses are considered through conventional column strength formulae but using the actual, rather than effective, length. Story out-of-plumbness is taken into account through the inclusion of notional lateral loads in the global structural analysis. The aim of this section of the report is to examine in detail the correspondence between a physical story out-of-plumbness imperfection ψ (Fig. 4.16(a)) and notional lateral loads N applied to a perfectly plumb story (Fig. 4.16(b)), as would be used in the elastic global structural analysis, and to calibrate the required magnitude of the notional load to give accurate predictions of the column strength.

4.5.1 Notional Load for Equivalent Elastic Behavior

The elastically restrained sway column of Fig. 4.17(a) has a (real) out-of-plumbness defined by ψ, and the column of Fig. 4.17(b) is identical to that analyzed theoretically in Appendix D, being perfectly plumb and subjected to a notional lateral load N. The problem at hand is to determine, if indeed it is possible, the value of N such that the elastic second-order behavior exhibited by the problems of Figs. 4.17(a) and (b) is identical. By solving the differential equations pertaining to the two sway columns of Fig. 4.17 (the column in Fig. 4.17(b) is analyzed in Appendix D), it can be shown that identical distributions of *second-order elastic bending moment* $M = M(z,P)$ are obtained if $N = \psi P$, a result which confirms intuition. It should be noted, however, that if $N = \psi P$, the *deflected shapes* $v = v(z,P)$, measured from the plumb line, do not coincide precisely for the two cases of Figs. 4.17(a) and (b). The bending moment distribution along the beam-column is given by Eq. (D.17) of Appendix D.

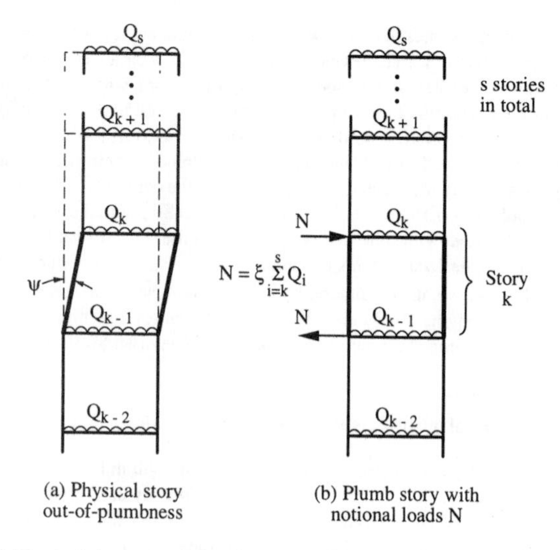

Fig. 4.16 Physical story out-of-plumbness ψ and equivalent notional lateral loads N

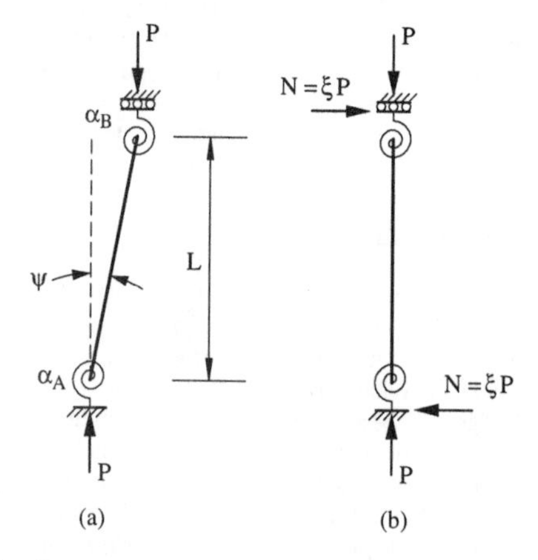

Fig. 4.17 Rotationally restrained sway column with (a) out-of-plumbness; and (b) notional lateral load

4.5.2 Notional Load for Equivalent Inelastic (Advanced) Behavior

As discussed in Section 4.5.1 above, the two cases of Figs. 4.17(a) and (b) have an identical elastic bending moment distribution $M = M(z,P)$ when $N = \psi P$. It can therefore be asserted that, for the purposes of determining the column strength using *advanced analysis*, the physical story out-of-plumbness imperfection of angular deviation ψ can be replaced by a notional lateral load of magnitude $N = \psi P$ and identical results achieved. This assertion has been confirmed by numerical experiments using advanced analysis [Clarke and Bridge, 1991, 1992; Clarke et al., 1992].

4.6 Theoretical Calibration of Notional Load for Column Strength

4.6.1 Calibration Rationale and Theoretical Development

In Section 4.4.3 and Fig. 4.11 it was shown that a close match to the AISC LRFD column curve could be achieved by conducting plastic zone advanced analyses of pin-ended columns with a defined initial out-of-straightness in the shape of a half-sine wave with maximum amplitude $\delta_0 = 0.001L$ at midspan, together with residual stresses based on the Galambos and Ketter [1959] distribution in Fig. 4.2(a) with a maximum compressive value of $\sigma_{rc} = 0.3F_y$ at the flange tips. For sway columns with a range of different elastic rotational end-restraints, it was also shown that the adoption of an out-of-plumbness imperfection of magnitude $\Delta_0 = 0.002L$ (in conjunction with the out-of-straightness and residual stresses defined above) results in advanced analysis column strength curves which are very close to the AISC LRFD column curve, but only provided the column effective length is based on the *inelastic* (rather than *elastic*) effective length factor.

For a sway column with out-of-plumbness designated by ψ, the required notional load N to be applied to a perfectly plumb member to produce an identical second-order elastic bending moment distribution is simply $N = \psi P$ (Section 4.5.1). The same notional load can also be used in advanced inelastic analysis to determine the ultimate strength as a column. However, when second-order elastic behavior is extrapolated to ultimate (nominal) strength through beam-column interaction equations, the required notional load will not necessarily be equal to ψP; generally speaking, although not always, the required value will be larger than ψP. This "artificially large" notional load is similar in concept to that of "equivalent" story out-of-plumbness imperfections which are larger than real (physical) values [Vogel, 1985; CEN, 1992].

The principle objective of this section is to calibrate the notional lateral load for use in design based on second-order elastic analysis. This calibration is undertaken for an elastically restrained sway member with end-restraint stiffnesses defined by $\overline{\alpha}_A, \overline{\alpha}_B$ subjected to a (real) axial load P, zero real lateral load H, and a notional lateral load N (Fig. 4.17(b)). This loading configuration corresponds to what, in the effective length approach, would be a sway column subjected to pure axial load. The omission of a real lateral load in the calibration exercise is appropriate as this is the case for which the sensitivity of the deduced member strength to the notional lateral load is greatest. The notional load approach to the design of axially loaded columns therefore considers a column to be a limiting case of a beam-column, with the moment in the member arising solely from "notional", rather than "real", loads.

It is assumed that the required notional lateral load is some proportion, say ξ, of the applied axial load, so that $N = \xi P$. The elastic second-order bending moment distribution in the sway column is given by Eq. (D.17), and it is assumed henceforth that both end-restraints do indeed provide zero or positive restraint (i.e., $\bar{\alpha}_A, \bar{\alpha}_B \geq 0$). If there is no translational restraint, the maximum moment will then occur at one of the ends (A or B), and is thus given by

$$M_u = N_u L\Omega = \xi P_u L\Omega \tag{4.5}$$

in which N_u and M_u define the required beam-column strength, and

$$\Omega = \max(\Omega_A, \Omega_B) \tag{4.6}$$

with Ω_A and Ω_B corresponding to parameters defined by Eq. (D.21) of Appendix D as a function of the end-restraints and the axial force.

Assuming compact section behavior for calibration purposes, the nominal section moment capacity M_n is equal to the plastic moment and is given by

$$M_n = ZF_y \tag{4.7}$$

in which Z is the plastic section modulus and F_y is the yield stress. Furthermore, by defining

S = elastic section modulus
η = shape factor
I = second-moment of area
A = area
D = overall section depth
r = radius of gyration
P_y = squash load ($= AF_y$)

and making the substitutions

$$Z = \eta S$$
$$S = \frac{2I}{D} \quad \text{(symmetrical cross-section assumed)} \tag{4.8}$$
$$r = \sqrt{\frac{I}{A}}$$

Eq. (4.7) can then be expressed in the form

$$M_n = 2\left(\frac{\eta r}{D}\right) r P_y \tag{4.9}$$

Using Eq. (4.5), the required-to-nominal moment ratio is then

$$\frac{M_u}{M_n} = \frac{1}{2}\xi\left(\frac{P_u}{P_y}\right)\left(\frac{L}{r}\right)\left(\frac{D}{\eta r}\right)\Omega \tag{4.10}$$

Finally, expressing the geometric slenderness L/r in terms of the modified slenderness parameter $\lambda_{c(L)}$, defined as

$$\lambda_{c(L)} = \frac{1}{\pi}\left(\frac{L}{r}\right)\sqrt{\frac{F_y}{E}} \tag{4.11}$$

and substituting into Eq. (4.10) furnishes

$$\frac{M_u}{M_n} = \frac{\pi}{2}\xi\lambda_{c(L)}\left(\frac{P_u}{P_y}\right)\sqrt{\frac{E}{F_y}}\left(\frac{D}{\eta r}\right)\Omega \tag{4.12}$$

Equation (4.12) above indicates that the moment ratio M_u/M_n is a function of the following characteristic parameters:

- ξ, which defines the magnitude of the notional lateral load
- $\lambda_{c(L)}$, a column slenderness (length) parameter
- E/F_y, a material parameter
- $D/(\eta r)$, a section geometric parameter
- P_u/P_y, which defines the column axial force at the required strength
- $\overline{\alpha}_A, \overline{\alpha}_B$, which define the rotational stiffnesses of the end-restraints.

For the correct calibration of ξ for a member subjected to pure axial compression, the maximum value of the *required* axial capacity P_u computed using the notional load approach in conjunction with the beam-column interaction equation must be equal to the *nominal* axial capacity P_n computed using the effective length approach (Eq. (D.7)), that is

$$P_u = 0.877\frac{\pi^2 EI\tau}{(KL)^2} = \frac{0.877\tau}{\left(K\lambda_{c(L)}\right)^2}P_y = \frac{0.877\tau}{\lambda_c^{\,2}}P_y \tag{4.13}$$

in which K is the (inelastic) effective length factor, τ is the inelastic stiffness reduction factor (see Chapter 2 and Appendix D), and $\lambda_c = K\lambda_{c(L)}$ is the modified effective member slenderness defined by Eq. (4.4). Procedures for the calculation of K are given in Chapter 2 and Appendix D.

In the notional load approach, the nominal axial capacity $P_{n(L)}$ which is used in the interaction equation relates to the actual member length ("$K = 1$") and is therefore given by

$$P_{n(L)} = \begin{cases} 0.658^{\lambda_{c(L)}^2} P_y & \text{for } \lambda_{c(L)} < 1.5 \\ 0.877/\lambda_{c(L)}^2 P_y & \text{for } \lambda_{c(L)} \geq 1.5 \end{cases} \tag{4.14}$$

By defining an inelastic stiffness reduction factor τ_L which relates to the column of actual length L rather than effective length KL (i.e., τ_L is given by Eq. (D.10) with P_n replaced by $P_{n(L)}$), Eq. (4.14) can be expressed in unified form as

$$P_{n(L)} = \frac{0.877\tau_L}{\lambda_{c(L)}^2} P_y \tag{4.15}$$

The required-to-nominal axial force ratio can then be determined from Eqs. (4.13) and (4.15) as

$$\frac{P_u}{P_{n(L)}} = \frac{1}{K^2}\left(\frac{\tau}{\tau_L}\right) \tag{4.16}$$

and the required-to-nominal moment ratio from Eqs. (4.12) and (4.13) as

$$\frac{M_u}{M_n} = \frac{\pi}{2}\frac{\xi}{\lambda_{c(L)}}\frac{0.877\tau}{K^2}\sqrt{\frac{E}{F_y}}\left(\frac{D}{\eta r}\right)\Omega \tag{4.17}$$

The ultimate strength interaction surface defined by the AISC LRFD beam-column interaction equation, neglecting ϕ-factors, is

$$\begin{aligned} \frac{P_u}{P_{n(L)}} + \frac{8}{9}\frac{M_u}{M_n} &= 1 \qquad \text{for } \frac{P_u}{P_{n(L)}} \geq 0.2 \\ \frac{P_u}{2P_{n(L)}} + \frac{M_u}{M_n} &= 1 \qquad \text{for } \frac{P_u}{P_{n(L)}} < 0.2 \end{aligned} \tag{4.18}$$

The neglect of ϕ-factors in Eq. (4.18) is believed to be appropriate since the calibration being undertaken is for the nominal strength rather than the design strength[3]. However, the use of ϕ-factors is included in design examples discussed in later sections of this chapter and in Chapter 6.

Assuming $P_u/P_{n(L)} \geq 0.2$ (as would overwhelmingly be the case since the only moment contribution to the interaction equation is from the notional loads) and combining Eqs. (4.16) and (4.17) with the first of the AISC LRFD beam-column interaction limits given in Eq. (4.18), and solving for ξ furnishes

$$\xi = \lambda_{c(L)}\left(K^2 - \frac{\tau}{\tau_L}\right)\frac{9}{4\pi}\frac{1}{0.877\tau}\sqrt{\frac{F_y}{E}}\left(\frac{\eta r}{D}\right)\frac{1}{\Omega} \tag{4.19}$$

If by chance the column is sufficiently slender such that $P_u/P_{n(L)} < 0.2$, then the appropriate beam-column interaction limit is the second equation in Eq. (4.18) from which the corresponding solution for the notional load parameter ξ is

$$\xi = \lambda_{c(L)}\left(2K^2 - \frac{\tau}{\tau_L}\right)\frac{1}{\pi}\frac{1}{0.877\tau}\sqrt{\frac{F_y}{E}}\left(\frac{\eta r}{D}\right)\frac{1}{\Omega} \tag{4.20}$$

For a column with a defined length, end-restraints, cross-sectional shape and material properties, the required notional lateral load parameter ξ can therefore be computed from Eqs. (4.19) or (4.20), although it should be recalled that calculation of the inelastic stiffness reduction factors τ and τ_L requires an iterative procedure. For sway columns of atypically high slenderness such that $\lambda_{c(L)} > 1.5$ (which therefore implies that $\lambda_c = K\lambda_{c(L)} > 1.5$), Eqs. (4.19) and (4.20) can be simplified by virtue of the fact that $\tau_L = \tau = 1$.

Although it appears from Eqs. (4.19) and (4.20) that the theoretically required notional load appears unbounded as the slenderness increases, it is proposed to perform the notional load calibration at a modified effective slenderness parameter λ_c of around unity since this is the length corresponding to the greatest sensitivity to imperfections and therefore to notional loads. Having performed the calibration in the vicinity of $\lambda_c = 1$, the ramifications of the calibration over the full slenderness range can be assessed through parametric studies. The results of the initial calibrations are given in Table 4.2. The results shown in Table 4.2(a) indicate that, in the intermediate slenderness range of $\lambda_c = 1$, the theoretically required notional load parameter varies from $\xi = 0$ for the extreme case of a sway column fully restrained against rotation at both ends ($G_A = G_B = 0$, effective length factor $K = 1$), to a value approaching $\xi = 0.01$ for a sway column which is lightly restrained against rotation by restraints of equal stiffness (giving an elastic

---

[3] As outlined in Section 1.2.2, the calibration of the AISC LRFD beam-column interaction equations themselves was also undertaken on the basis of nominal strengths. Appropriate ϕ-factors were only incorporated at the final stage to convert the formulae into predictions of design strength rather than nominal strength.

effective length factor $K > 7$). Therefore, the increase in K has an effect on the theoretically required notional load parameter ξ. For sway columns with end-restraints (at one or both ends) of practical stiffnesses likely to be achieved in steel frames, however, the majority of the ξ values, when calibrated for the material and cross-sectional parameters of $D/(\eta r) = 2.07$ and $E/F_y = 800$, do not exceed 0.005. As shown in Table 4.2(b), slightly higher bounds are inferred for high strength steels ($E/F_y = 444$); in this instance $\xi = 0.005$ would correspond more to a "mean" value than an "upper bound".

As suggested by the form of Eqs. (4.19) and (4.20), and as also demonstrated by results in Table 4.2, it is apparent that the following factors contribute to an increase in the notional lateral load required for precise calibration:

- an increase in column slenderness, as defined by $\lambda_{c(L)}$
- an increase in the effective length factor K
- an increase in the yield stress F_y.

From an analysis of the $D/\eta r$ term, it would also appear from the calibration that higher notional loads are required for major-axis behavior (low $D/\eta r$) than for minor-axis behavior (high $D/\eta r$) of typical I-section columns as indicated in Table 4.2(c).

It should be borne in mind that in the AISC LRFD Specification no distinction in the form of the interaction equation is made between strong and weak axis bending, nor between cross-sectional strength and in-plane strength. Many other specifications, such as the Canadian Standard CSA-S16.1-M94 [CSA, 1994], the Australian Standard AS4100-1990 [SA, 1990] and the Eurocode 3 [CEN, 1992], employ separate interaction equations for in-plane and cross-sectional strength and, furthermore, differentiate between strong and weak axis behavior. In these circumstances, larger notional loads may be required for weak axis compared to strong axis bending in order to minimize the potential unconservatism in "stability critical" frames.

In the following, three "tiered" notional load calibrations are presented. These are termed a "simple" calibration, a "modified" calibration, and a "refined" calibration, and are described in Sections 4.6.2 to 4.6.4 hereafter.

4.6.2 "Simple" Notional Load Calibration

The first objective of this calibration exercise is to derive a single "simple" or "basic" value of $\xi = \xi_0$ for "universal" use in beam-column design which is reasonably effective throughout the range of end-restraints likely to be achieved in practice, which is applicable for both major and minor axis behavior, and which can be applied without due regard to the yield stress F_y of the section (provided $F_y \leq 65$ ksi (450 MPa)). With this intention, it is proposed that

$$\xi = \xi_0 = 0.005 \tag{4.21}$$

represents the best possible calibration of the required notional lateral load $N = \xi P$ under a wide range of practical conditions.

Table 4.2(a) Calibration of Notional Lateral Load for $D/\eta r = 2.07$, $E/F_y = 800$, and Various End-Restraint Conditions G_A, G_B

$\lambda_{c(L)}$	G_A	G_B	K_e[a]	K_i[b]	τ	$\lambda_c = K_i\lambda_{c(L)}$	$P_{n(L)}/P_y$	P_u/P_y	ξ
0.114	60	∞	10.100	8.825	0.755	1.006	0.995	0.655	0.00741
0.189	20	∞	6.021	5.302	0.752	1.002	0.985	0.657	0.00492
0.303	6	∞	3.652	3.305	0.751	1.001	0.962	0.657	0.00385
0.465	0.6	∞	2.199	2.149	0.750	0.999	0.913	0.658	0.00380
0.500	0	∞	2.000	2.000	0.750	1.000	0.901	0.658	0.00390
0.163	60	60	7.083	6.174	0.756	1.006	0.989	0.654	0.00991
0.469	6	6	2.405	2.136	0.752	1.002	0.912	0.657	0.00367
0.871	0.6	0.6	1.196	1.148	0.750	1.000	0.728	0.658	0.00125
1.000	0	0	1.000	1.000	0.750	1.000	0.658	0.658	0.00000

Table 4.2(b) Calibration of Notional Lateral Load for $D/\eta r = 2.07$, $G_A = 0$, $G_B = \infty$ and Various Material Parameters E/F_y

$\lambda_{c(L)}$	E/F_y	K_e[a]	K_i[b]	τ	$\lambda_c = K_i\lambda_{c(L)}$	$P_{n(L)}/P_y$	P_u/P_y	ξ
0.500	800	2.000	2.000	0.750	1.000	0.901	0.658	0.00390
0.500	571	2.000	2.000	0.750	1.000	0.901	0.658	0.00462
0.500	444	2.000	2.000	0.750	1.000	0.901	0.658	0.00523

Table 4.2(c) Calibration of Notional Lateral Load for $E/F_y = 800$, $G_A = 0$, $G_B = \infty$ and Various Cross-Sectional Parameters $D/\eta r$

$\lambda_{c(L)}$	$D/\eta r$	K_e[a]	K_i[b]	τ	$\lambda_c = K_i\lambda_{c(L)}$	$P_{n(L)}/P_y$	P_u/P_y	ξ
0.500	2.07	2.000	2.000	0.750	1.000	0.901	0.658	0.00390
0.500	2.17	2.000	2.000	0.750	1.000	0.901	0.658	0.00372
0.500	2.58	2.000	2.000	0.750	1.000	0.901	0.658	0.00313
0.500	3.07	2.000	2.000	0.750	1.000	0.901	0.658	0.00263

[a] Elastic K-factor

[b] Inelastic K-factor

4.6.3 "Modified" Notional Load Calibration; Investigation of Accuracy

The results shown in Table 4.2(b) indicate that, with all other parameters being constant, the material parameter E/F_y has a direct effect on the theoretically required notional load. Since the elastic modulus E for steel is essentially constant at 29000 ksi (200000 MPa), variations in E/F_y are due to variations in the yield stress F_y. The yield stress itself is well-defined and practically always known at the commencement of the analysis/design process. It is therefore considered appropriate to modify the basic notional load parameter of $\xi_0 = 0.005$ in an elementary manner for the effects of different yield stresses. The notional load parameter so derived is termed the "modified" notional load parameter ξ_M, and is expressed in terms of a yield stress modification factor k_y as

$$\xi_M = \xi_0 k_y \tag{4.22}$$

It is hinted by Eqs. (4.19) and (4.20) that such a modification factor should be proportional to $\sqrt{F_y/E}$, and combining this notion with the results given in Table 4.2(b) furnishes

$$k_y = 22\sqrt{\frac{F_y}{E}} \tag{4.23}$$

as a simple and effective formula which modifies the notional load appropriately for the effects of different yield stresses[4]. Furthermore, the specified yield stress of structural steel typically corresponds to just a few standard values, such as 36 ksi (250 MPa), 50 ksi (350 MPa) and 65 ksi (450 MPa). For these standard yield stress values, it is proposed that the notional load parameters which can be universally applied under all practical conditions of end-restraint stiffness and slenderness (which also reflect the values determined using the more "precise" Eq. (4.22) to within 0.0002) are:

- $F_y = 36$ ksi (250 MPa): $\xi = 0.004$
- $F_y = 50$ ksi (350 MPa): $\xi = 0.0045$
- $F_y = 65$ ksi (450 MPa): $\xi = 0.005$.

It should be pointed out that although the above values of ξ have been proposed as the "best possible" unique calibrations of the notional load parameters for the given commonly specified yield stresses, in many instances the predicted strength of the column or beam-column may not be overly sensitive to the precise value of ξ assumed. The sensitivity of the member strength to ξ is likely to be greatest in the intermediate slenderness range for which the differences in P_n and

---

[4] The yield stress modification factor defined by Eq. (4.23) is one of four modification factors employed in the "refined" notional load calibration outlined in Section 4.6.4 and described in detail in Appendix F.

$P_{n(L)}$ are greatest (where the modified effective slenderness λ_c is approximately unity). The sensitivity of the member strength to other imperfections, such as member out-of-straightness and residual stresses, is also greatest in the intermediate slenderness region. The member strength will be somewhat less sensitive to ξ in regions of high slenderness where elastic buckling becomes important. As the column slenderness decreases, the sensitivity of the column strength P_n to the effective length K diminishes (P_n tends towards $P_{n(L)}$, which approaches P_y) and the second-order elastic amplification effects become less significant. For such stocky columns, the effect of the notional load on the design bending moment is approximately first-order.

For any prescribed value of ξ, the AISC LRFD beam-column interaction equation can be solved (iteratively) to deduce the nominal strength P_n of a sway column, where the axial resistance term $P_{n(L)}$ is based on the *actual* column length. In this manner "column curves" based on the notional load approach can be derived. The deduced column curves shown in Figs. 4.18, 4.19 and 4.21, and discussed hereafter, have been computed assuming $E = 29000$ ksi (200000 MPa), $F_y = 36$ ksi (250 MPa) (i.e., $E/F_y = 800$) and $\xi = 0.004$. For sway columns with a single end-restraint at end A (i.e., $G_B = \infty$), which therefore experience single-curvature bending ($\beta = 0$) due to the notional load, the effect of the end-restraint stiffness is shown in Fig. 4.18. Generally, the deduced column strength curves for all end-restraint stiffnesses are close to the AISC LRFD column curve, although it can be seen that an increase in the stiffness of the end-restraint (corresponding to a decrease in the G-factor) produces a slight lowering of the deduced column curve. For sway columns with end-restraints of equal stiffness ($G_A = G_B$), which are therefore bent in double-curvature ($\beta = 1$) by the notional load, the influence of the restraint stiffness is shown in Fig. 4.19. In common with Fig. 4.18, higher restraint stiffnesses result in lower column curves. The case of $G_A = G_B = 0$ is a theoretical lower limit since in this instance the column *effective* length is equal to the *actual* length ($K = 1$) and consequently the theoretically required notional load is zero. Accordingly, enforcing the assumption of $\xi = 0.004$ necessarily produces a significantly lower column curve than that defined in the AISC LRFD Specification. It is relevant to note in passing that the refined notional load calibration outlined in Section 4.6.4 infers $\xi = 0$ for the extreme case of $G_A = G_B = 0$.

For a cantilever column with a rigid base ($G_A = 0$, $G_B = \infty$) and $E = 29000$ ksi (200000 MPa), the effectiveness of the assumptions of $\xi = 0.004$, 0.0045 and 0.005, which correspond to yield stresses $F_y = 36$, 50 and 65 ksi (250, 350 and 450 MPa), respectively (i.e. $E/F_y = 800$, 571 and 444), is shown in Fig. 4.20. It can be seen in this figure that the deduced column curves are effectively coincident, thus vindicating the rationality of the yield stress adjustment factor k_y defined by Eq. (4.23).

Also for a cantilever column with a rigid base, the sensitivity of the deduced column curves to the axis of bending, as reflected in the cross-sectional parameter $D/(\eta r)$, is shown in Fig. 4.21. The parameter $D/(\eta r) = 2.07$ corresponds to the *minimum* value for *major-axis* bending of Australian Universal Column (UC) sections [BHP Steel, 1991]. The parameter $D/(\eta r) = 3.07$ corresponds to the *maximum* value applicable to *minor-axis* bending of Australian Universal Beam (UB) sections [BHP Steel, 1991]. These limits are similar to those for American W-shapes. It can be seen in Fig. 4.21 that the $D/(\eta r) = 3.07$ (minor-axis) curve is lower than the $D/(\eta r) = 2.07$ (major-axis) curve, particularly in the intermediate slenderness range. It is pertinent to note, however, that column strength curves derived by *advanced analysis* of imperfect pin-ended columns (with the same W8×31 cross-sectional shape and imperfection parameters $\delta_0 = 0.001L$ and $\sigma_{rc} = 0.3F_y$) also indicate that columns failing about the minor axis are noticeably weaker than the columns of corresponding slenderness failing about the major axis. There is therefore a degree of consistency between the column curves derived using the notional load analysis, and the column curves derived using advanced analysis, despite the fact that in both cases the minor axis curve is somewhat below the (unique) AISC LRFD curve. In Eurocode 3 [CEN, 1992], which employs multiple column curves, a lower column curve is employed for rolled I-sections buckling about the weak axis than for the corresponding sections buckling about the strong axis.

4.6.4 "Refined" Notional Load Calibration

The objective of the "refined" notional load calibration is to improve the overall accuracy of the "simple" and "modified" calibrations described above. Although in many cases the simple and modified calibrations produce accurate estimates of column strength when compared with the corresponding advanced analysis results, in other cases the results are somewhat conservative. It is apparent from the results shown in Figs. 4.18 and 4.19, and from the studies of beam-column and frame strength presented in later sections of this chapter, that the conservatism is greatest in frames comprising columns of low to intermediate slenderness which are bent in reverse-curvature, have significant rotational end-restraints, and are subjected to significant axial force relative to bending moment. It is likely that such columns are prevalent in multistory frames under gravity load, especially in the lower stories. This section summarizes the refined calibration of the notional load parameter ξ which will significantly reduce or eliminate the conservatism of the simple and modified approaches; full details are given in Appendix F. Such a refined calibration will be particularly advantageous for the application of the notional load approach to the design of multistory frames (see Sections 4.10 to 4.13).

Motivation and guidance for the refined notional load calibration stems from the following observations:

- As the yield stress of the member increases, so too does the notional load required for precise calibration against the AISC LRFD column curve[5].
- Irrespective of the length of the column, as the lateral stiffness of the sway member approaches that of one restrained rigidly against rotation at both ends (i.e., as the effective length factor tends to unity), the required notional load tends to zero.
- Irrespective of the magnitude of the end restraints (and hence the effective length factor), as the actual length of the member approaches zero, so too does the required notional load.

---

[5] This effect alone was the basis of the modified calibration described previously.

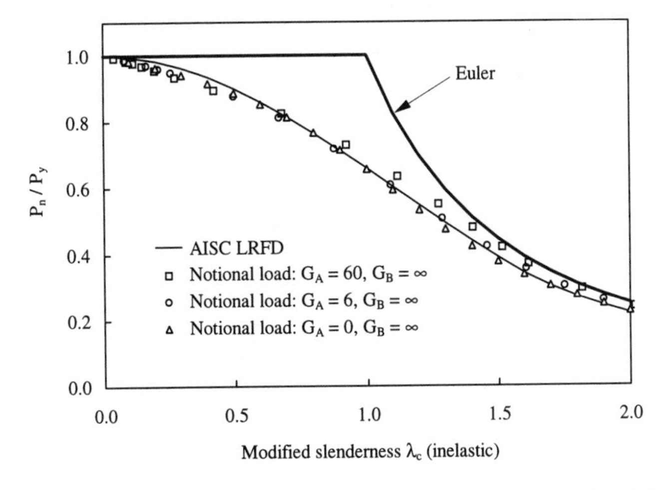

Fig. 4.18 Strength curves for sway columns in single-curvature bending derived using notional load approach with $\xi = 0.004$

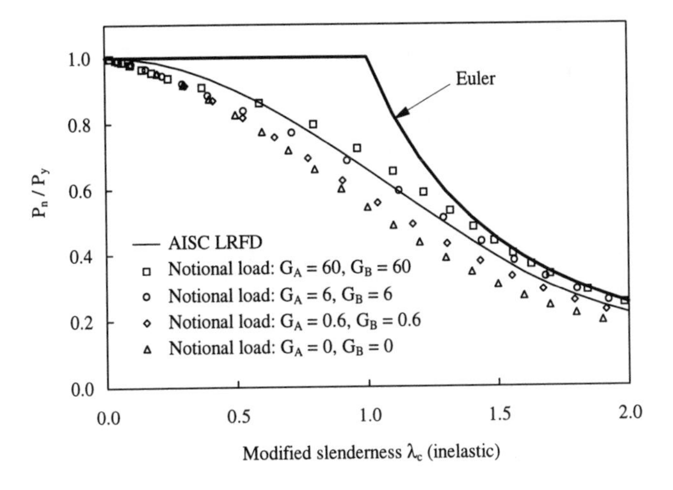

Fig. 4.19 Strength curves for sway columns in reverse-curvature bending derived using notional load approach with $\xi = 0.004$

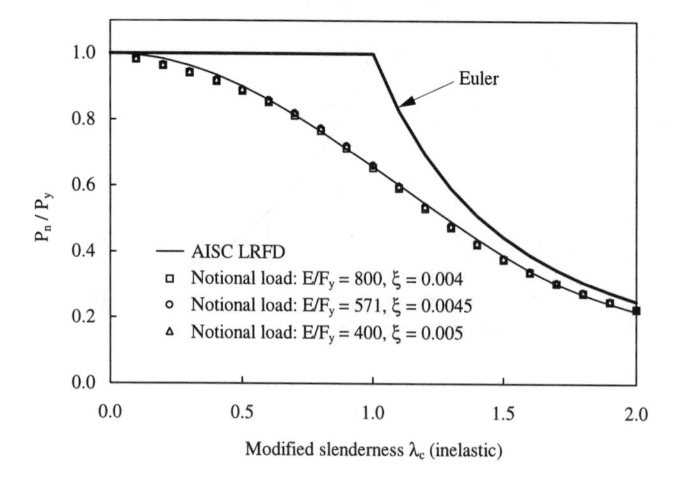

Fig. 4.20 Strength curves for cantilever column derived using notional load approach with $D/(\eta r) = 2.07$—effect of material parameter E/F_y

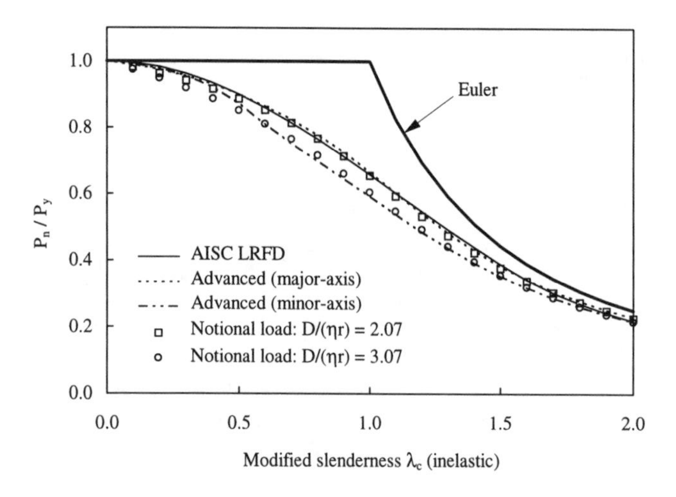

Fig. 4.21 Strength curves for cantilever column derived using notional load approach with $\xi = 0.004$—effect of cross-sectional shape parameter $D/(\eta r)$

As well as the above facts, it may be surmised from probabilistic arguments that the notional load parameter ξ applicable to multistory frames should reduce as the number of bays in a story increases. The justification for this supposition, which has parallels with the question of how the number of bays in a story should influence the story out-of-plumbness imperfection included in advanced analysis (see Section 4.11.2), is outlined in Section F.4 of Appendix F.

Proceeding directly from the above observations, the refined notional load parameter $\xi = \xi_R$ which is generally applicable to a complete story is expressed in the form

$$\xi = \xi_R = \xi_0 k_y k_S k_\lambda k_c \tag{4.24}$$

in which ξ_0 is the "basic" or "simple" notional load parameter of 0.005 employed in the simple notional load approach, and k_y, k_S, k_λ and k_c are adjustment or refinement factors accounting for the yield stress of the columns in the story, the lateral stiffness of the story, the (effective) slenderness of the story, and the number of columns in the story, respectively. Full details on each of these refinement factors are given in Appendix F.

4.7 Application to Individual Beam-Column Members

4.7.1 General

The preceding sections of this chapter have dealt almost exclusively with the assessment of the strength of axially loaded *columns* using the notional load approach. This same approach can clearly be extended to assess the strength of *beam-columns* subjected to combined compression and bending, where the bending action effects arise from both *real* and *notional* lateral loads. In this section, the ultimate strength of beam-columns which are free to sway laterally is investigated. Comparisons are made between the strength interaction curves determined using the following four methods:

(a) advanced analyses of physically imperfect members (Fig. 4.22(a)), the results of which are assumed to define the "exact" strength interaction;

(b) direct second-order elastic analysis of the geometrically perfect beam-column (Fig. 4.22(b)) to determine M_u for a given value of P_u and set of real lateral loads (no notional loads), the use of the equivalent pin-ended column concept with *elastic* effective length factors to determine P_n, and the assessment of the strength interaction through the AISC LRFD beam-column equations (Eq. (1.1) of Chapter 1, simplified for in-plane behavior and ignoring capacity factors for comparison of the nominal strength with the "exact" strength);

(c) as in (b) above, but with the column strength P_n based on the inelastic, rather than elastic, effective length;

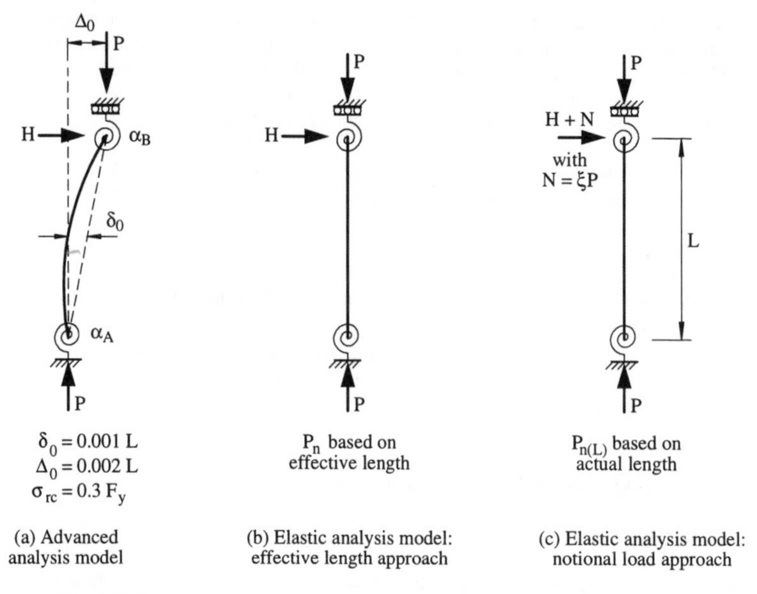

Fig. 4.22 Beam-column models for determination of strength interaction curves

(d) determination of M_u and P_u using direct second-order elastic analysis of the geometrically perfect beam-column subjected to both real lateral loads H and notional lateral loads $N = \xi P$ (Fig. 4.22(c)), with the column strength $P_{n(L)}$ based on the actual member length ($K = 1$), and evaluation of the interaction using the AISC LRFD beam-column equations, again ignoring capacity factors for comparison purposes.

The accuracy of (d) above is the prime focus of this chapter.

4.7.2 Computational Procedures for Calculation of Interaction Curves Including the Effect of Notional Lateral Loads

The differences between the notional load and other approaches for the assessment of beam-column strength are most easily comprehended when the interaction diagrams are drawn in terms of real loads only (not including notional loads) or first-order moment quantities arising from real loads, rather than second-order moments. On the other hand, the beam-column interaction equation of the AISC LRFD Specification is expressed directly in terms of M_u which by definition is the maximum second-order moment in the member. In the notional load approach, the notional as well as the real lateral loads contribute to M_u.

In the figures following, beam-column strengths are represented as strength interaction diagrams of axial load P_u and real lateral load H_u. The procedure used to compute these curves, taking into account the notional load $N_u = \xi P_u$, is described in the following.

For a given value of P_u, the AISC LRFD beam-column design equations can be used to determine the maximum required *second-order* moment M_u in the member. This moment can be expressed as the sum of two components

$$M_u = M_{uR} + M_{uN} \tag{4.25}$$

in which M_{uR} is the moment due to the real lateral load H_u, and M_{uN} is the moment due to notional load $N_u = \xi P_u$.

The maximum required second-order moment M_u can be written in the form

$$M_u = \delta_R M_{fR} + \delta_N M_{fN} \tag{4.26}$$

in which M_{fR} and M_{fN} are the first-order moments due to real and notional loads, respectively, and $\delta_R = \delta_R(P_u)$ and $\delta_N = \delta_N(P_u)$ are the corresponding elastic amplification factors. The elastic amplification factors δ_R and δ_N are determined from elastic second-order analyses of the perfect beam-column subjected to axial and real lateral loads (for δ_R), and axial and notional lateral loads (for δ_N). Where the real and notional lateral loads are identical in direction and point of action, as is often the case, the two amplification factors δ_R and δ_N are equal. The amplification factor δ for a sway beam-column with zero translational restraint, elastic rotational end-restraints, and a single imposed lateral load at the top is defined by Eq. (D.22) of Appendix D.

From Eq. (4.26), M_{fR} can thus be determined as

$$M_{fR} = \frac{M_u - \delta_N M_{fN}}{\delta_R} \tag{4.27}$$

in which it should be noted that M_{fN} is defined directly by the notional load $N_u = \xi P_u$ (see Eq. (D.13)) as

$$M_{fN} = \xi P_u L \Phi \tag{4.28}$$

in which $\Phi = \max(\Phi_A, \Phi_B)$, with Φ_A and Φ_B defined by Eq. (D.14) of Appendix D. The real lateral load H_u acting on the beam-column at the strength limit state, expressed as a fraction of the simple plastic collapse load H_p, is thus

$$\frac{H_u}{H_p} = \frac{M_{fR}}{M_p} \tag{4.29}$$

in which M_p is the full plastic moment.

4.7.3 Interaction Curves for Sway Beam-Columns

The examples presented in this section have been computed for beam-columns comprising a W8×31 section subjected to major-axis bending. Material behavior was assumed elastic-perfectly plastic with elastic modulus $E = 29000$ ksi (200000 MPa) and yield stress $F_y = 36$ ksi (250 MPa). For these material parameters, the relevant modified notional load parameter proposed in Section 4.6.3 is $\xi = 0.004$; this value has been used to derive all the notional load results shown in this section. All the advanced analysis results were computed for the physically imperfect member, with imperfection parameters $\delta_0/L = 0.001$, $\Delta_0/L = 0.002$ and $\sigma_{rc}/F_y = 0.3$.

Interaction curves for a cantilever beam-column with $G_A = 0$ and $G_B = \infty$, which is therefore in single-curvature bending ($\beta = 0$), are shown in Figs. 4.23 and 4.24 for the two cases of $\lambda_{c(L)} = 0.5$ and 1.0, respectively. Both elastic and inelastic effective length factors K are equal to 2.0 for this problem. In Fig. 4.23 ($\lambda_{c(L)} = 0.5$, $\lambda_c = K\lambda_{c(L)} = 1.0$), it can be seen that there is negligible difference between the strength interaction curves defined by the advanced analysis, effective length and notional load (with $\xi = 0.004$) approaches. For pure axial load ($H_u = 0$), the value of P_u given by the notional load ($\xi = 0.004$) curve is virtually identical to the value from the effective length curve (from Table 4.2(a), the theoretically required notional load parameter is $\xi = 0.0039$). The highest curve in Fig. 4.23 corresponds to the strength interaction consequent on basing the column strength P_n on the actual length, but ignoring the notional lateral load ($\xi = 0.0$). The unconservatism of this method, especially in the high axial force-low moment regime, is clearly evident and serves to enforce the irrationality of blindly basing the nominal column strength term in the beam-column interaction equation on the actual member length ($K = 1$) without due consideration of notional lateral loads applied additively with the real lateral loads. The more slender case ($\lambda_{c(L)} = 1.0$, $\lambda_c = K\lambda_{c(L)} = 2.0$) is shown in Fig. 4.24. It is evident in this figure that the strengths predicted by the notional load approach with $\xi = 0.004$ agree closely with those of advanced analysis, with the greatest degree of unconservatism, as measured radially from the origin, being less than 4 %. On the other hand, the effective length approach appears somewhat conservative (up to about 18 %) for intermediate values of lateral and axial load. Thus it appears—for these cantilever column examples at least—that while the effective length approach seems to become more conservative in the intermediate range of axial force and moment as the member slenderness increases, the notional load approach consistently matches the advanced analysis strengths across the slenderness range.

Strength interaction curves for a cantilever beam-column of slenderness $\lambda_{c(L)} = 0.167$ and with an elastically restrained base defined by $G_A = 20$ are given in Fig. 4.25. The elastic and inelastic effective length factors for this problem are equal to 6.0 and 4.41, respectively, thus emphasizing the low degree of rotational restraint present. As for the rigid-base cantilever, the interaction curves corresponding to the advanced analysis results, the *inelastic* effective length approach and the notional load approach (with $\xi = 0.004$) are all in quite close agreement over the complete range of axial force and moment. For the effective length approach, the importance of utilizing the *inelastic* rather than the *elastic* effective length factor for the determination of the column strength P_n is clearly evident in Fig. 4.25. The nominal column strength P_n based on the elastic effective length is 17.5 % conservative compared to the strength based on the inelastic effective length. The latter column strength is virtually identical to that determined from an advanced analysis of the physically imperfect member. If the actual column length is used to determine P_n, the unconservative consequences of ignoring the notional load can also be seen in Fig. 4.25.

The strength interaction curves for a beam-column fixed rigidly against rotation at both ends, but free to sway (zero translational restraint) are given in Figs. 4.26 and 4.27 for the two cases of $\lambda_{c(L)} = \lambda_c = 1.0$ and 2.0, respectively. As identified in Table 4.2(a) and Fig. 4.19, the theoretically required notional load for this problem is zero since the actual and effective lengths (both elastic and inelastic) are equal ($K = 1$). Using the notional load approach in conjunction with $\xi = 0.004$ therefore furnishes strength interaction curves which are quite conservative compared to the effective length based approach; for pure axial load the conservatism compared to advanced analysis results is of the order of 20% for the case of $\lambda_{c(L)} = 1.0$ (Fig. 4.26) and 30 % for the more slender case of $\lambda_{c(L)} = 2.0$ (Fig. 4.27). It can be seen however, especially in Fig. 4.27, that the differences between the notional load and effective length approaches diminish as the lateral load increases. *Among all possible conditions of elastic restraint, this sway beam-column with complete rotational fixity at both ends is believed to produce the greatest degree of conservatism when compared to the interaction curve deduced from the (inelastic) effective length approach.* The conservatism associated with the modified notional load approach ($\xi = 0.004$ for $F_y = 36$ ksi (250 MPa)) for beam-columns bent in double curvature with stiff end-restraints (as evidenced by the results in Figs. 4.26 and 4.27) can be alleviated if the refined expression for the notional load parameter ξ is employed (see Section 4.6.4 and Appendix F). For perfectly rigid rotational restraints, the refined notional load calibration infers $\xi = 0$; under these circumstances the strength interaction curves deduced from the notional load and effective length approaches for the problems elucidated in Figs. 4.26 and 4.27 are identical.

It is important to comment that if geometric imperfections are either ignored or assumed in the buckling mode (see Fig. 4.4), the elastic and inelastic behavior pertaining to the beam-column problems of Figs. 4.23 and 4.26 (and also Figs. 4.24 and 4.27) is theoretically identical. This identical behavior is not reproduced precisely in the advanced analysis results due to the slightly different effects of the (physical) out-of-straightness and out-of-plumbness imperfections in each case. Since the notional loads, in effect, account for the out-of-plumbness imperfection, the notional load approach also does not predict identical behavior. In contrast, the effective length approach implicitly assumes the imperfections are in the buckling mode and thus predicts identical interaction curves for all the beam-columns shown in Fig. 4.4.

Rather than being rigid, if the rotational end-restraints of the sway beam-column are assumed to be elastic and defined by $G_A = G_B = 6.0$ (implying elastic and inelastic effective length factors of 2.40 and 1.88, respectively) it can be seen in Fig. 4.28 that the notional load approach with $\xi = 0.004$ yields an interaction curve which is in very close agreement with the results of the advanced analysis and inelastic effective length methods for all ratios of axial load to lateral load. As in Fig. 4.25, evaluation of the nominal column strength P_n using the elastic rather than inelastic effective length produces significant conservatism in the deduced interaction curve;

conversely, the use of the actual column length but without a notional lateral load yields some unconservatism where the axial force is high.

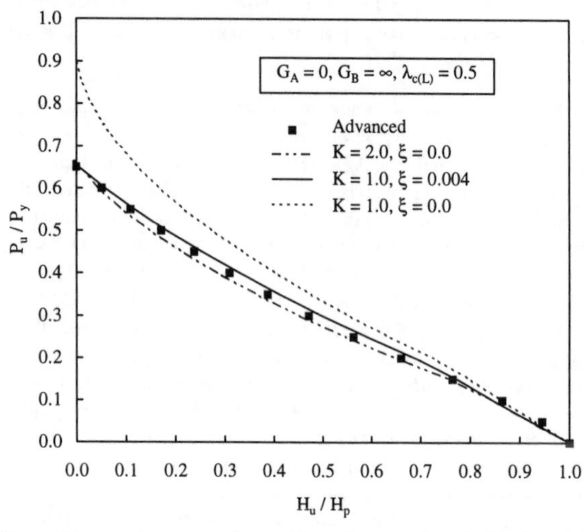

Fig. 4.23 Strength interaction curves for cantilever beam-column with $G_A = 0$, $\lambda_{c(L)} = 0.5$

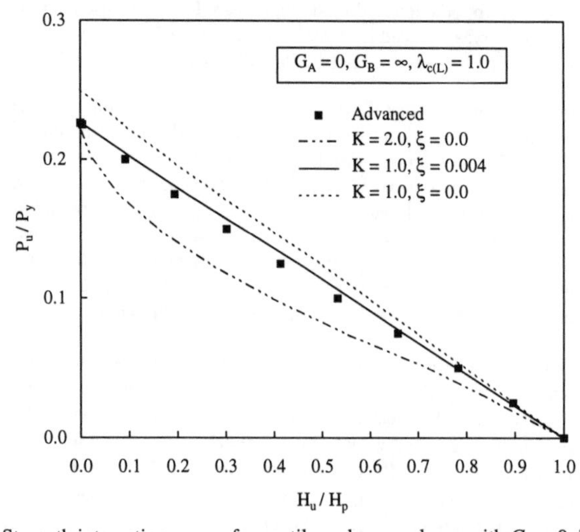

Fig. 4.24 Strength interaction curves for cantilever beam-column with $G_A = 0$, $\lambda_{c(L)} = 1.0$

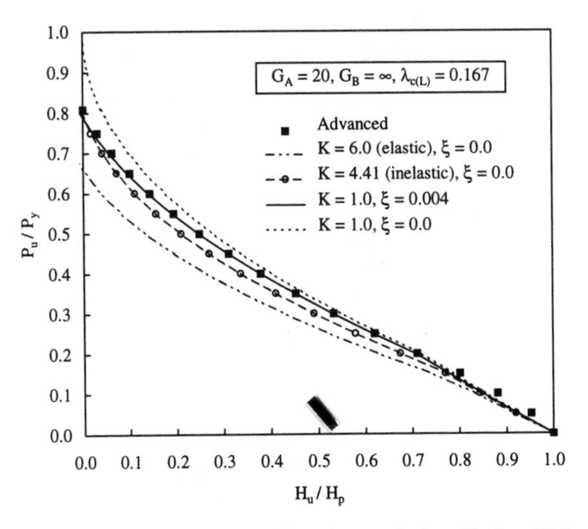

Fig. 4.25 Strength interaction curves for cantilever beam-column with $G_A = 20$, $\lambda_{c(L)} = 0.167$

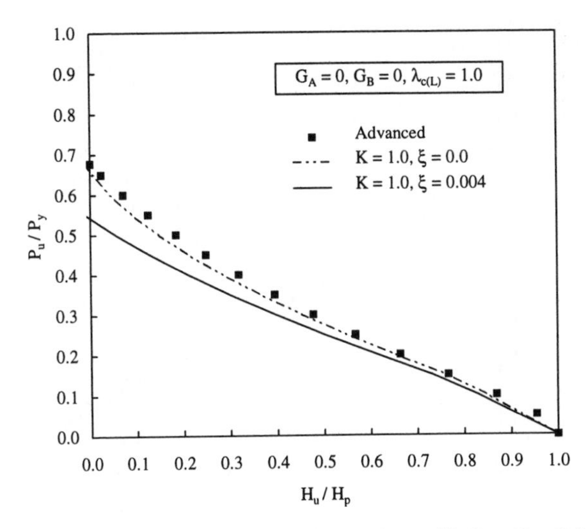

Fig. 4.26 Strength interaction curves for sway beam-column with $G_A = G_B = 0$, $\lambda_{c(L)} = 1.0$

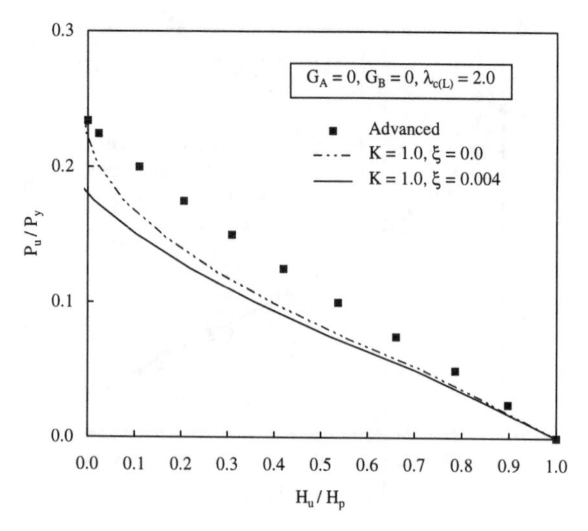

Fig. 4.27 Strength interaction curves for sway beam-column with $G_A = G_B = 0$, $\lambda_{c(L)} = 2.0$

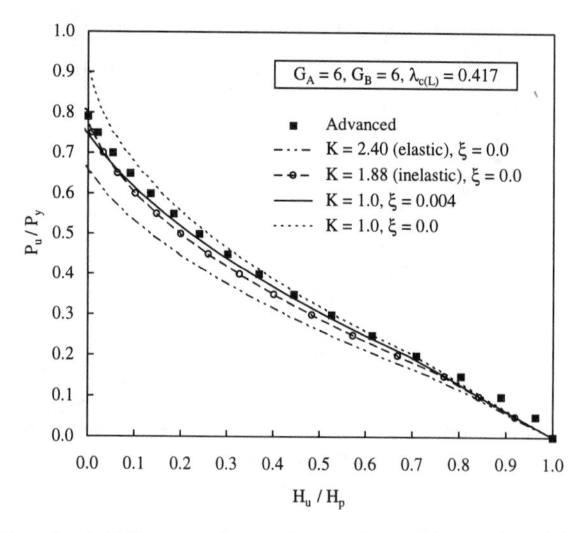

Fig. 4.28 Strength interaction curves for sway beam-column with $G_A = G_B = 6$, $\lambda_{c(L)} = 0.417$

4.8 Application to Single-Story Frames

Previous work in this chapter has related to the calibration and use of the notional load approach to assess the strength of *individual* column and beam-column members. In this section, the notional load approach is applied to a complete story; this process is therefore akin to the concept of story (as opposed to member) buckling elucidated in Section 2.4 of Chapter 2.

For the notional load approach applied to a story, every member in the story may be treated as a beam-column. Leaning columns pinned at both ends may be interpreted as a special case of a beam-column in which there is no bending. From the viewpoint of design based on second-order elastic analysis, the capacity of the story is assumed to be limited to the strength associated with the first achievement of a limit state, as defined by the AISC LRFD beam-column interaction equations, within any of the members of the story. Any "reserve capacity", associated with the inelastic redistribution of forces and moments from the most critically loaded members, is not considered. Of course, in statically indeterminate structures, the effects of such inelastic redistribution are automatically included in strength estimates computed using advanced analysis.

4.8.1 Frames with a Leaning Column

The leaning column frame illustrated in Fig. 4.29 is an extreme case for the story buckling load concept (see Section 2.4) and is therefore investigated here using the notional load approach. For simplicity, it is assumed that both columns are of equal length, although for generality their cross-sectional properties, yield stresses and applied gravity loads may differ. In the following, all quantities relating to the left-hand (restrained) column, henceforth termed column 1, are

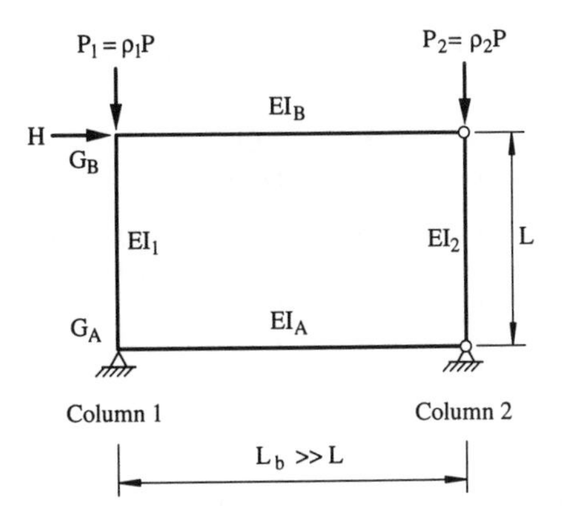

Fig. 4.29 Leaning column frame

indicated by the subscript "1". Similarly, all quantities relating to the right-hand (leaning) column, hereafter termed column 2, are indicated by the subscript "2". All lateral restraint for the frame is thus provided by column 1, which is assumed to be restrained rotationally by girders such that the G-factors pertaining to the base and top of the column are given by

$$G_A = 2\frac{EI_1/L}{EI_A/L_b} \qquad \text{and} \qquad G_B = 2\frac{EI_1/L}{EI_B/L_b} \tag{4.30}$$

respectively, in which EI_1, EI_A and EI_B are the flexural rigidities of column 1, the lower girder and the upper girder, respectively, L is the length of the columns, and L_b is the length of the girders. These expressions for the G-factors include the effect of the modified moment distribution in the girders arising from the fact that they are pinned to the right-hand column.

The relevant theoretical analysis for the leaning column frame, including elastic second-order behavior, elastic and inelastic buckling analysis, and correct application of the beam-column interaction equations for the notional load and effective length approaches, is presented in Appendix E.

The aim of the present exercise is to deduce or compute the "strength" of the frame under gravity loads alone, or combined gravity and lateral loads, using the following methods:

- advanced analysis of the physically imperfect frame, with the imperfection parameters $\Delta_0/L = 0.002$, $\delta_0/L = 0.001$ and $\sigma_{rc}/F_y = 0.3$;
- elastic analysis/design incorporating the modified simple notional load approach with the axial resistance term $P_{n(L)}$ based on the actual column length;
- elastic analysis/design, with the column strengths P_n based on effective length factors determined from an *elastic* system buckling analysis;
- elastic analysis/design, with the column strengths P_n based on effective length factors determined from an *inelastic* system buckling analysis.

The frame can therefore be visualized as being loaded *proportionally* by the gravity loads P_1 and P_2 and the lateral load H until the ultimate strength ("failure") of the "critical member" or system occurs. For convenience, the applied gravity loads may be written in terms of a single load parameter P as

$$\begin{aligned} P_1 &= \rho_1 P \\ P_2 &= \rho_2 P \end{aligned} \tag{4.31}$$

with $\rho_1+\rho_2 = 1$ and therefore $P_1+P_2 = P$. An axial strength parameter for the frame can be defined by the ratio

$$\gamma = \frac{P_{u1} + P_{u2}}{P_{y1} + P_{y2}} = \frac{\sum P_{ui}}{\sum P_{yi}} \tag{4.32}$$

in which P_{ui} is the applied gravity load acting on column i at the ultimate strength of the frame, and P_{yi} is the squash load of column i, with the summation extending over both columns. The slenderness of the frame is characterized by the slenderness parameter $\lambda_{c(L)i}$ pertaining to column i, which is defined by

$$\lambda_{c(L)i} = \frac{1}{\pi}\left(\frac{L}{r_i}\right)\sqrt{\frac{F_{yi}}{E}} \tag{4.33}$$

In this section, results are reported for several configurations of the leaning column frame. In each case, the "strength" of the frame under gravity load alone ($H = 0$) is evaluated for various axial load ratios P_1:P_2 for the different methods of analysis/design. An example illustrating the strength interaction between axial and lateral load for a particular ratio P_1:P_2 is reported in Section 4.8.3.

4.8.2 Leaning Column Frames Under Gravity Loading

Frames 1 and 2

Column 1: W8×31 (major axis); $E = 29000$ ksi (200000 MPa); $F_y = 36$ ksi (250 MPa).

Column 2: As for column 1.

Frame 1: $L/r_1 = L/r_2 = 20$ ($\lambda_{c(L)1} = \lambda_{c(L)2} = 0.2243$); $G_A = \infty$, $G_B = 4$.

Frame 2: $L/r_1 = L/r_2 = 40$ ($\lambda_{c(L)1} = \lambda_{c(L)2} = 0.4486$); $G_A = 2$, $G_B = 2$.

The frames 1 and 2 specified above correspond to the identically named frames shown in Fig. 1.6 and discussed in Section 1.4. If the mode of failure is one of sway, and if imperfections are assumed consistently in the buckling mode for both frames, the inelastic strength and behavior of the frames is identical, as are the strength predictions of the design approach based on elastic or inelastic effective length factors. However, the independent inclusion of out-of-straightness and out-of-plumbness imperfections based on physical considerations will result in the advanced analysis results for frames 1 and 2 being slightly different when the frame strength is governed by sway failure of the restrained column. Similarly, the strength predictions of the notional load approach will also differ marginally for these two so-called "equivalent" frames. The strength of frames 1 and 2, as computed from advanced analysis and application of the effective length and notional load approaches, is plotted in Figs. 4.30(a) and 4.30(b), respectively. It can be seen in these figures that, when the frame strength is governed by the restrained column, the effective length results are identical for frame 1 and frame 2, and the advanced analysis and notional load results are nearly so. However, the point at which the transition from the "sway" mode of failure of the restrained column to the "braced" mode of failure of the leaning column occurs differs between frames 1 and 2 because of the fact that the leaning column in frame 1 is only half the length of and thus has a higher strength than the corresponding column in frame 2. The advanced analysis results are predicted quite well by the notional load approach for both frame 1 (Fig. 4.30(a)) and frame 2 (Fig. 4.30(b)), although slightly conservatively in the latter case.

The switch in failure mode from sway failure of the restrained column to braced failure of the leaning column is captured well by the notional load approach. This aspect, together with the ability to accurately represent the frame strength for all ratios of column axial loads, is quite remarkable and serves to further confirm the validity of the notional load approach as a rational and consistent means of assessing in-plane frame stability when elastic second-order analysis is employed. The procedures outlined in Chapter 2 for the computation of story buckling loads and the associated effective length factors are only relevant to the sway buckling mode, and separate limits need to be checked to determine the propensity of the frame to buckle in a braced mode. This transition from the sway to the braced mode of failure appears to be able to be detected automatically by the notional load approach.

When dealing with frames with leaning columns, it may be thought that the ratio (termed R_L in Chapter 2) of the gravity load resisted by the leaning columns to the total story gravity load may influence the value of the notional load parameter which should be employed in order to achieve strength estimates which exhibit a consistent trend with those of advanced analysis. However, the results shown in Figs. 4.30(a) and 4.30(b), together with the results plotted in Figs. 4.31, 4.32 and 4.33 pertaining to frames 3, 4 and 5, respectively (see following sections), indicate that it is appropriate to adopt a single value of the notional load parameter ξ (say $\xi = 0.004$ for $F_y = 36$ ksi (250 MPa)) based on the *total* gravity load acting on the story since the effect of the proportion of load resisted by the leaning columns is relatively insignificant.

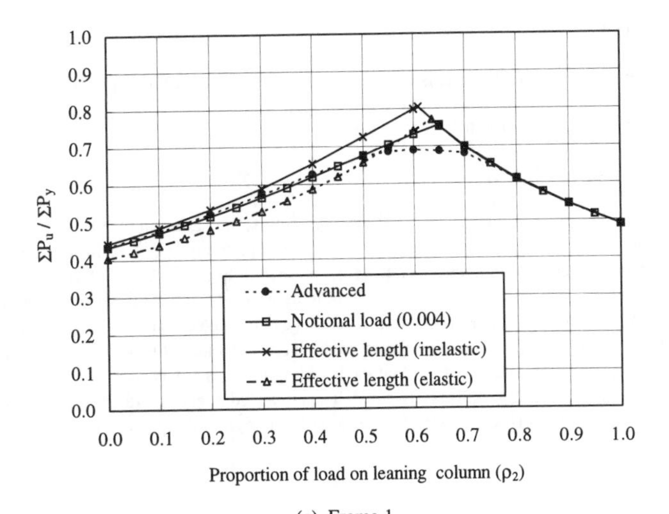

(a) Frame 1

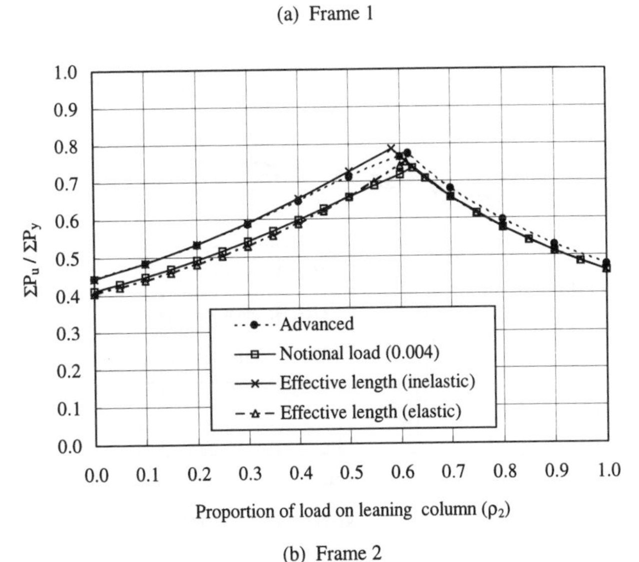

(b) Frame 2

Fig. 4.30 Strength of "equivalent" leaning column frames 1 and 2 under gravity loading

Frame 3

Column 1: W8×31 (major axis); $E = 29000$ ksi (200000 MPa); $F_y = 36$ ksi (250 MPa).

Column 2: As for column 1.

$\lambda_{c(L)1} = \lambda_{c(L)2} = 1.0$; $G_A = 0$, $G_B = \infty$.

For the frame described by the above parameters, the variation of the "strength" with the proportion of axial load on the leaning column is depicted in Fig. 4.31 for the advanced, notional load and effective length approaches. It can be seen in Fig. 4.31 that, in terms of the total gravity load the frame can support, there is a slight increase in strength as the proportion of load on the leaning column (ρ_2) increases, and this trend is captured effectively by both the notional load and effective length approaches. In relation to the latter approach, it is worth noting that the slenderness of this frame is such that the buckling mode is always one of sway, with the elastic and inelastic buckling loads being equal. The advanced analysis and notional load results also indicate that the "critical member" is column 1 (the rigid column) for all ratios P_1:P_2.

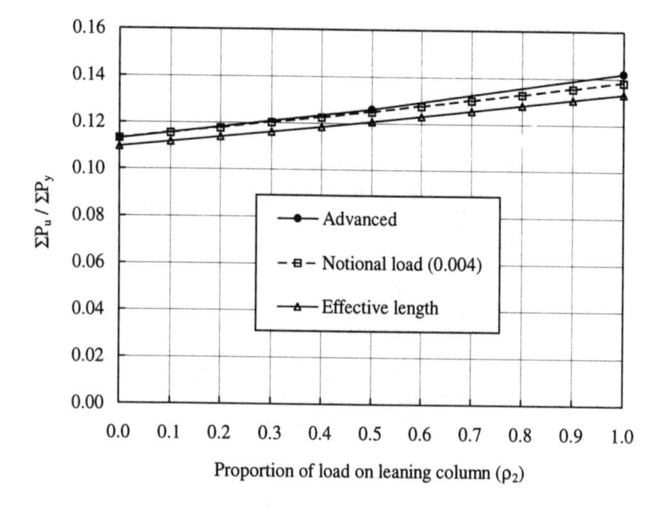

Fig. 4.31 Strength of leaning column frame 3 under gravity loading

Frame 4

Column 1: W8×31 (major axis); $E = 29000$ ksi (200000 MPa); $F_y = 36$ ksi (250 MPa).

Column 2: As for column 1.

$\lambda_{c(L)1} = \lambda_{c(L)2} = 0.5$; $G_A = 0$, $G_B = \infty$.

Results for this frame, which is stockier than but otherwise identical to frame 3, are shown in Fig. 4.32. It is interesting to observe in this figure that, in common with Fig. 4.30(a) and (b), there are two modes of failure depending on the proportion of load on the leaning column (ρ_2). With reference to the advanced analysis results, when $0 \le \rho_2 < 0.88$ the frame fails in a sway mode with column 1 critical. The effective length results appear to be somewhat optimistic for this mode, while the notional load strength predictions are quite accurate. For $0.88 < \rho_2 \le 1$, the frame fails in a braced mode with the leaning column (column 2) critical.

An interesting result for the frame studied in Fig. 4.32 is that when the proportion of gravity load on the leaning column is high ($\rho_2 \ge 0.88$), the advanced and inelastic buckling analyses predict a *braced* mode of failure, but the elastic buckling mode is one of *sway* for all ρ_2 (even though the leaning column is "critical" from a strength viewpoint).

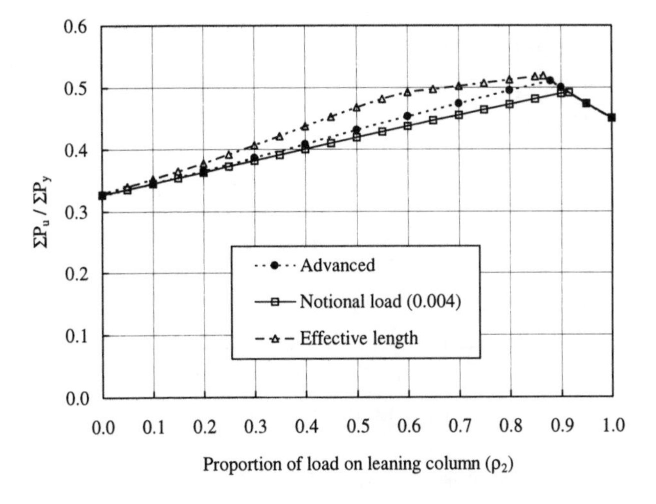

Fig. 4.32 Strength of leaning column frame 4 under gravity loading

Frame 5

Column 1: W8×31 (major axis); $E = 29000$ ksi (200000 MPa); $F_y = 36$ ksi (250 MPa).

Column 2: As for column 1.

$\lambda_{c(L)1} = \lambda_{c(L)2} = 0.1667$; $G_A = 20$, $G_B = \infty$.

This frame buckles in the inelastic range and, due to the finite flexibility of the rotational restraint G_A, the elastic and inelastic effective length factors for the restrained column are not equal (for sufficiently small ρ_2). Comparisons of the "strength" of the frame are given in Fig. 4.33, and again it can be seen that both the notional load and *inelastic* effective length approaches accurately estimate the advanced analysis results, while the use of elastic effective lengths is somewhat conservative. All analysis methods imply that the mode of failure and buckling is one of sway for all ratios of gravity load, and the critical member is always the restrained column.

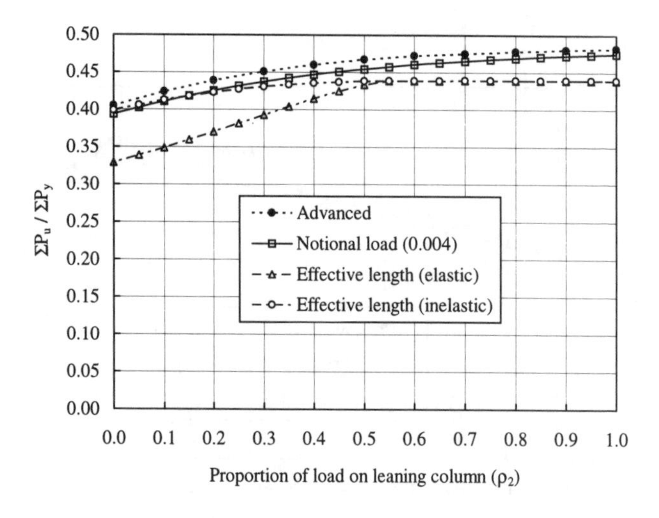

Fig. 4.33 Strength of leaning column frame 5 under gravity loading

4.8.3 Leaning Column Frames Under Combined Axial and Lateral Loading

The single-story frame examples presented in the previous section were all assumed to be subjected to gravity loads only. In this section, the effectiveness of the notional load approach for assessing the strength interaction between gravity and lateral loads acting on a single story with leaning columns is studied. The frames investigated are the "equivalent" frames 1 and 2 specified above, which correspond to the identically named frames presented as illustrative examples in Fig. 1.6 and Section 1.4.

Strength interaction curves for frame 1 based on advanced analysis, effective length and notional load approaches are shown in Fig. 4.34(a); corresponding curves for frame 2 are plotted in Fig. 4.34(b). The application of various effective length approaches to this frame has been studied in detail in Chapter 1, and so the effective length results included in Fig. 4.34 for comparison purposes are therefore based only on elastic and inelastic K-factors computed using an exact frame buckling analysis, as described in Appendix E. The advanced analysis results for frame 1 differ from those pertaining to frame 2 due to the fact that member out-of-straightness and story out-of-plumbness geometric imperfections were modeled independently. Slight differences also exist in the notional load solutions for frames 1 and 2, but in both cases there is excellent agreement between the notional load and advanced analysis results—agreement across the majority of the interaction which is as good as or better than the corresponding results based on the exact inelastic effective length. It can also be seen in Fig. 4.34 that it is clearly inappropriate to ignore both effective length factors and notional loads.

4.9 Practical Design Considerations in the Use of the Notional Load Approach

This section illustrates the practical application of the notional load approach to frame 2 shown in Fig. 1.6. The relevant assumptions for the analysis and design of this frame, and the calculation of story-based effective length factors, were elucidated in Section 2.4.5. In common with Section 2.4.5, it is assumed here that the loads applied to the frame as shown are factored and have values of $P_u = 430$ kips and $H_u = 9.2$ kips. In the following, the modified notional load approach with $\xi = 0.004$ is employed. Theoretical analysis of leaning column frames of the type considered here is detailed in Appendix E.

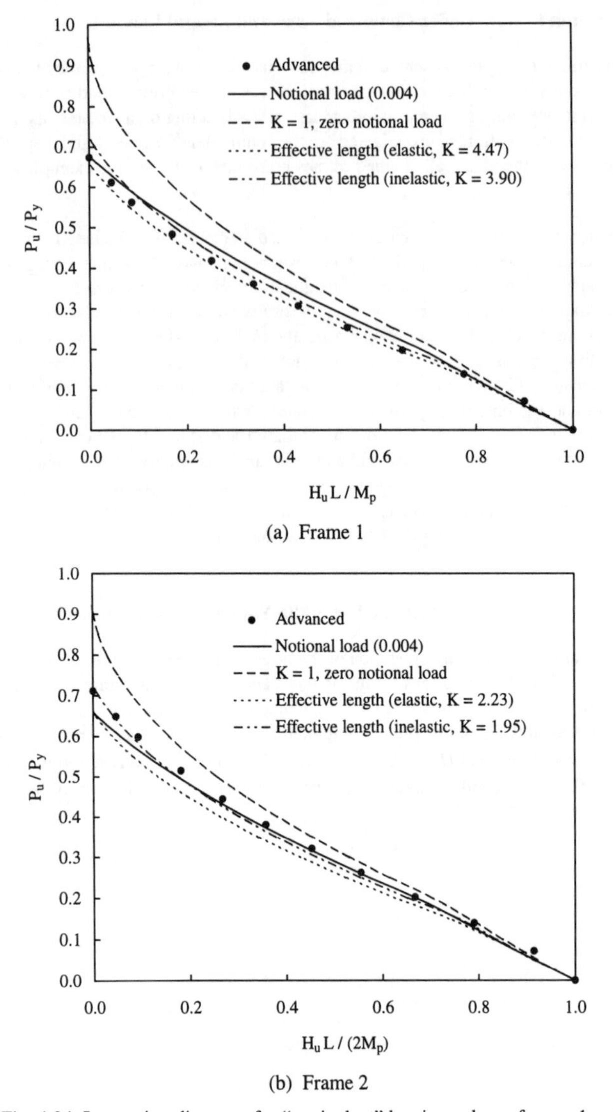

(a) Frame 1

(b) Frame 2

Fig. 4.34 Interaction diagrams for "equivalent" leaning column frames 1 and 2

Frame 2

The restrained column in the frame depicted in Fig. 1.6 has end-restraints which are of equal stiffness defined by $G_A = G_B = G = 2.0$ (i.e., $\overline{\alpha}_A = \overline{\alpha}_B = \overline{\alpha} = 3.0$). Consequently, the maximum moment M_u in the member coincides to the end-moments M_A and M_B defined by Eq. (E.10), and can thus be expressed in the form

$$M_{ux} = \frac{(H_u + N_u)\rho_1 L\Omega}{1 - 2\rho_2\Omega} \tag{4.34}$$

in which N_u is the notional horizontal load, $\mu L = L\sqrt{P_u / EI_x}$ is an axial force parameter, and ρ_1 and ρ_2 are the proportions of the total story gravity load acting on the rigidly connected column and leaning column, respectively. Since the end-restraints are of equal stiffness, the term Ω is defined either by Ω_A or Ω_B of Eq. (D.21). For this example:

$$P_u = 430 \text{ kips}, \; H_u = 9.2 \text{ kips}$$

$$N_u = \xi(P_u + P_u) = 0.004 \times (430 + 430) = 3.44 \text{ kips}$$

$$\rho_1 = 0.5, \; \rho_2 = 0.5$$

$$L = 245.5 \text{ in}$$

$$\mu L = L\sqrt{\frac{P_u}{EI_x}} = 245.5\sqrt{\frac{430}{29000 \times 999}} = 0.9458$$

$$\Omega = \frac{\overline{\alpha}(\mu L \sin \mu L - \overline{\alpha} \cos \mu L + \overline{\alpha})}{\mu L\left[2\overline{\alpha}\mu L \cos \mu L + \left(\overline{\alpha}^2 - (\mu L)^2\right)\sin \mu L\right]} = 0.6450$$

thus furnishing[6]

$$M_{ux} = \frac{(9.2 + 3.44) \times 0.5 \times 245.5 \times 0.6450}{1 - 2 \times 0.5 \times 0.6450} = 2819 \text{ kip in}$$

The nominal flexural capacity M_{nx} can calculated as

$$M_{nx} = Z_x F_y = 157 \times 36 = 5652 \text{ kip in}$$

---

[6] Rather than use Eq. (4.34), M_{ux} could have been computed by amplifying the first-order moment by the AISC LRFD B_2 factor, which, as shown in Section 2.4.5, equals 1.81 for this example. In this case

$$M_{ux} = B_2(H_u + N_u)L\Phi = 1.81 \times (9.2 + 3.44) \times 245.5 \times 0.5 = 2803 \text{ kip in}$$

which is in very close agreement with the accurate value given by Eq. (4.34).

The normalized column slenderness parameter $\lambda_{c(L)x}$ for the rigidly connected column, based on its actual length, is

$$\lambda_{c(L)x} = \frac{1}{\pi}\left(\frac{L}{r_x}\right)\sqrt{\frac{F_y}{E}} = \frac{1}{\pi} \times 40 \times \sqrt{\frac{36}{29000}} = 0.4486$$

The column axial strength term $P_{n(L)x}$, also based on the column effective length being equal to its actual length, is therefore

$$P_{n(L)x} = 0.658^{\lambda_{c(L)x}^2} A_g F_y = 0.658^{0.4486^2} \times 26.5 \times 36 = 876.9 \text{ kips}$$

The AISC LRFD interaction can now be checked as follows:

$$\frac{P_u}{\phi_c P_{n(L)x}} = \frac{430}{0.85 \times 876.9} = 0.577 > 0.2$$

AISC LRFD Eq. (H1-1a) (Eq. (1.1a)) therefore applies, yielding

$$\frac{P_u}{\phi_c P_{n(L)x}} + \frac{8}{9}\frac{M_{ux}}{\phi_b M_{nx}} = \frac{430}{0.85 \times 876.9} + \frac{8}{9} \times \frac{2819}{0.9 \times 5652} = 1.07 > 1.0 \qquad \text{No good}$$

It can thus be concluded that, according to the notional load approach, the rigidly connected column (W14×90 section) of frame 2 of Fig. 1.6 is *not* satisfactorily designed. For comparison with conventional design procedures based on effective lengths, it is interesting to note that the AISC LRFD interaction equation was also *not* satisfied when calculations were founded on either the elastic or inelastic nomograph-based story effective length factor (K_{K_n}) (see Section 2.4.1) or the elastic or inelastic lateral stiffness-based story effective length factor (K_{R_L}) (see Section 2.4.2). In the former case, the interaction equation evaluated to 1.09, while in the latter case it evaluated to 1.10 (see Section 2.4.5). In contrast, however, the interaction equation *was* satisfied if the nomograph member effective length factor (elastic, with uncorrected G factors) of $K_n = 1.32$ was used. It is also of interest to report the results of a plastic zone advanced analysis (including residual stresses and geometric imperfections) of the frame when subjected to the design ultimate loads of $P_u = 430$ kips and $H_u = 9.2$ kips, applied proportionally. With no allowance for capacity (ϕ) factors, the frame nominal ultimate strength corresponded to $\lambda_n = 1.08$ times the design loads. Once some allowance for ϕ-factors is made, say by adopting a system ϕ-factor of $\phi_s = 0.9$, the design frame strength of $\phi_s\lambda_n = 0.97$ would be deemed inadequate according to advanced analysis.

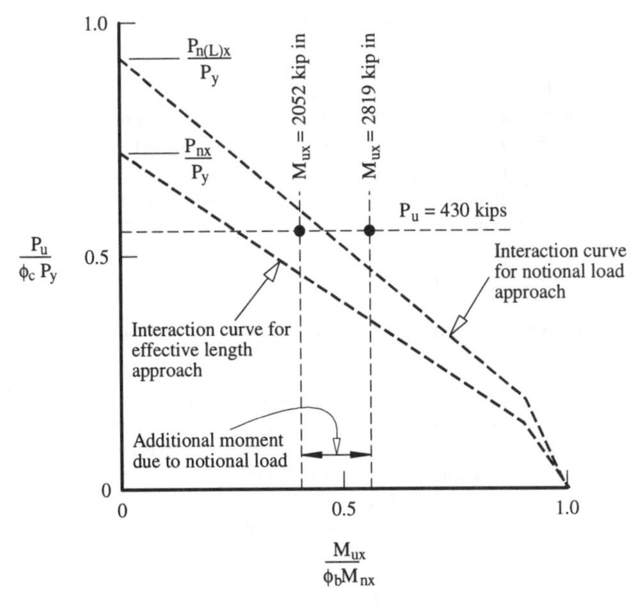

Fig. 4.35 Interaction diagrams for notional load example

The mechanism of the notional load approach compared to the effective length approach can be understood more clearly if the relevant quantities are plotted on an interaction diagram as shown in Fig. 4.35. For a given value of axial load P_u, the maximum permissible moment M_{ux} is greater for the notional load approach than the effective length approach. This is logical since in the effective length approach, it is only *real* loads acting on the structure that contribute to the design moment M_{ux}; alternatively, in the notional load approach, the design moment M_{ux} is increased due to the additional effect of the *notional* loads which, insofar as the structural analysis is concerned, act in conjunction with the real loads. The notional load approach attempts to balance the resulting increase in the design bending moment by increasing the axial limit from P_{nx} (computed using an effective length) to $P_{n(L)x}$ (computed using the actual length), and hence expanding the interaction curve.

It is saliently evident, even for this simple example, that the *calculation effort* required for the notional load approach is much less than for methods requiring the calculation of story-based effective length factors. This *vast reduction of design effort*, often with an improvement in or no loss of design accuracy (as assessed by comparison with advanced analysis results), and occasionally at the cost of some minor design conservatism, is emphasized here as a major advantage of the notional load philosophy compared to traditional effective length based methods.

As outlined above, the strength of the rigidly connected column (W14×90) is not influenced by the size of the leaning column, since the latter column provides no contribution to the lateral stiffness of the frame. The nominal axial strength P_{nx} plotted on Fig. 4.35 was computed based on an *inelastic* effective length factor of $K_x = 1.95$ as determined from the solution of Eq. (E.15) of Appendix E. It is rational to base the design of leaning columns on a K-factor of 1.0. For the example being conducted here, the leaning column should be designed to carry its axial load of $P_u = 430$ kips.

4.10 Introductory Comments on the Application of the Notional Load Approach to Multistory Frames

Previous work in this chapter has focused on the calibration and application of the notional load approach to individual unbraced beam-columns and to one-story one-bay portal frames, the latter also considering the effects of leaning columns. The notional load approach employed in the majority of these calibration and verification studies is referred to as the "modified" notional load approach since, corresponding to the assumption of $F_y = 36$ ksi (250 MPa), the notional lateral load deemed to act at each story level was universally assumed to have a magnitude equal to 0.004 times the total gravity load acting on the corresponding story.

This section, and Sections 4.11 to 4.13 following, explore some of the issues involved in extending the realm of the notional load approach to multistory frames. Unfortunately, a direct extension of the simple and modified notional load approaches, as developed and applied to one-story frames, to multistory frames may produce an over-conservative design. Particular concerns include the following:

1. It is acknowledged that the simple and modified notional load approaches may be unduly conservative for unbraced frames comprising columns of low to intermediate slenderness which are bent in reverse-curvature and have appreciable rotational restraints at their ends (see Figs. 4.19, 4.26 and 4.27). Columns with these characteristics are commonly encountered in sidesway uninhibited multistory frames, particularly in the lower stories.
2. As the height-to-width aspect ratio of the frames increases, the additional axial forces in the columns generated by notional loads alone constitute an increasing proportion of the total axial forces in the columns due to real gravity and lateral loads. This effect will be more severe in the lower level leeward columns and will contribute to additional conservatism in the design. However, to some extent the additional axial forces generated by the "overturning" effect of the notional loads is a real phenomenon arising from the actual out-of-plumbness of a "real" frame.

Of importance to the analysis of multistory frames using the notional load approach is the calibration of a "refined" notional load parameter ξ; this is outlined in Section 4.6.4 and described fully in Appendix F. Prior to a detailed consideration of the assessment of the stability of multistory frames using the notional load approach, however, several aspects relating to the *advanced analysis* of multistory frames are elucidated in Section 4.11. Such an exposition is important because there are parallels between the advanced analysis and notional load approaches with respect to the effects of out-of-plumbness imperfections, and ultimately the effectiveness of the latter approach will be judged by how well it reproduces the advanced analysis strengths of individual stories in multistory frames. In Section 4.12, a rationale is developed for applying the notional load approach to multistory frames which maintains the simplicity of the simple and modified notional load approaches, but avoids the introduction of an unreasonable degree of conservatism into the design. One of the fundamental questions addressed by the work described in Section 4.12 and Appendix F is what constitutes an appropriate ("refined") distribution of notional loads for use in the design of multistory frames. The exposition on the application of the notional load approach to multistory frames concludes in Section 4.13 with a detailed example comprising several closely related five-story one-bay frames subjected to gravity loading. The "design strengths" of these frames are determined using advanced analysis, effective length and notional load approaches, and the results compared.

As in previous sections, the scope of the work described in Sections 4.11 to 4.13 is limited to considerations of in-plane behavior and strength only. Although the notional load approach, including examples, is discussed primarily with reference to the beam-column interaction equations of the AISC LRFD Specification, the general principles expounded are universally applicable. With respect to the exposition on advanced analysis, guidance on geometric imperfections has been obtained primarily from AS4100-1990 and Eurocode 3.

4.11 Advanced Analysis of Multistory Frames

4.11.1 General Considerations; Imperfections

Prior to discussing the notional load approach in the context of multistory frames, it is pertinent to consider how the strength of such frames may be rationally evaluated using advanced analysis. In the example presented later in Section 4.13, the advanced analysis result is assumed to be the benchmark solution to which the notional load and effective length based design approaches are compared. In the context of the present work, the advanced analyses were conducted using a program based on a plastic zone (distributed plasticity) model for material nonlinearity [Clarke et al., 1992; Clarke, 1994] as described briefly in Section 4.4.1. Residual stresses in the pattern proposed by Galambos and Ketter [1959] (Fig. 4.2) were modeled explicitly in the advanced analysis, as were initial geometric imperfections of both the story out-of-plumbness and member out-of-straightness types [Clarke et al., 1992] (Fig. 4.3).

A controversial issue in the practical application of plastic zone advanced analysis to multistory frames is the determination of a "suitable" (in terms of type, orientation and magnitude) distribution of geometric imperfections. In plastic zone analysis, the emphasis is on the explicit (physical) modeling of the geometric imperfections in a manner which is as consistent as possible with fabrication tolerances (in the case of member out-of-straightness) and erection tolerances (in the case of story out-of-plumbness and global frame non-verticality) specified in design standards and manuals. A survey of the fabrication and erection tolerances prescribed in several design specifications was presented previously in Section 4.3.4. A key point, however, is that for unbraced rectangular frames, story or frame out-of-plumbness imperfections are generally more significant than member out-of-straightness imperfections [Clarke et al., 1992]; discussion hereafter is restricted to imperfections of the former type. As outlined in Section 4.3.4 and illustrated in Fig. 4.3, design standards specify erection tolerances relating to both story out-of-plumbness and global frame non-verticality. In AS4100-1990, for example, the erection tolerance for any specific story of height h is given by $\psi_s = \Delta_0/h = 0.002$, in which Δ_0 is the lateral deviation from the vertical of the top of the story relative to the base (Fig. 4.3(b)). AS4100-1990 also prescribes an envelope defining the maximum permissible overall non-verticality of the frame, shown schematically in Fig. 4.3(c). In a real frame, it would be expected that the actual distribution of story out-of-plumbness imperfections would be somewhat random, both in direction and magnitude, the only restrictions being compliance with the local story out-of-plumbness erection tolerance, and the global non-verticality envelope.

The concept of "equivalent" frame imperfections may be employed as an alternative to prescribing member and frame geometric imperfections on an independent "physical" basis [De Luca and De Stefano, 1994]. An equivalent imperfection is defined as an enlarged frame imperfection of uniform angular deviation from the vertical which accounts for the combined effects of member and frame imperfections prescribed in design specifications. With reference to the Eurocode 3 provisions, De Luca and De Stefano [1994] proposed a simple expression for an equivalent frame imperfection which can be used to simplify imperfection-related issues in the advanced analysis of multistory frames. However, such equivalent imperfections have not been considered in the present work.

4.11.2 Influence of the Number of Bays

Although it may be reasonable for the advanced analysis of a *one-bay* frame to assume that both columns and hence the whole story is out-of-plumb by an amount corresponding to the erection tolerance specified in design standards, this assumption of maximum out-of-plumbness for all columns is not statistically justifiable as the number of bays, or columns per plane, increases. It would thus seem plausible from a statistical viewpoint for the level of story out-of-plumbness modeled in plastic zone advanced analysis to decline as the number of bays increases. Similar comments also apply to the magnitude of the notional load parameter ξ used in the notional load approach (see Appendix F).

While no guidance on the influence of the number of bays on the story out-of-plumbness which should be prescribed in advanced analysis is given in AS4100-1990, Eurocode 3 does have a relevant provision which, in the lack of any better information, has been adopted here. Accordingly, the story out-of-plumbness imperfection ψ_s assumed in advanced analysis for any particular story with multiple bays is of the form

$$\psi_s = \frac{\Delta_0}{h} = \psi_0 k_c \tag{4.35}$$

in which $\psi_0 = 0.002$, $k_c = \sqrt{0.5 + 1/c}$ (but $k_c \leq 1.0$), and c is the number of columns per plane. Presumably, the preceding definition for k_c has a rational statistical basis founded on a particular mean imperfection magnitude and a corresponding standard deviation. As discussed in Appendix F, an identical adjustment parameter k_c has been adopted in the refined notional load approach.

4.11.3 Concept of Critical Story; Practical Determination of Frame Strength

For a given multistory frame subjected to a given distribution of design loads, it is inevitable that the design strength of the overall frame will be "governed" largely by the strength of one particular story; this story is termed the "critical story" in this chapter. If the frame were subjected to gravity loads only, the critical story is most likely the one which has the lowest load factor at inelastic story buckling. It is interesting to note, however, that the results of a system (eigenvalue) buckling analysis may not facilitate the identification of the critical story since all members of the frame are assumed to buckle "simultaneously" at the same critical load. However, knowledge of the buckled shape derived from the eigenvector may reveal that the sidesway mode of buckling is largely confined to one story; for all practical considerations, this story is then the critical one.

In the following discussion it will be convenient to think in terms of a multistory frame with s stories, with the height of the ith story ($1 \leq i \leq s$) denoted h_i. The overall height of the frame is then given by $H = \sum_{i=1}^{s} h_i$. In the context of plastic zone advanced analysis, the critical story of the frame is defined as the one which furnishes the minimum value of λ_{ni} ($1 \leq i \leq s$), where λ_{ni} corresponds to the advanced analysis strength of the frame with a story out-of-plumbness imperfection localized in the ith story[7].

While a large number of random permutations of geometric imperfections could have been considered, some simple practical guidelines are proposed instead to aid in the evaluation of the strength of multistory frames using plastic zone advanced analysis. Some distribution of member out-of-straightness imperfections should probably be assumed, although for sway members bent in double-curvature, the orientation of the out-of-straightness as shown in Fig. 4.3(a) is not particularly significant. Hence, the asymmetric pattern illustrated in Fig. 4.36(a) has been adopted for all advanced analysis studies reported herein. Also, the precise nature of the pattern of member imperfections is not of great importance to the advanced analysis of unbraced frames since it is the frame and story out-of-plumbness imperfections that dominate. Indeed, in many practical cases it may be prudent to ignore member imperfections altogether, possibly with the aid of the concept of "equivalent" frame or story imperfections [De Luca and De Stefano, 1994].

---

[7] A (fixed) distribution of member out-of-straightness imperfections in all columns is assumed concurrently with any distribution (localized or otherwise) of story out-of-plumbness imperfections.

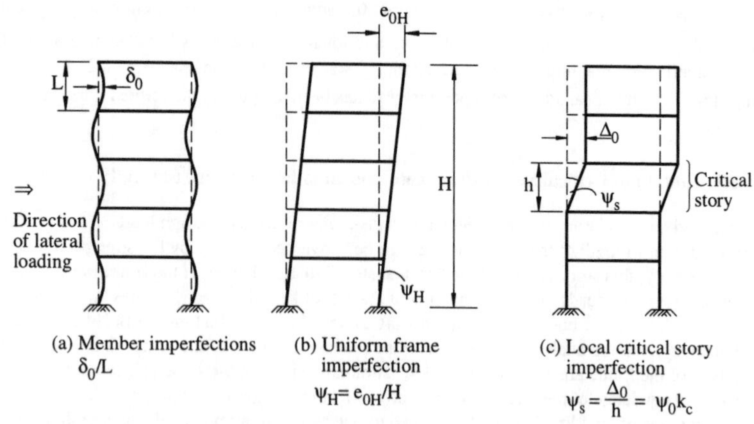

Fig. 4.36 Member, frame and story out-of-plumbness imperfections considered in advanced analysis of multistory frames

With regard to out-of-plumbness imperfections, it is proposed here that the only two patterns which need to be considered for the practical evaluation of the strength of a multistory frame using advanced analysis are:

1. a global initial non-verticality of the whole frame which is in the direction of the lateral loading and of uniform angular tilt $\psi_H = e_{0H}/H$, in which e_{0H} is the maximum permissible initial lateral deviation of the top of the frame relative to the base (see Fig. 4.36(b)); and
2. a local story initial out-of-plumbness in the critical story which is in the direction of the lateral loads and of angular deviation from the vertical $\psi_s = \psi_0 k_c$ as given by Eq. (4.35) (see Fig. 4.36(c)); evidently $\psi_H \leq \psi_s$.

The practical advanced analysis frame strength is then taken as the lower of the two results computed using the two out-of-plumbness imperfection patterns described above. In the majority of cases (see Section 4.13.2), the imposition of a localized imperfection in the critical story will result in a lower advanced analysis strength than the frame imperfection of lower magnitude extending over the entire height of the frame. As an aside, it is interesting to surmise that this brings into question the appropriateness of the out-of-plumbness imperfection provisions of Eurocode 3 (see Section 4.3.3) which enforce only a global envelope requirement without explicitly acknowledging the possibility of an imperfection of higher magnitude localized in a single story.

4.12 Modified Notional Load Analysis Procedures

4.12.1 Axial Forces Induced by Notional Loads

As well as the obvious effect of amplifying the bending moments, the explicit modeling of story out-of-plumbness imperfections in the *advanced analysis* of frames results in a slightly different distribution of axial forces in the columns compared to a corresponding analysis of a perfectly plumb frame. The out-of-plumbness under gravity loads will result in superimposed tension in the "windward" columns and superimposed compression in the "leeward" columns.

Notional lateral loads, acting in conjunction with real gravity and lateral loads, were originally contrived as a device for representing story out-of-plumbness imperfections within the context of a design procedure based on elastic analysis. Although the primary role of the notional loads—and indeed the only effect considered in the calibrations described in Section 4.6 and Appendix F—is to increase the design *bending moments* M_u in the members of a frame so as to counterbalance the increase in the beam-column interaction equation axial resistance term from P_n (computed using the effective length) to $P_{n(L)}$ (computed using the actual length), they will also induce changes in the axial forces P_u in the columns. Akin to the effects of story out-of-plumbness imperfections included in advanced analysis, the notional lateral loads—like real lateral loads—will induce superimposed tension in the "windward" columns, and superimposed compression in the "leeward" columns.

Fundamental to the notional load approach is the correspondence (determined through calibration) between a physical story out-of-plumbness imperfection ψ_s modeled in advanced analysis, and the notional load parameter ξ employed in elastic analysis. In the advanced analysis of a multistory frame of any reasonable height, it is inappropriate to assume the *whole frame* is out-of-plumb by an amount corresponding to the local story out-of-plumbness erection tolerance ψ_s; the frame must merely remain inside the global non-verticality envelope specified in design standards. Analogously, for the purpose of determining the axial force distribution in the columns of the frame using the notional load approach, it is also inappropriate to accept the cumulative superimposed effect resulting from notional loads acting at every story level. On the other hand, the additional moments arising from these notional loads acting at every story *are* needed in order to fulfill the dominant philosophical intention of the notional load approach, which is to increase the design moments M_u in the members to offset the increase in the beam-column interaction equation axial resistance term from P_n to $P_{n(L)}$.

It is therefore evident that although the superimposed axial force effect is clearly a real phenomenon in the context of advanced analysis, the equivalent effect in multistory frames with a notional load at every floor level will be unrealistically exaggerated and thus may contribute to conservatism in the resulting design. The effect will be most severe in the lower-level leeward columns of multistory frames which have a large height-to-width aspect ratio.

At the expense of ignoring a real but relatively insignificant physical phenomenon, the inducement of superimposed axial forces from the action of notional loads alone is not considered to be of central importance to the notional load approach. It is thus appropriate to ignore the effect if possible, and indeed such an assumption will help reduce the design conservatism associated with the application of the notional load approach to multistory frames. It is consequently of some import to devise a modified analysis procedure which preserves the basic philosophy of the notional load approach—to increase the design moments in the members to offset the increase in the axial resistance term from P_n to $P_{n(L)}$—while avoiding the "side-effect" of increasing the compression unrealistically in the leeward columns. The modified notional load analysis procedure so devised is termed the *dual R-N analysis procedure*, and is described in Section 4.12.2 following.

4.12.2 Dual R-N Analysis Procedure

Due to its complexity, the dual R-N analysis procedure is predominantly of theoretical interest and is not intended to be of practical design significance under usual circumstances.

Consider the problem of a multistory frame with s stories for which each story i is acted on by some general distribution q_i of real gravity loads (beam loading), as well as real lateral loads H_i and notional lateral loads N_i, as shown in Fig. 4.37(a) for the case of $s = 5$. In the conventional notional load analysis procedure, a second-order elastic analysis of the geometrically perfect frame subjected to both real and notional loads (termed an "integrated analysis" here), as depicted in Fig. 4.37(a), would be conducted to determine the design axial force P_u and moment distribution M_u for each member. The effects of the notional loads on the axial force P_u (and of course the moment distribution M_u) in each member would be included implicitly. The modified notional load analysis procedure described here avoids this axial force side-effect by decomposing the aforementioned single integrated analysis into two analyses—termed an "R-analysis" and an "N-analysis"—as described below:

- In the R-analysis, illustrated in Fig. 4.37(b), a second-order elastic analysis of the frame under the action of real loads q_i and H_i only is conducted. For each member of the frame, the results of this analysis yield the design axial force P_u and the moment distribution M_{uR} due to real loads. An analysis of this kind is exactly the same as would be undertaken in the effective length approach to the assessment of frame stability.
- In the N-analysis, illustrated in Fig. 4.37(c), the basic idea is to compute the additional moments in the members solely due to notional lateral loads. It should be borne in mind that these moments must be second-order moments which have been amplified for the effects of the column axial forces arising from the (real) gravity loads. This enables the notional load moments to be isolated from the moments generated in the columns from the beam loading. In order for these requirements to be met, the distributed gravity loads $q_i = q_i(x)$ acting across each story i must first be replaced by equivalent nodal (column) gravity loads Q_{ij}, such that

$$Q_i = \sum_{j=1}^{c} Q_{ij} = \int_0^{W_i} q_i \, dx \tag{4.36}$$

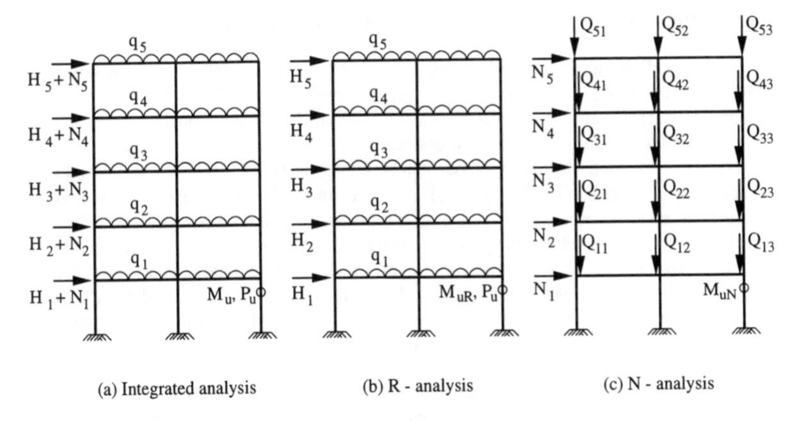

Fig. 4.37 Dual R–N analysis procedure

in which Q_i is the total story gravity load, c is the number of columns per plane, and W_i is the overall width of the story. Strictly speaking, the Q_{ij} should be chosen so that the resulting column axial forces are identical to those obtained from the R-analysis described above. However, due to the relative dominance of "P-Δ" second-order effects (associated with relative lateral translation of the ends of the member) over "P-δ" second-order effects (associated with member deformations relative to the chord joining the two ends) in sidesway uninhibited frames, for all practical purposes the Q_{ij} can be taken as the energy equivalent nodal shears pertaining to the distributed beam loading q_i. The N-analysis thus yields the (second-order) distribution of bending moments M_{uN} in the members due to notional loads alone.

For each frame member, the resultant distribution of bending moment due to both real and notional loads is given by the sum of the M_{uR} and M_{uN} distributions. For assessment of the member's design adequacy through the beam-column interaction equations, the appropriate design moment M_u to be used is given by the maximum value of $(M_{uR} + M_{uN})$ over the length of the member. It should be kept in mind, however, that the position along the member at which $(M_{uR} + M_{uN})$ is a maximum need not necessarily correspond to the position at which M_{uR} is a maximum, which again may differ from the location of maximum M_{uN}. In many practical columns, though, the positions of maximum M_{uR} and maximum M_{uN} will be identical, and will correspond to one of the ends of the member. Finally, it is worth reiterating that the design axial force employed in the beam-column strength check is obtained directly from the R-analysis of the geometrically perfect frame and therefore does not include the unwanted contribution of the notional loads.

4.12.3 Rigorous Story-Based Notional Load Analysis Procedure

The dual R-N analysis procedure described in the previous section, in conjunction with refined notional loads as described in Section 4.6.4 and Appendix F, is one means through which additional accuracy (less conservatism) can be achieved from the notional load approach. The aim of this section of the report is to describe another means through which the accuracy and rationality of the notional load approach can be improved even further for cases where the notional load parameter ξ differs for each story. The method to be described is essentially a *story-based* notional load procedure which aims to represent the localized story imperfection effect, and it may be contended that the procedure is the most rigorous possible within the realm of notional load analysis. Unfortunately, however, the story-based notional load procedure is impractical for routine design use because of the large number of second-order elastic analyses required (one per story for each load combination); it is nevertheless interesting from a theoretical perspective to explore the concept.

Focusing attention on the strength of a *particular story k* in a frame with *s* stories in total, the most rational and rigorous way of representing the "imperfection" effect local to this story through notional loads is to prescribe the notional load N_k as some fraction ξ_k (determined using the refined approach) of all the gravity load acting at that story level *k*, *and all stories above* so that

$$N_k = \xi_k \sum_{i=k}^{s} Q_i \tag{4.37}$$

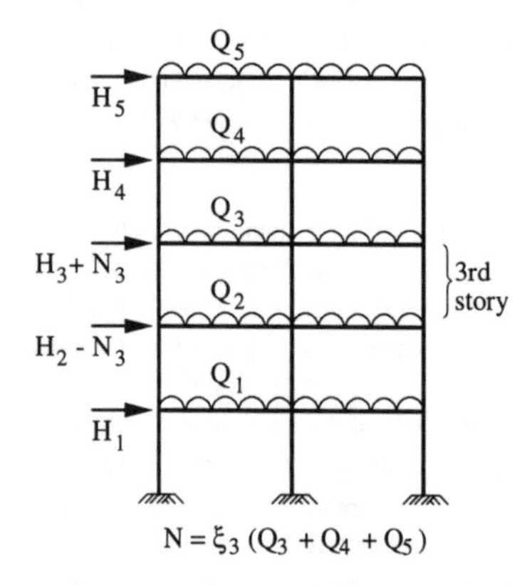

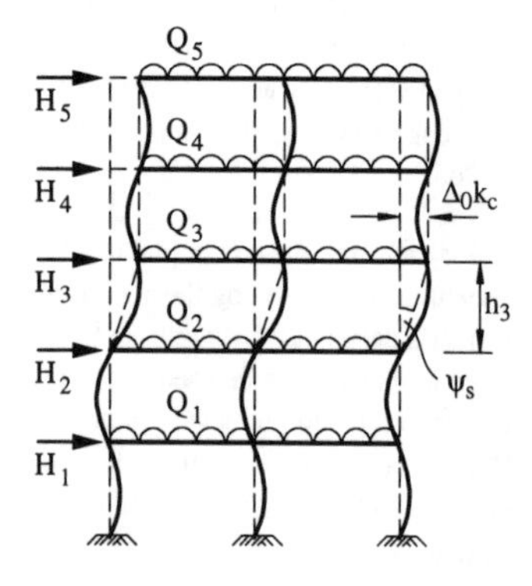

(a) Rigorous story-based notional load analysis model

(b) Advanced analysis model

Fig. 4.38 Rigorous story-based analysis notional load procedure

To restrict the notional load effects to the story k under consideration, the notional load N_k defined by Eq. (4.37) is imposed equally in magnitude but opposite in direction to the upper and lower levels of story k (i.e. story levels k and k–1), thus creating the couple $N_k h_k$. This "story based" notional load concept is illustrated in Fig. 4.38(a) for the case of $s = 5$ and $c = 3$. The procedure is akin to conducting an advanced analysis of the frame where a local story out-of-plumbness imperfection is prescribed in the kth story only, based on erection tolerances in steel design standards and specifications (see Fig. 4.38(b)). If the kth story were the "critical" one, then the adoption of notional loads at this story alone would be sufficient to assess the design strength of the entire frame.

It should be noted, however, that when the one value of ξ is assumed for all stories (as it would be in the "simple" notional load approach, for example), the loading configuration shown in Fig. 4.38(a) is exactly equivalent, as far as the generated columns moments in the third story are concerned, to the configuration shown in Fig. 4.37(a) in which a notional load $N_i = \xi Q_i$ acts at each story level i. This is not the case when the notional load parameter ξ is not identical for every story of the frame.

Following the determination of the notional load for each story of a multistory frame using the refined procedure outlined in Appendix F, it would of course be excessively tedious to then undertake s separate notional load analyses of the frame (one for each story i, $1 \le i \le s$), as inferred by the preceding discussion, merely to verify the design adequacy of every story of the frame for a particular load case. It is much more practical to simply assume one distribution of notional loads for the frame, with $N_i = \xi_i Q_i$ (where in general $\xi_i \ne \xi_j$, $i \ne j$) acting at each story level i, as is the case in the dual R-N analysis procedure described in Section 4.12.2. For practical cases, this simplified "frame based" analysis procedure will seldom be on the unconservative side since generally $\xi_i > \xi_j$ for $i > j$, and therefore $\sum_{i=k}^{s} \xi_i Q_i > \xi_k \sum_{i=k}^{s} Q_i$ for any particular story k. This last result comes about largely through the notional load adjustment parameter k_λ (see Appendix F), which tends to increase as the columns become more slender with increasing story number.

4.13 Multistory Frame Example—Five-story One-bay Frame

4.13.1 Problem Description

The effectiveness of the recommendations encompassed in Appendix F and Section 4.12 for a somewhat practical multistory frame have been investigated using variants of the five-story one-bay frame shown in Fig. 4.39[8] [Clarke and Bridge, 1995]. Attention is initially focused on

---

[8] This example was originally formulated using metric units and Australian Universal Beam (UB) and Universal Column (UC) sections [BHP, 1991; Clarke and Bridge, 1995]. For the sake of consistency with the rest of this report, in the description of the example and in the presentation of the results pertaining to it, all sections have been converted to the closest corresponding U.S. W-section, and all dimensions and

the strength and behavior of this frame when subjected to gravity loads only ($H = 0$), since this is the loading condition which will elucidate to the greatest degree the differences in the various analysis/design philosophies. If the notional load approach is shown to be effective for the case of pure gravity load, then one could be confident that it will also work well for any combination of gravity and real lateral loads (combined axial force and bending in the members) [Clarke and Bridge, 1992]. Combined gravity and lateral loading is investigated in Section 4.13.4.

The frame 0 listed in the table accompanying Fig. 4.39 corresponds to an "optimum design" under pure gravity loading. The term "optimum" is used here in the sense that none of the stories in the frame are "over-designed" or "under-designed" relative to the others. The remaining frames 1 to 5 as listed in the table accompanying Fig. 4.39 are identical to frame 0 except that, for each frame i ($1 \leq i \leq 5$), the W-section used for the columns in story i have been reduced slightly from the corresponding section designation employed in the "optimum" frame 0. The frames 1 to 5 have therefore been deliberately manufactured for analytical purposes so it can be categorically stated that for frame i, story i is the critical one under gravity loading. The beam section sizes were chosen such that the plastic section modulus of the beam in any particular story exceeds the sum of the plastic section moduli of the columns above and below. It was thus ensured that the strength of the frame was dictated by inadequate column strength rather than insufficient beam strength. Material parameters of $E = 29000$ ksi (200000 MPa) and $F_y = 36$ ksi (250 MPa) have been assumed for all members of all frames, and all cross-sections were tacitly assumed to be compact. For the advanced analyses, residual stresses were incorporated as described in Section 4.11.1 with $\sigma_{rc}/F_y = 0.3$.

---

loads have been directly converted to U.S. units. The inexactness of the conversions from Australian to U.S. section shapes does not alter the tenor of the example nor the conclusions drawn from it. However, if the example had been recomputed using the actual U.S. sections, the results would be slightly different from those shown in Tables 4.3 to 4.6 and Figs. 4.40 to 4.42 due to the approximate nature of the conversions between Australian and U.S. sections.

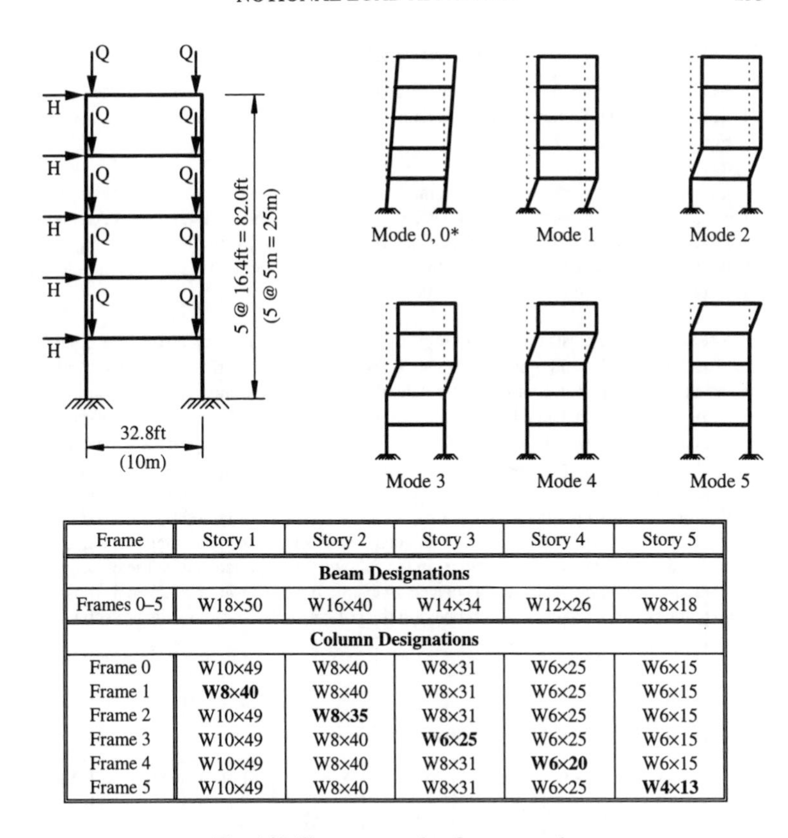

Frame	Story 1	Story 2	Story 3	Story 4	Story 5
Beam Designations					
Frames 0–5	W18×50	W16×40	W14×34	W12×26	W8×18
Column Designations					
Frame 0	W10×49	W8×40	W8×31	W6×25	W6×15
Frame 1	**W8×40**	W8×40	W8×31	W6×25	W6×15
Frame 2	W10×49	**W8×35**	W8×31	W6×25	W6×15
Frame 3	W10×49	W8×40	**W6×25**	W6×25	W6×15
Frame 4	W10×49	W8×40	W8×31	**W6×20**	W6×15
Frame 5	W10×49	W8×40	W8×31	W6×25	**W4×13**

Fig. 4.39 Five-story one-bay frame example

4.13.2 Advanced Analysis Results for Gravity Loaded Frame

Both member out-of-straightness and story out-of-plumbness imperfections were modeled in the plastic zone advanced analyses of the five-story frames shown in Fig. 4.39. The pattern of member out-of-straightness imperfections assumed in all advanced analyses is as depicted in Fig. 4.36(a). As far as story out-of-plumbness and global frame non-verticality are concerned, the distributions assumed in the various advanced analyses undertaken were:

1. a uniform global non-verticality (see Figs. 4.36(b) and 4.39) of $\psi_H = e_{0H}/H = 0.002$; in subsequent discussion this imperfection mode is termed mode 0;

2. a uniform global non-verticality of $\psi_H = e_{0H}/H = 0.001$, hereafter termed mode 0*; and
3. localized story imperfections of magnitude $\psi_s = \Delta_0/h_i = 0.002$, for which mode i ($1 \le i \le 5$) denotes the imperfection mode comprising a local story out-of-plumbness in story i (see Figs. 4.36(c) and 4.39).

A total of seven geometric imperfection modes (modes 0, 0*, and 1 to 5) were thus considered for each of the six frames studied (frames 0 to 5). It should be noted that mode 0 does not comply with the erection tolerance envelope specified in AS4100-1990, while mode 0* is right on the boundary at a height of 82.0ft (25m). Modes 1 to 5 are consistent with the erection tolerances specified for an individual story in AS4100-1990.

The advanced analysis ultimate strengths Q_u for frames 0 to 5 subjected to gravity load only are given in Table 4.3 and plotted in Fig. 4.40 for the various imperfection modes. Although the lowest advanced analysis strengths are obtained for imperfection mode 0 for all frames, these strengths are closely matched by those pertaining to the case of a local story out-of-plumbness in the critical story (frame i, mode i, $1 \le i \le 5$)[9]. This important result appears to vindicate the "critical story" concept proposed in Section 4.11.3, but simultaneously provides support for the application of a global out-of-plumbness (as in frame 0) of the same magnitude as that used for a single story. It is also pertinent to note that the frame strengths obtained for mode 0*, which corresponds to a uniform angular tilt of reduced magnitude, exceed the strengths corresponding to localized critical-story imperfections. In the following comparisons with the effective length and notional load approaches, the advanced analysis strengths have been taken as those corresponding to a local out-of-plumbness in the critical story. These results are shown in bold underlined typeface in Table 4.3.

Table 4.3 Advanced Analysis Strengths Q_u (kips) for Gravity-Loaded Five-Story Frames

Frame	Mode 0	Mode 0*	Mode 1	Mode 2	Mode 3	Mode 4	Mode 5
Frame 0	88.1	91.9	93.5	**89.7**	91.2	95.9	98.9
Frame 1	72.4	75.3	**72.8**	79.8	79.1	79.1	78.9
Frame 2	77.5	80.9	85.4	**78.4**	84.9	86.3	86.7
Frame 3	65.2	69.2	76.2	74.2	**66.3**	74.2	76.4
Frame 4	77.1	81.8	90.3	89.2	87.0	**78.9**	87.9
Frame 5	44.0	46.5	50.1	50.3	50.3	49.9	**44.3**

[9]Of course, the optimum frame 0 also has a critical story, although it is less conspicuous than the critical stories of frames 1 to 5; advanced analysis results indicate that the second story is critical for frame 0.

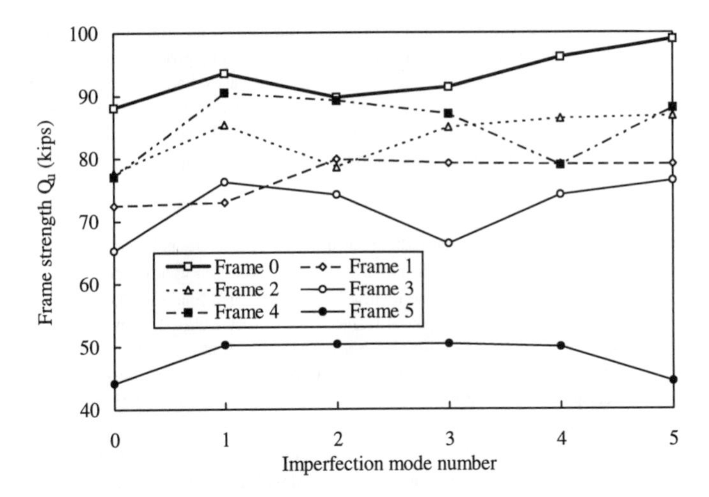

Fig. 4.40 Advanced analysis results for five-story frames with different imperfection modes

4.13.3 Effective Length and Notional Load Results for Gravity Loaded Frame

The advanced analysis strengths and the maximum design loads deduced from the effective length approach are compared in Table 4.4 and Fig. 4.41. In the effective length based design calculations, the elastic and inelastic K-factors were computed using a system (eigenvalue) buckling analysis, and the AISC LRFD beam-column interaction equations were applied assuming ϕ-factors of unity. The results shown in Table 4.4 and Fig. 4.41 reveal that the design strengths predicted using the inelastic effective length approach are comparable in accuracy to those given by advanced analysis, and are superior to those based on elastic effective lengths.

To assess the merits of the notional load approach for multistory frames, four variants of the notional load/$K = 1$ procedure have been applied to the present example. They are:

1. the use of $K = 1$ in conjunction with zero notional load ($\xi = 0$);
2. the use of $K = 1$ in conjunction with notional lateral loads defined by $\xi = \xi_0 = 0.005$ acting at each story level;
3. the use of $K = 1$ in conjunction with notional lateral loads at each story level given by the refined values of $\xi = \xi_R$ computed in accordance with the procedure detailed in Appendix F;
4. as in 3 above, but using the "rigorous" story-based analysis procedure described in Section 4.12.3.

Although the effect of the additional axial forces induced by the notional loads is only minor for the five-story frames considered here, the "dual R-N" analysis procedure described in Section 4.12.2 was nonetheless employed for all cases. For cases 3 and 4 above, which utilize "refined" notional loads, the relevant values of ξ computed in accordance with Appendix F are shown in Table 4.5. It is interesting to note from the figures in Table 4.5, which range between 0.00072 and 0.00277, that significant reductions in ξ from the "simple" or "basic" value of $\xi_0 = 0.005$ result from the refined approach.

The design strengths deduced using the four notional load variants listed above are given in Table 4.6, together with the corresponding "errors" e_{adv} compared to advanced analysis. The results are also plotted in Fig. 4.41. The results for zero notional load ($\xi = 0.0$) can be seen to be surprisingly accurate with the exception of frame 5 (fifth story critical) for which the unconservative error is 9.5 %. The "simple" notional load procedure based on universal adoption of $\xi = \xi_0 = 0.005$ results in estimates of design strength which are fairly conservative, with the degree of conservatism ranging from 15.7 % to 21.4 %. The adoption of refined notional loads $\xi = \xi_R$ (Table 4.5) for each story in a "frame-based" analysis procedure can be seen to approximately halve the unconservative error (maximum 11.7 %) for each frame compared to the case of $\xi = \xi_0 = 0.005$. The unconservatism associated with the frame-based analysis procedure (Fig. 4.37) employed in conjunction with refined notional loads ($\xi = \xi_R$) can be reduced by using the "rigorous" story-based notional load approach, as described in Section 4.12.3 and Fig. 4.38. It can be seen in Table 4.6 that the maximum conservative error associated with the story-based notional load approach of this kind is 8.6 %.

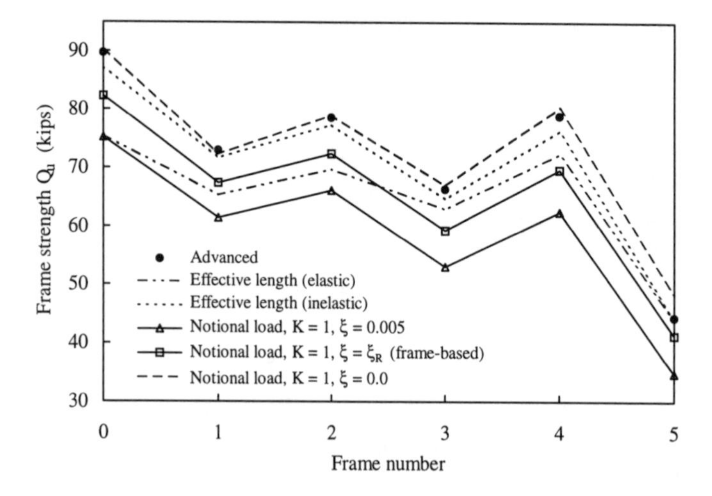

Fig. 4.41 Comparison of advanced analysis, effective length and notional load results for five-story frame under gravity load

Table 4.4 Comparison of Effective Length and Advanced Analysis Results

Frame	Advanced Analysis	Elastic Effective Length		Inelastic Effective Length		
	Q_u (kips)	Q_u (kips)	e_{adv}[a] (%)	Q_u (kips)	e_{adv}[a] (%)	K[b]
Frame 0	89.7	75.3	−16.0	87.2	−2.8	1.10
Frame 1	72.8	65.4	−10.2	71.7	−1.5	1.03
Frame 2	78.4	69.7	−11.1	77.3	−1.4	1.06
Frame 3	66.3	62.9	−5.1	64.7	−2.4	1.06
Frame 4	78.9	72.4	−8.2	76.2	−3.4	1.08
Frame 5	44.3	44.0	−0.7	44.0	−0.7	1.06

[a] "Error" compared to advanced analysis result.
[b] Inelastic K-factor in critical story.

Table 4.5 Refined Notional Load Parameters for Five-Story Frames

Frame	Notional Load Parameter $\xi = \xi_R = \xi_0\, k_c\, k_S\, k_\lambda\, k_y$				
	Story 1	Story 2	Story 3	Story 4	Story 5
Frame 0	0.00100	0.00180	0.00212	0.00218	0.00276
Frame 1	0.00072	0.00184	0.00213	0.00219	0.00277
Frame 2	0.00099	0.00160	0.00213	0.00218	0.00276
Frame 3	0.00100	0.00180	0.00145	0.00219	0.00276
Frame 4	0.00100	0.00180	0.00212	0.00178	0.00276
Frame 5	0.00100	0.00180	0.00212	0.00218	0.00127

Table 4.6 Comparison of Notional Load and Advanced Analysis Results

Frame	Notional Load $\xi = 0.0$ (Frame-Based)		Notional Load $\xi = 0.005$ (Frame-Based)		Notional Load $\xi = \xi_R$ (refined) (Frame-Based)		Notional Load $\xi = \xi_R$ (refined) (Story-Based)	
	Q_u (kips)	e_{adv} (%)	Q_u (kips)	e_{adv} (%)	Q_u (kips)	e_{adv} (%)	Q_u (kips)	e_{adv} (%)
Frame 0	90.3	+0.7	75.0	−16.4	82.2	−8.4	84.5	−5.8
Frame 1	72.1	−1.0	61.3	−15.8	67.4	−7.4	70.6	−3.0
Frame 2	78.9	+0.6	66.1	−15.7	72.4	−7.7	74.4	−5.1
Frame 3	67.0	+1.1	53.0	−20.1	59.3	−10.6	61.8	−6.8
Frame 4	80.2	+1.6	62.5	−20.8	69.7	−11.7	72.1	−8.6
Frame 5	48.5	+9.5	34.8	−21.4	41.3	−6.8	41.6	−6.1

The important conclusion from the preceding results is that the conservative error of the refined notional load approach is predominantly within 10 % of the corresponding advanced analysis strengths. Although this may be regarded by some as a touch on the high side, it should be borne in mind that the multistory frames considered here constitute a particularly severe test of the notional load approach since, in all cases, the frames are subjected to pure gravity loading and the columns in the critical story are in double-curvature at buckling and are heavily restrained by the beams. Consequently, the inelastic effective length factors in the critical story are not much greater than unity, $K = 1$ being the theoretical limit for which zero notional load is required. Also, the notional load adjustment parameters for lateral stiffness (k_S) and slenderness (k_λ) (see Appendix F) were calibrated so as to be applicable to columns in both single-curvature bending (where the G-factor at one end equals infinity, for instance) and double-curvature bending (where the most severe case is equal G-factors at both ends) without the end-moment ratio being explicitly included in the definitions. The calibrations were undertaken to provide a balance between the competing objectives of simplicity, unconservatism of a few percent for some single-curvature bending cases and double-curvature cases at high slendernesses, and conservatism of a few percent typical of double-curvature columns in the low and intermediate slenderness range. It is conceivable that the explicit inclusion of the end-moment ratio in the expressions for k_S and k_λ may lead to more accurate (less conservative) calibrations for columns which are bent in double curvature.

Despite the significant effort which has been expended on the calibration of refined notional loads, the results for frames 0 to 4 shown in Table 4.6 and Fig. 4.42 indicate that in many practical cases, both effective length factors and notional loads can be comfortably ignored. The result for frame 5, on the other hand, is unacceptably unconservative (9.5 %) under the dual assumptions of $K = 1$ and $\xi = 0$. The advantage of the refined notional load approach is the assurance that the predicted frame strengths are either accurate or slightly conservative under all practical conditions.

4.13.4 Frame Strength Under Combined Gravity and Lateral Loading

The results discussed in Sections 4.13.2 and 4.13.3 pertained to the five-story frame of Fig. 4.39 acted on by gravity loads Q only—the most severe case for the notional load procedure. It is also of interest to generate strength interaction curves which show the deduced strength of the frame when subjected to any ratio of lateral loads H to gravity loads Q. Such interaction curves have been computed for frame 0 using analysis procedures based on advanced analysis (assuming a mode 2 imperfection), elastic and inelastic effective lengths (system based), and the first three notional load approaches listed in Section 4.13.3.

The resulting strength interaction curves are shown in Fig. 4.42. In this figure, the applied lateral load H_u has been normalized by H_p, this latter quantity corresponding to the ultimate strength, determined using advanced analysis, of the frame when acted on by lateral loads H only ($Q = 0$). As for the case of pure gravity loading, the advanced analysis results indicate that the critical story of frame 0 (the "optimum" frame) under pure lateral loading is the second story, although this is not clear-cut. This value for H_p determined by advanced analysis corresponds very closely to the plastic collapse load of the frame computed using simple plastic hinge

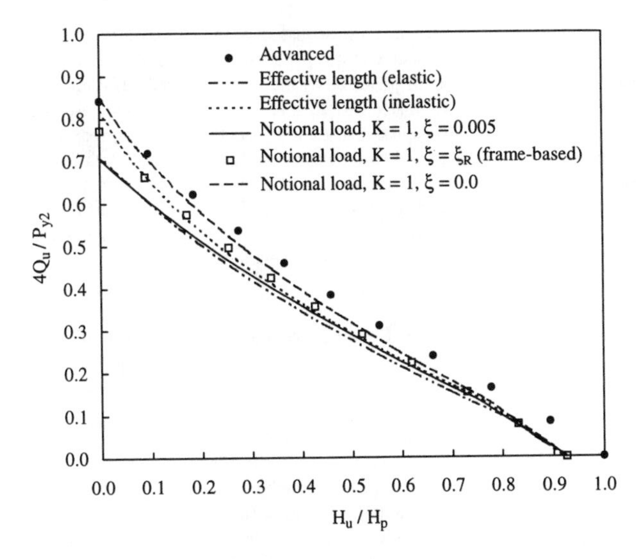

Fig. 4.42 Strength interaction diagram for frame 0

analysis. The quantity $4Q_u/P_{y2}$ on the vertical axis of Fig. 4.42 corresponds to the average axial force in the two second-story columns, expressed as a proportion of the squash load P_{y2} of the same columns.

Perusal of the results shown in Fig. 4.42 for all the effective length and notional load approaches reveals that there is universal conservatism of about 7 % for the case of pure lateral load ($Q = 0$). The main reason for this conservatism stems from the reduction in the permissible moment in the presence of axial force implied by the AISC LRFD beam-column interaction equation.

A design approach based on elastic effective lengths appears to be the most conservative of all the engineering procedures over the full range of the interaction—even more so than the notional load approach with $\xi = 0.005$. A design approach based on the use of $K = 1$ with zero notional load ($\xi = 0$) is the most accurate procedure over the complete range of axial force and moment. Although slightly more conservative than strengths based on $\xi = 0$, the performance of the refined notional load approach is marginally superior to an inelastic effective length based procedure over the whole range of the interaction diagram, with the exception of the very high axial force regime.

The above results and discussion indicate that all of the effective length and notional load analysis/design procedures considered in this paper are satisfactory for this particular frame. The differences in the design strengths predicted by the various procedures diminish as the behavior of the frame is increasingly dominated by lateral (wind) loads at the expense of gravity loads. Broadly speaking, it is therefore quite realistic to expect that a design procedure based on the use of $K = 1$, but without notional loads, will be appropriate for a large percentage of practical design scenarios. Limits within which the use of $K = 1$ in conjunction with zero notional loads is "acceptably" (less than 5 %) unconservative compared to a corresponding effective length based design result are derived in the following chapter. However, there is scope to undertake additional work in this area using more rigorous benchmark solutions.

4.14 Three-Dimensional Beam-Column Strength Under In-Plane Bending

4.14.1 Overview

Discussion in this chapter up to this point has focused on the calibration and application of the notional load approach for the assessment of the *in-plane* strength of beam-columns and frames. It has been shown that the in-plane strength of both individual sway beam-columns, and complete stories in a sidesway uninhibited frame (with or without leaning columns) can be predicted either accurately or conservatively, using the notional load approach as described, in conjunction with the AISC LRFD interaction equation formulae. The refined notional load approach produces results which are superior to those of the simple and modified notional load approaches, but at the expense of some additional effort.

In practice, steel frames are three-dimensional and the three-dimensional strength (which may be governed by "in-plane" or "out-of-plane" failure) of the beam-column members must therefore be determined. Three-dimensional steel frames rectangular in plan are commonly designed as a series of parallel two-dimensional moment-resisting sway frames spanning the shorter plan dimension, with these frames interconnected by a simple bracing system in the orthogonal direction. In such a structural system, the primary load carrying steel frames are subjected to *in-plane bending only* and are free to sway in-plane. The frames are braced out-of-plane and it is assumed here that there are no out-of-plane design bending moments. Although subjected to in-plane flexure only, the possibility of out-of-plane failure of the members exists. Under these circumstances, the AISC LRFD beam-column interaction equations are simply applied as follows:

$$\begin{aligned} &\frac{P_u}{\phi_c P_n} + \frac{8}{9}\frac{M_{ux}}{\phi_b M_{nx}} \le 1.0 \qquad \text{for } \frac{P_u}{\phi_c P_n} \ge 0.2 \\ &\frac{P_u}{2\phi_c P_n} + \frac{M_{ux}}{\phi_b M_{nx}} \le 1.0 \qquad \text{for } \frac{P_u}{\phi_c P_n} < 0.2 \end{aligned} \tag{4.38}$$

in which P_n is the lower of the in-plane and out-of-plane nominal column strengths, and M_{nx} is the lower of the in-plane and out-of-plane nominal flexural strengths. The AISC LRFD approach therefore adopts what is essentially a single equation check encompassing both in-plane and out-of-plane failure. This is in contrast to other steel design standards around the world, such as the Canadian Standard CSA-S16.1-M94, the Eurocode 3 and the Australian Standard AS4100-1990, all of which prescribe independent checks of in-plane and out-of-plane strength, with the three-dimensional strength defined as the lower of the two.

If a notional load/$K = 1$ approach was used to check the *in-plane strength*, the term P_n in Eq. (4.38) would be defined as

$$P_n = \min(P_{n(L)x}, P_{ny}) \tag{4.39}$$

in which $P_{n(L)x}$ is the nominal column strength about the major axis, based on the actual length L of the member, and P_{ny} is the minor-axis nominal column strength based on the minor-axis effective length $K_y L$ [10]. In the vast majority of practical cases, in fact whenever $r_x/r_y > 1/K_y$, the column strength P_n from Eq. (4.39) will be governed by the out-of-plane strength P_{ny}. This may lead to some confusion as to the role of the notional load approach in assessing the three-dimensional beam-column strength, since, as currently calibrated and verified, the approach only relates to in-plane strength.

4.14.2 Three-Dimensional Advanced Analysis Studies

Before considering a rational design procedure for assessing the three-dimensional strength of beam-columns, it is appropriate to conduct some finite element studies to determine the "true" strength and behavior. These numerical studies were conducted using a three-dimensional plastic zone advanced analysis developed at the University of Sydney [Pi and Trahair, 1994a, 1994b]. The beam-column problems investigated using the three-dimensional advanced analysis, including the in-plane and out-of-plane restraints and geometric imperfections, are shown in Fig. 4.43. Two types of loading were considered: axial load P in conjunction with a tip lateral load H, and axial load P in conjunction with a tip moment C. Two cross-sections were employed: a W10×25 beam-type section for which $P_{ny} < P_{nx}$, and a W8×31 column-type section for which $P_{nx} < P_{ny}$. The length of the members studied is characterized by the in-plane normalized effective slenderness parameter

$$\lambda_{cx} = \frac{1}{\pi}\left(\frac{K_x L}{r_x}\right)\sqrt{\frac{F_y}{E}} \tag{4.40}$$

in which $K_x = 2.0$ for the rigid-base cantilever columns of Fig. 4.43. The out-of-plane normalized effective slenderness can similarly be defined

---

[10] For a frame which is braced out-of-plane, the out-of-plane effective length factors K_y are relatively simple to compute compared to the in-plane effective length factors K_x for sway members. It is often appropriate to assume $K_y = 1.0$.

$$\lambda_{cy} = \frac{1}{\pi}\left(\frac{K_y L}{r_y}\right)\sqrt{\frac{F_y}{E}} \tag{4.41}$$

in which the theoretical value of the out-of-plane effective length factor for the beam-column illustrated in Fig. 4.43 is $K_y = 0.699 \approx 0.7$.

From the viewpoint of *column* strengths (under pure axial load) determined using three-dimensional advanced analysis, it is possible to compute:

- the in-plane strength P_{nx} by assuming in-plane geometric imperfections only, and restraining the column fully in the out-of-plane direction;
- the out-of-plane strength P_{ny} by assuming out-of-plane geometric imperfections only, and restraining the column fully against in-plane deformations; and
- the three-dimensional strength P_n by assuming both in-plane and out-of-plane geometric imperfections, and allowing the column to deform freely in space.

If P_{nx} is similar in magnitude to P_{ny}, it may be expected that some interaction between in-plane and out-of-plane modes of failure will occur, resulting in the three dimensional strength P_n being lower again than the smaller of P_{nx} and P_{ny}. Although this phenomenon was detected in the plastic zone studies described here for which $P_{ny} < P_{nx}$, the reduction from the out-of-plane strength P_{ny} to the three-dimensional strength P_n was not significant. Practically therefore, when $P_{ny} < P_{nx}$, it is appropriate, at least for all the examples considered here, to assume that the three-dimensional strength P_n is equal to the out-of-plane strength P_{ny}. In Figs. 4.44 and 4.45, both the out-of-plane strength P_{ny} and the three-dimensional strength P_n have been plotted for comparison purposes.

Both the three-dimensional and in-plane[11] advanced analysis results for the cantilever beam-column comprising a beam-type section ($P_{ny} < P_{nx}$) and subjected to a tip lateral load H are plotted in Fig. 4.44 for slendernesses λ_{cx} of 1.0, 1.5 and 2.0. The curves shown in Fig. 4.44(a) are expressed in terms of first-order loads while those in Fig. 4.44(b) are plotted in terms of the corresponding maximum elastic second-order moment for the geometrically perfect member. The curves of Fig. 4.44(b) have therefore been obtained by amplifying those of Fig. 4.44(a) along the moment axis by the theoretical elastic amplification factor. For the cantilever beam-column comprising a beam-type section ($P_{ny} < P_{nx}$) and subjected to a tip moment C, the advanced analysis results are shown in Figs. 4.45(a) and (b). For a corresponding cantilever but with a column-type section for which the in-plane column strength is lower than the out-of-plane column strength ($P_{nx} < P_{ny}$), relevant advanced analysis results are given in Fig. 4.46.

---

[11]All in-plane advanced analysis results were computed by ignoring out-of-plane geometric imperfections and assuming full lateral restraint.

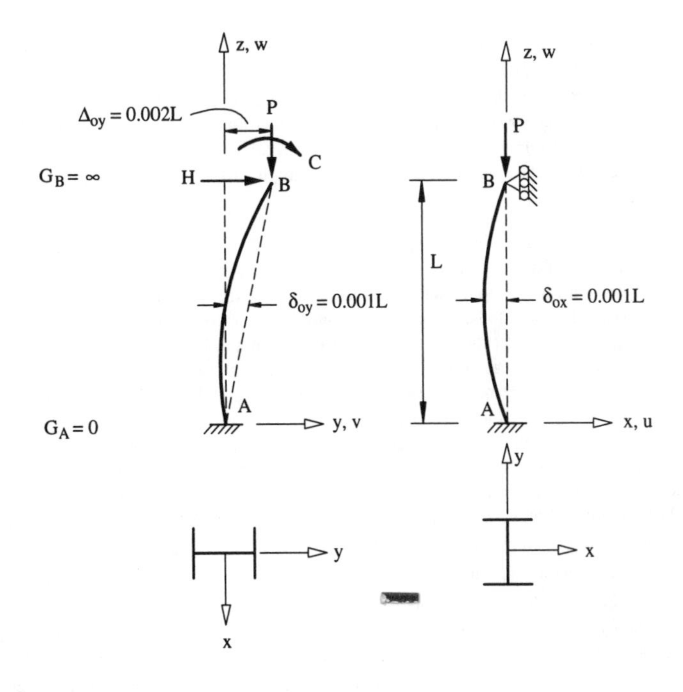

Boundary conditions

At A: $u = u' = v = v' = w = \phi = \phi' = 0$

At B: $u = \phi = 0$ (ϕ = twist angle)

Material properties and residual stresses

Elastic-perfectly plastic

$E = 29000$ ksi (200000 MPa), $F_y = 36$ ksi (250 MPa)

$\sigma_{rc} = 0.3\, F_y$ [Galambos and Ketter, 1959]

Fig. 4.43 Cantilever column problems studied with three-dimensional plastic zone advanced analysis

A key observation from the advanced analysis results shown in Figs. 4.44 to 4.46 is that a transition from "in-plane failure" to "out-of-plane failure" can occur along a given strength interaction curve (see the λ_{cx} = 1.0 curves in Figs. 4.44 and 4.45 for example). *If $P_{ny} < P_{nx}$ it is therefore not a necessary consequence that the three-dimensional strength curve lies below the in-plane strength curve over the full range of axial force and moment.* Of course, if the member is sufficiently slender, the three-dimensional strength will be consistently lower than the in-plane strength.

From the preceding three-dimensional plastic zone advanced analysis results, it is apparent that the mode of failure (in-plane or out-of-plane) depends on many factors including the column effective lengths about both axes, the bending moment distribution, the axial force distribution, the ratio of axial force to bending moment, and the cross-sectional shape. It would therefore appear appropriate and rational from a design viewpoint to have separate interaction equations corresponding to the various possible modes of failure, all of which must be satisfied for the beam-column to be deemed satisfactorily designed. Although it is beyond the scope of this report to investigate specific interaction equations beyond those currently in use by the AISC, it is interesting to note that CSA-S16.1-M94, AS4100-1990 and Eurocode 3 contain separate interaction equations for in-plane section strength, in-plane member strength, and out-of-plane member strength. The possible interactions between the different modes of failure are shown schematically in Fig. 4.47[12], with the (true) three-dimensional strength corresponding to the greatest lower bound of all curves. In this figure, M_{sx} denotes the in-plane strength in pure flexure, M_{bx} denotes the out-of-plane member strength for in-plane flexure, with M_{nx} being the lower of M_{sx} and M_{bx}. The correspondence between the failure modes depicted schematically in Figs. 4.47(a) to (e) and the three-dimensional plastic zone advanced analysis results given in Figs. 4.44(b), 4.45(b) and 4.46(b) can be easily seen. For example, the three-dimensional strength for the case of λ_{cx} = 1.0 shown in Fig. 4.44(b) corresponds to the schematic representation shown in Fig. 4.47(a). Similarly, the strength curve shown for λ_{cx} = 1.0 in Fig. 4.45(b) corresponds to that depicted schematically in Fig. 4.47(b).

The adoption of separate interaction formulae to check the in-plane and out-of-plane strengths of members in combined compression and flexure about the major axis is of no particular consequence for the notional load approach. It has been shown in this chapter that the in-plane strength can be assessed, either accurately or conservatively, using the notional load approach. Therefore the in-plane strength of a member which can potentially fail in an out-of-plane mode can be assessed essentially equally well using either using the effective length method (preferably using *inelastic* effective length factors), or the notional load method. It is emphasized here, however, that the calculation procedures associated with the notional load approach are vastly simpler than those required for the effective length approach since the calculation of in-plane (inelastic) effective length factors is avoided entirely.

---

[12]The curves shown in Fig. 4.47 are intended to represent "true strengths" as determined from advanced analysis (where this is possible), rather than the predictions of any particular interaction equations.

It would appear from the advanced analysis results shown in this section that it is rational to adopt a separate interaction equation for checking out-of-plane strength. It is beyond the scope of this chapter, however, to assess the merit of the various such equations prescribed in steel design specifications, or to propose a suitable alternative interaction equation for assessing out-of-plane strength.

4.14.3 Recommendations

Since this report addresses design only within the framework of the existing AISC LRFD interaction equations, a rigorous assessment of the three-dimensional strength of a member within the philosophy of the notional load approach generally requires independent checks for both in-plane and out-of-plane strength.

In-plane strength

$$\begin{aligned} &\frac{P_u}{\phi_c P_{n(L)x}} + \frac{8}{9}\frac{M_{ux}}{\phi_b M_{sx}} \leq 1.0 \qquad \text{for} \quad \frac{P_u}{\phi_c P_{n(L)x}} \geq 0.2 \\ &\frac{P_u}{2\phi_c P_{n(L)x}} + \frac{M_{ux}}{\phi_b M_{sx}} \leq 1.0 \qquad \text{for} \quad \frac{P_u}{\phi_c P_{n(L)x}} < 0.2 \end{aligned} \tag{4.42}$$

in which $P_{n(L)x}$ is the axial resistance term for buckling about the major principal x-axis, based on the actual member length ($K = 1$), and M_{sx} is the nominal section flexural strength about the major principal x-axis.

Out-of-plane strength

$$\begin{aligned} &\frac{P_u}{\phi_c P_{ny}} + \frac{8}{9}\frac{M_{ux}}{\phi_b M_{bx}} \leq 1.0 \qquad \text{for} \quad \frac{P_u}{\phi_c P_{ny}} \geq 0.2 \\ &\frac{P_u}{2\phi_c P_{ny}} + \frac{M_{ux}}{\phi_b M_{bx}} \leq 1.0 \qquad \text{for} \quad \frac{P_u}{\phi_c P_{ny}} < 0.2 \end{aligned} \tag{4.43}$$

in which P_{ny} is the axial resistance term for buckling about the minor principal y-axis, based on the minor axis effective length K_yL, and M_{bx} ($\leq M_{sx}$) is the nominal member out-of-plane flexural strength for in-plane bending about the major principal x-axis.

For assessing in-plane strength (Eq. (4.42)), notional loads must be included in the second-order elastic global analysis employed to determine the moments M_{ux}. However, if it is desired to maintain consistency with the existing AISC LRFD procedure for checking member strength when out-of-plane strength governs, the moments M_{ux} in Eq. (4.43) should be determined from second-order elastic global analysis *without* notional loads. Alternately, from the viewpoint of the effects of physical imperfections, it could be argued that it is actually more appropriate to include the notional loads in the assessment of out-of-plane strength. The latter approach also simplifies the notional load design procedure but at the cost of some minor conservatism compared to current AISC LRFD practice.

Prior to undertaking the beam-column interaction equation checks described above, the following characteristic axial and flexural strength quantities are required: $P_{n(L)x}$, P_{ny}, M_{sx} and M_{bx}. Based on the relative magnitudes of these quantities, it can be determined whether an in-plane strength check only is sufficient, or if both in-plane and out-of-plane checks are obligatory. The possible scenarios and required actions are:

- $P_{n(L)x} < P_{ny}$, $M_{bx} = M_{sx}$: An in-plane strength check alone is sufficient.
- $P_{n(L)x} < P_{ny}$, $M_{bx} < M_{sx}$: Both in-plane and out-of-plane strength checks should be performed.
- $P_{n(L)x} > P_{ny}$: Both in-plane and out-of-plane strength checks should be performed.

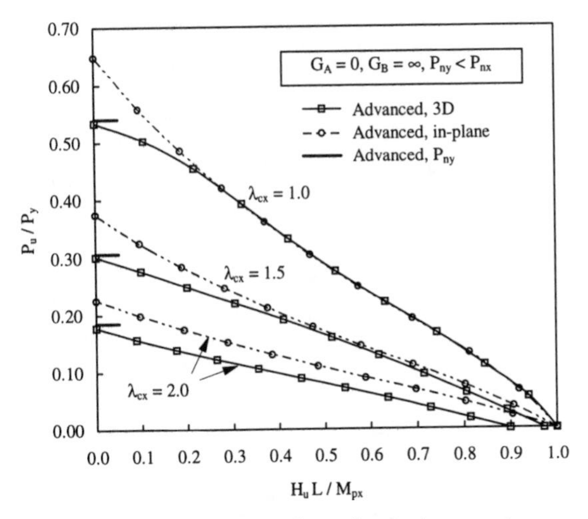

(a) Axial load versus first-order elastic moment

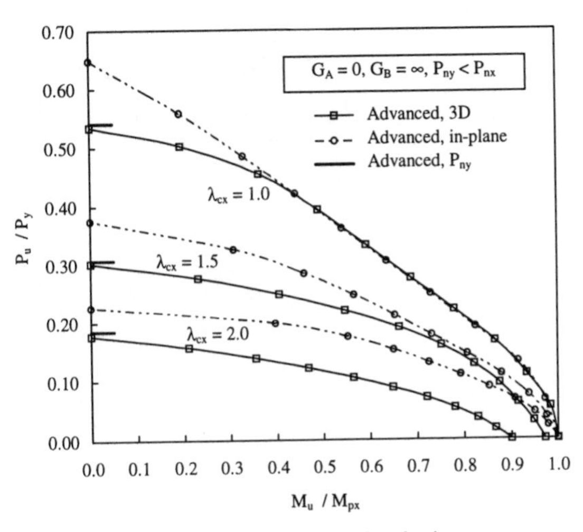

(b) Axial load versus second-order elastic moment

Fig. 4.44 Advanced analysis strength curves for cantilever beam-column with tip lateral load (beam-type section)

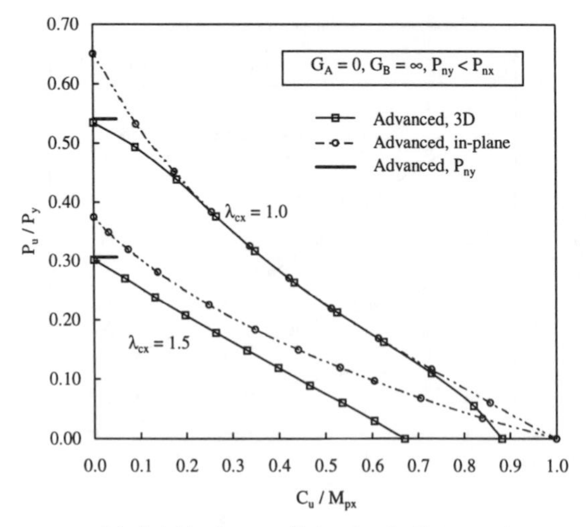

(a) Axial load versus first-order elastic moment

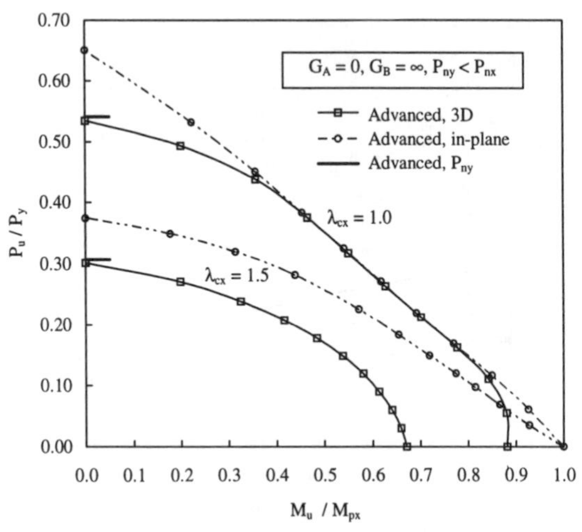

(b) Axial load versus second-order elastic moment

Fig. 4.45 Advanced analysis strength curves for cantilever beam-column with tip moment (beam-type section)

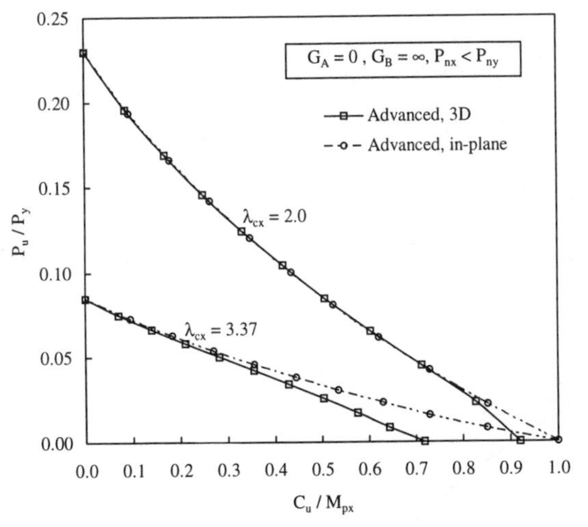

(a) Axial load versus first-order elastic moment

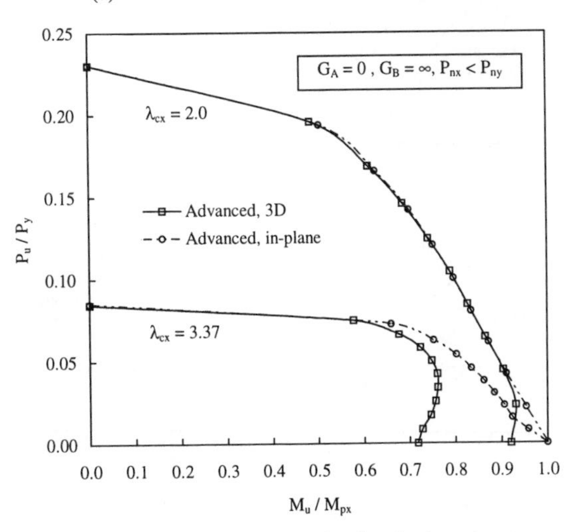

(b) Axial load versus second-order elastic moment

Fig. 4.46 Advanced analysis strength curves for cantilever beam-column with tip moment (column-type section)

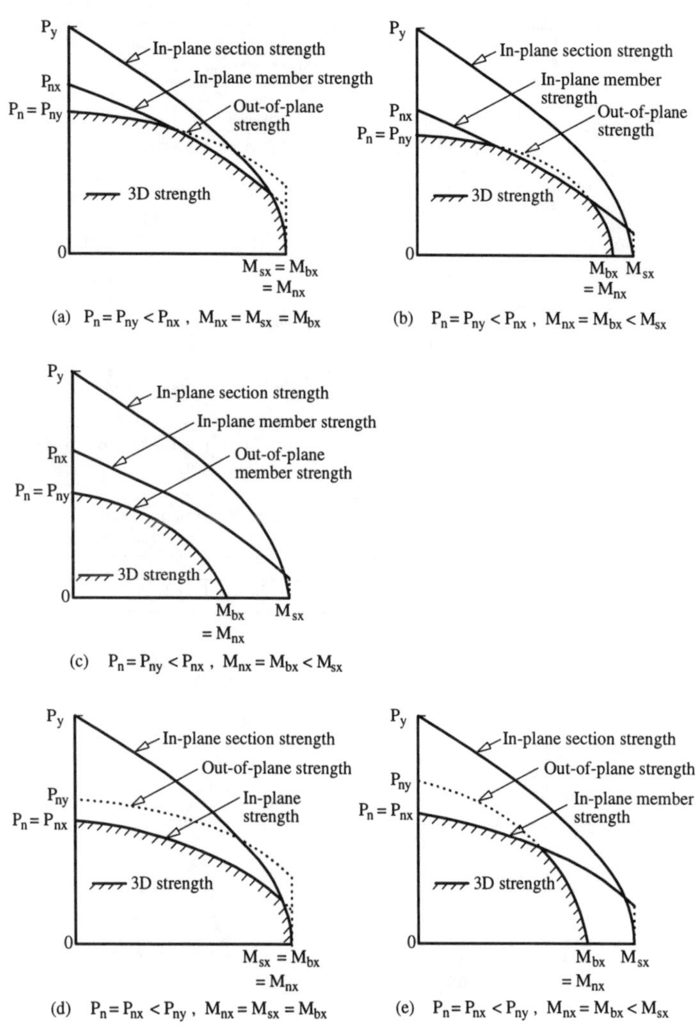

Fig. 4.47 Schematic representation of in-plane section strength, in-plane member strength, out-of-plane member strength and three-dimensional member strength of steel beam-columns subjected to in-plane flexure about the major (x) axis

4.15 Three-Dimensional Beam-Column Strength Under Biaxial Bending

The rigorous application of the notional load approach to frames which are unbraced in both orthogonal directions is outside the scope of this chapter. To facilitate its rational calibration and verification, any such exposition on the notional load approach would first need to investigate the advanced analysis of three-dimensional sway frames, an important aspect of which is the determination of a suitable philosophy for the inclusion of out-of-plumbness (sway) imperfections in both orthogonal directions. The authors are not aware of any readily available finite element software which can perform refined plastic zone analyses of three-dimensional frames including the effects of geometric imperfections and residual stresses[13].

Central to the adoption of a fair and reasonable approach to the inclusion of the effects of story and frame out-of-plumbness imperfections in three-dimensional frames is the correct interpretation of the erection tolerances given in steel design specifications. If the erection tolerance for a story is prescribed as $\psi_s = \Delta_0/h$, for example, then it would be quite reasonable to consider the possibility that the story is out-of-plumb by this amount in either (but not both) of the principal orthogonal directions of the frame. The statistical realities of the erection procedure dictate that it is unreasonable to conceive the possibility of the story being out-of-plumb by the maximum amount ψ_s *simultaneously* in both orthogonal directions.

The provisions of Eurocode 3 with regard to an appropriate philosophy for considering the effects of initial out-of-plumbness (sway) imperfections (or the equivalent notional horizontal loads) are basically consistent with the reasoning posed above since it is stated that the initial sway imperfections should "apply in all horizontal directions, but need only be considered in one direction at a time." (Eurocode 3, Section 5.2.4.3, paragraph 4 [CEN, 1992]). In relation to the preceding quotation from Eurocode 3, the interpretation adopted here is that the term "horizontal directions" refers to the two principal orthogonal directions of the three-dimensional frame.

Due to their close relationship, the discussion above in the context of out-of-plumbness imperfections is also directly applicable to notional loads. Thus, in terms of being a direct reflection of the physical realities of out-of-plumbness imperfections arising from the erection procedure, and with the aim of minimizing design conservatism, an "optimum" notional load procedure may entail the use of notional loads in all horizontal directions, but only in one principal orthogonal direction at a time. It is clear that strict application of this procedure will lead to a large number of load cases requiring consideration. For a gravity load case, for example (no real lateral load), notional loads will need to be considered independently in four directions if the frame is asymmetric in both principal orthogonal directions.

In view of the current lack of results for the advanced analysis of three-dimensional frames which can be used for comparison purposes, and because of the relative immaturity of the notional load approach relative to the effective length based philosophy, it is considered that full endorsement of the notional load guidelines outlined in the previous paragraph is inappropriate at

---

[13] The analysis employed to obtain the results reported in Section 4.14.2 is restricted in application to individual *members* subjected to axial load, biaxial bending and torsion.

the present time. Alternately, it is recommended that notional loads be applied *simultaneously (at full magnitude) in both principal orthogonal directions of the frame*. Although this may lead to some design conservatism compared to an approach in which notional loads are considered only in one direction at a time, this conservatism is not anticipated to be appreciable nor inconsistent with that inherent in an effective length based procedure. The simultaneous application of notional loads in both orthogonal directions will also simplify the mechanics of the notional load procedure in terms of reducing the number of load cases which need be considered. The practical application of the notional load approach according to the guidelines proposed above is illustrated for a three-dimensional unbraced frame in Section 6.3.

4.16 Summary and Conclusions

This chapter has described the calibration and verification of the notional load approach for the assessment of frame stability. As for the traditional effective length approach described in Chapters 2 and 3, the notional load approach is an engineering procedure intended to be applied in conjunction with second-order elastic analysis of the geometrically perfect structure.

The chapter began with an exposition on the influence of residual stresses and geometric imperfections on column strength. Based on "physical" considerations of imperfect members or frames, as would be modeled in plastic zone advanced analysis, the following distinct types of geometric imperfections were identified:

- member initial out-of-straightness, the magnitude of which should be consistent with fabrication tolerances prescribed in steel design specifications; and
- story initial out-of-plumbness and global frame non-verticality, the magnitudes and distributions of which should be consistent with erection tolerances prescribed in steel design specifications.

In the effective length approach to beam-column design, all of the above imperfection effects are included implicitly in the column strength curve using the effective length of the member. In particular, the effective length approach cannot embrace the notion of independent consideration of member out-of-straightness and story out-of-plumbness imperfections (the "physically imperfect member" concept) which, being a direct reflection of the fabrication and erection tolerances prescribed in steel design specifications, is an integral component of the rational plastic zone advanced analysis of frames.

On the other hand, the notional load approach attempts to include the various imperfection effects in a more direct manner which has parallels with the "physically imperfect member" concept employed in advanced analysis. The effects of residual stresses and member out-of-straightness imperfections alone can be regarded as being included in the column strength curve using the actual member length L ("$K = 1$"). The effects of story out-of-plumbness can be considered independently through notional lateral loads acting at each story level; conceptually, these notional loads account for the effect of story out-of-plumbness under gravity loads.

One of the main objectives of this chapter was to calibrate rationally the magnitude of the notional lateral load such that the deduced beam-column strengths are consistent with those determined using advanced analysis. This calibration was undertaken for columns subjected to axial load and notional lateral load only (no real lateral load). It should be noted that under such a loading regime, what would simply be an axially loaded column in the effective length approach to design becomes the limiting case of a beam-column according to the notional load philosophy. Three variants of the notional load approach were developed and validated in this chapter: a "simple" approach, a "modified" approach and a "refined" approach (the latter predominantly described in Appendix F).

When applied to a single sidesway uninhibited member with applied axial load P, the notional lateral load was assumed to be given by $N = \xi P$, where ξ is termed the notional load parameter. When applied to a frame with one or more stories, the notional load deemed to act laterally at each story level was assumed to be $N = \xi \Sigma P$, in which ΣP represents the sum of all gravity loads acting on the corresponding story. These notional lateral loads act in conjunction with real lateral loads and so the design moments M_u employed in the beam-column interaction equation strength checks include contributions from both real and notional lateral loads.

The simple notional load calibration indicated that the universal adoption of $\xi = 0.005$ infers nominal member and frame strengths which are either accurate or conservative compared to the corresponding advanced analysis strengths for the vast majority of practical cases. The modified notional load approach improves on the simple calibration of $\xi = 0.005$ by incorporating a modifying factor $k_y = 22\sqrt{F_y/E}$ accounting for the effect of steel yield stress. As strictly defined, the modified notional load parameter is expressed as $\xi_M = \xi_0 k_y$ (with $\xi_0 = 0.005$), although as a convenience it may be assumed with sufficient accuracy that $\xi_0 = 0.004$, 0.0045 and 0.005 for the "standard" yield stress values of 36 ksi (250 MPa), 50 ksi (350 MPa) and 65 ksi (450 MPa), respectively. The results of the modified notional load approach are particularly accurate for beam-columns in single curvature bending, but are more conservative for members bent in double-curvature with significant elastic end-restraints. The strength of a *single story* in a sidesway uninhibited frame with or without leaning columns and with varying end-restraints, load patterns and member slendernesses can also be accurately or conservatively reproduced through the use of a notional lateral load of magnitude $N = \xi \Sigma P$.

The refined notional load calibration ($\xi = \xi_R$, see Appendix F) was undertaken in order to improve the general accuracy of the simple and modified notional load procedures, principally by reducing the inherent conservatism associated with the latter procedures for columns of low to intermediate slenderness bent in double-curvature. Columns with these characteristics are common in multistory frames. The refined expression for the notional load parameter $\xi = \xi_R$ considers the influence of yield stress, story slenderness, story lateral stiffness, and the number of columns in the story.

For the example five-story frames studied, the refined notional load parameters $\xi = \xi_R$ are considerably lower than the simple value of $\xi = 0.005$. The unconservatism associated with the refined notional load approach was approximately half that pertaining to the simple approach with $\xi = 0.005$. The five-story frame examples also demonstrated that the use of $K = 1$ but *without notional loads* may also be appropriate in some circumstances (even for pure gravity loading), particularly when the end-restraints provided by the attached beams and supports are fairly stiff and the story is not atypically slender. However, further work in this area is required.

In the advanced analysis of multistory frames, the story out-of-plumbness imperfections under gravity loads result in a small amount of superimposed tension in the "windward" columns and superimposed compression in the "leeward" columns. The advanced analysis also predicts an increase in the beam moments due to real (physical) imperfections and inelasticity both of which are modeled within the analysis. Although similar phenomena will arise from a conventional notional load analysis procedure whereby notional loads are applied at all floor levels, the effect is unrealistically exaggerated to the extent that unwarranted conservatism may result in the design. It is thus considered appropriate, if desired, to neglect the axial force effects resulting from notional loads; this has been achieved through the introduction of a so-called "dual R-N analysis procedure". This modified analysis procedure entails separate second-order analyses for real loads and notional lateral loads so that the moments associated with the former are computed independently of those associated with the latter. In such a dual analysis procedure, the notional loads fulfill their original primary philosophical intention, which is to account for the effects of story out-of-plumbness imperfections by adjusting the required moment (M_u) used in the beam-column interaction equation to counter the increase in the nominal axial resistance term from P_n (based on the effective member length) to $P_{n(L)}$ (based on the actual member length), while leaving the required axial force P_u unaltered.

A feature of the notional load approach demonstrated using the single-story gravity-loaded frame examples is the automatic detection of the transition from a "sway" to a "non-sway" mode of failure (as determined through the interaction equation applied to the critical member). There is no need to consider whether the frame has sway or non-sway buckling modes as is the case for the effective length approach. In fact, if *elastic* effective length factors are used instead of the *inelastic* values, the wrong mode can be inferred for the critical member.

A principal advantage of the notional load approach is its simplicity. As the actual length ($K = 1$) is used in the notional load approach to determine the nominal strength $P_{n(L)}$ for use in the interaction equation, there is no need to calculate values of K and hence G-factors. If the simple or modified notional load approach is used, the need to conduct a special lateral load analysis for the determination of modified G-factors which account for the distribution of moment in the restraining members is also obviated. The refined notional load approach, however, does entail a first-order sidesway analysis to assess the story lateral stiffness, itself a contributing factor to the refined notional load parameter $\xi = \xi_R$. Furthermore, as the notional load approach has been calibrated against the results of plastic zone advanced analysis, which includes the effects of material nonlinearity and imperfections, there is no need to consider inelastic stiffnesses in the first-order sidesway analysis as is required in the calculation of the inelastic G-factors employed in the effective length approach.

4.17 References

AISC [1978], *Manual of Steel Construction - Allowable Stress Design and Plastic Design*, American Institute of Steel Construction, Inc., Chicago, Illinois.

AISC [1993a], *Code of Standard Practice for Steel Buildings and Bridges*, Manual of Steel Construction - Load and Resistance Factor Design, Second Edition, American Institute of Steel Construction, Inc., Chicago, Illinois.

AISC [1993b], *Load and Resistance Factor Design Specification for Structural Steel Buildings*, Second Edition, American Institute of Steel Construction, Inc., Chicago, Illinois.

BHP [1991], *Hot Rolled and Structural Products*, 1991 edition, BHP Steel, Australia.

Bild, S. and Trahair, N. S. [1989], "In-plane Strengths of Steel Columns and Beam-Columns, *Journal of Constructional Steel Research*, 13, 1-22.

Bradford, M. A. and Trahair, N. S. [1985], "Inelastic Buckling of Beam-Columns with Unequal End Moments", *Journal of Constructional Steel Research*, 5(4), 195-212.

BSI [1990], *BS5950; Part 1:1990 Structural Use of Steelwork in Building - Part 1. Code of practice in simple and continuous construction: hot rolled sections,* British Standards Institution, U.K.

CEN [1992], *ENV 1993-1-1 Eurocode 3, Design of Steel Structures, Part 1.1 - General Rules and Rules for Buildings*, European Committee for Standardization, Brussels.

Clarke, M. J. [1994], "Plastic-Zone Analysis of Frames", in *Advanced Analysis of Steel Frames: Theory, Software and Applications*, Chen, W.F. and Toma, S., Eds., CRC Press, Boca Raton, Florida, 259-319.

Clarke, M. J. and Bridge, R. Q. [1992], "The Inclusion of Imperfections in the Design of Beam-Columns", *Proceedings, 1992 Annual Technical Session*, Structural Stability Research Council, U.S.A., Pittsburgh, Pennsylvania, 327-346.

Clarke, M. J. and Bridge, R. Q. [1995], "Application of the Notional Load Approach to the Design of Multistory Steel Frames", *Proceedings, 1995 Annual Technical Session*, Structural Stability Research Council, U.S.A., Kansas City, Missouri, 191-211.

Clarke, M. J., Bridge, R. Q., Hancock, G. J. and Trahair, N. S. [1991], "Design Using Advanced Analysis", *Proceedings, Structural Stability Research Council Annual Technical Session*, Chicago, Illinois, 27-40.

Clarke, M. J., Bridge, R. Q., Hancock, G. J. and Trahair, N. S. [1992], "Advanced Analysis of Steel Building Frames", *Journal of Constructional Steel Research*, 23, 1-29.

Clarke, M. J. and Hancock, G. J. [1991], "Finite-Element Nonlinear Analysis of Stressed-Arch Frames", *Journal of Structural Engineering, ASCE*, 117(10), 2819-2837.

CSA [1994], *Limit States Design of Steel Structures, CAN/CSA-S16.1-M94*, Canadian Standards Association, Rexdale, Ontario.

Davids, A. J. and Hancock, G. J. [1987], "Nonlinear Elastic Response of Locally Buckled Thin-Walled Beam-Columns", *Thin-Walled Structures*, 5, 211-226.

De Luca, A. and De Stefano, M. [1994], "A Proposal for Introduction of Equivalent Frame Imperfections into Eurocode 3 Provisions", *Proceedings, Structural Stability Research Council Annual Task Group Technical Session*, Lehigh University, 457-466.

Galambos, T. V. and Ketter, R. L. [1959], "Columns Under Combined Bending and Thrust", *Journal of the Engineering Mechanics Division, ASCE*, 85(EM2), 1-30.

Kanchanalai, T. [1977], "The Design and Behavior of Beam-Columns in Unbraced Steel Frames", AISI Project No. 189, Report No. 2, Civil Engineering/Structures Research Laboratory, University of Texas, Austin, Texas.

Ketter, R. L. [1961], "Further Studies on the Strength of Beam-Columns", *Journal of the Structural Division, ASCE*, 87(ST6), 135-152.

Ketter, R. L., Kaminsky, E. L. and Beedle, L. S. [1955], "Plastic Deformation of Wide-Flange Beam-Columns", *Transactions of the ASCE*, 120, 1028.

Pi, Y. L. and Trahair, N. S., [1994a], "Nonlinear Inelastic Analysis of Steel Beam-Columns. I: Theory", *Journal of Structural Engineering*, American Society of Civil Engineers, 120(7), 2041-2061.

Pi, Y. L. and Trahair, N. S., [1994b], "Nonlinear Inelastic Analysis of Steel Beam-Columns. II: Applications", *Journal of Structural Engineering*, American Society of Civil Engineers, 120(7), 2062-2085.

SA [1990], *AS4100-1990 Steel Structures*, Standards Australia, Sydney.

SSRC [1976], *Guide to Stability Design Criteria for Metal Structures*, 3rd Edition, editor B.G. Johnston, Structural Stability Research Council, Bethlehem, U.S.A..

SSRC [1988], *Guide to Stability Design Criteria for Metal Structures*, 4th Edition, editor T.V. Galambos, Structural Stability Research Council, Bethlehem, U.S.A.

SSRC [1991], *Stability of Metal Structures - A World View*, 2nd Edition, editor L.S. Beedle, Structural Stability Research Council, Bethlehem, U.S.A.

SSRC [1993], *Plastic Hinge Based Methods for Advanced Analysis and Design of Steel Frames: An Assessment of the State-of-the Art*, White, D.W. and Chen, W.F., Eds., Structural Stability Research Council, Lehigh University, Bethlehem, Pennsylvania.

Trahair, N. S. [1983], "Inelastic Lateral Buckling of Beams", Chapter 2 of *Developments in the Stability and Strength of Structures*, Vol. 2, Beams and Beam-Columns, Applied Science Publishers, London.

Trahair, N. S. and Bradford, M. A. [1991], *The Behavior and Design of Steel Structures,* Chapman and Hall, Second Edition, London and New York.

Vogel, U. [1985], "Calibrating frames", Stahlbau, 54(10), 295-301.

CHAPTER 5

FRAMES THAT CAN BE DESIGNED WITHOUT CONSIDERATION OF EFFECTIVE LENGTH OR NOTIONAL LOAD

5.1 Introduction

Chapter 5 discusses conditions for which frames can be designed with negligible error using $K = 1$ in the context of the current AISC LRFD Specification [1993]. It is assumed that the design is based on second-order elastic forces, calculated by use of approximate amplifiers or by direct analysis, but without consideration of equivalent geometric imperfections or notional horizontal loads. In Section 5.2, limits that have been suggested in various design standards and specifications for definition of frames as "non-stability critical" are reviewed within the context of LRFD practice [AISC 1994]. The error in the AISC LRFD *beam-column interaction equations*, caused by use of $K = 1$ without the application of notional loads, is investigated in Section 5.3[1]. This error is derived directly from the LRFD equations, using one of the elastic story-based effective length procedures presented in Chapter 2 as the buckling model. By inspection of the error relationships, general limits on the story sidesway amplification factor, B_2, are deduced for which neither effective length nor notional load need to be considered in the design. Since these relations are based on elastic effective lengths, they are conservative for many design cases.

In certain situations substantial economy can be gained by the use of inelastic effective lengths. Therefore, simple relationships based on the inelastic buckling behavior are desirable. Equations for approximate inelastic effective length factors recently proposed by LeMessurier [1995] are outlined and explained in Section 5.4. These equations are derived with a goal of maintaining the error in the *axial strength ratio*, $P_u / \phi_c P_n$, to less than three percent for frames that meet a reasonably liberal criterion on their sidesway stiffness. The development and use of these equations provides several insights into when the design can be based on $K = 1$.

In Section 5.5, examples are provided to illustrate the types of frames associated with different ranges of error, as defined by the relationships outlined in Sections 5.3 and 5.4. Also, these examples are used to show the "actual" errors compared to strength predictions based on "exact" inelastic buckling analysis. Specific design recommendations are proposed in Section 5.6. The discussions and recommendations

---

[1] As noted in Chapter 1, it is assumed throughout the report that the error in calculating the second-order elastic design forces is negligible. If the design is based on $K = 1$ and first-order analysis, the unconservative error can be substantially larger than that discussed in this chapter.

of this chapter are valid for assessing when "braced" frames are sufficiently braced such that their beam-column members can be designed using $K = 1$, and for assessing when "unbraced" frames are "non-stability critical" such that they can be designed using $K = 1$ without applying notional horizontal loads. The chapter applies to any FR and PR frame construction for which story-based stability analysis and design is appropriate[2]. The base recommendations restrict the maximum unconservative error in the evaluation of the beam-column interaction equations to six percent by limiting B_2 to 1.11 (for many design situations, the error is substantially less). If desired, a conservative design can be ensured by limiting the beam-column interaction values based on $K = 1$ to $1/1.06 = 0.94$. It is explained that the use of $K = 1$ for larger B_2 values can be permitted by further limiting the interaction equation values based on upper-bound errors.

As discussed previously in Chapters 2 and 4, the use of $K = 1$ without applying notional loads can in general lead to significant unconservative errors, and thus it is generally not acceptable. However, it is believed that the proposed limits on B_2 are satisfied for a large number of cases in practice. The application of these limits to justify the use of $K = 1$ can greatly simplify frame design procedures.

5.2 Prior Recommendations and Provisions

Research by Lu, et al. [1975], and Liapunov [1973 & 74] demonstrated that within the context of AISC Allowable Stress Design [AISC 1989], certain classes of rigidly-connected unbraced frames can be designed without considering P-Δ or P-δ effects in the calculation of system forces, and with the column axial strength computed based on $K = 1$. Limits that define when these conditions exist were developed based on parametric studies of *17* frames. These studies considered a representative sample of practical frame geometries and loadings. Several elastic designs of these frames were performed, with one set of designs based on first-order moments and $K = 1$. All the designs were subjected to second-order elastic-plastic and/or plastic zone analyses to ascertain whether their ultimate capacities were sufficient. Some of the key results from these studies are summarized in [SSRC 1976]. The limits proposed based on this research are [SSRC 1976]:

- The maximum column axial-load ratios f_a / F_a and $f_a / 0.6F_y$ are not to exceed *0.75*.
- The maximum in-plane column slenderness ratio L/r is not to exceed 35.
- The bare-frame first-order drift index, $\frac{\Delta_{oh}}{L}$, is to be controlled such that

---

[2] Cases for which story-based equations may not be appropriate are discussed in Chapter 2, Sections 2.3.3 and 2.4.4.

$$\frac{\Delta_{oh}}{L} \le \frac{1}{7}\frac{\sum\limits_{non-leaner} H}{\sum\limits_{all} P_s} \tag{5.1}$$

where L is the story height, $\sum\limits_{non-leaner} H$ is the total story shear due to service lateral loads, Δ_{oh} is the first-order drift of the story due to $\sum\limits_{non-leaner} H$, and $\sum\limits_{all} P_s$ is the total *service level* gravity load on the story. As a result, the sidesway amplification at service load levels,

$$B_{2s} = \frac{1}{1 - \dfrac{\sum\limits_{all} P_s}{\sum\limits_{non-leaner} P_L}} = \frac{1}{1 - \dfrac{\sum\limits_{all} P_s}{\sum\limits_{non-leaner} \dfrac{HL}{\Delta_{oh}}}} \tag{5.2}$$

is limited to a maximum value of 1.17.

The above three requirements are simply the maximum values encountered in the cited design studies, and it is suggested in [SSRC 1976] that these recommendations are only tentative.

The above limits may be placed in the context of AISC LRFD as follows. As observed in Section 2.2, if the traditional allowable stress column equations [AISC 1989] are multiplied by $1.67\,\phi_c A_g$, the LRFD column strength is closely approximated [Salmon and Johnson 1996]. In fact, the LRFD column equations were developed as an approximate fit to the ASD equations at a live to dead load ratio of 1.1 and $\lambda_c = 1$ [Tide 1985]. Therefore, the first restriction in the above list may be expressed as[3]

$$1.67\phi_c A_g f_a \le 0.75\,[1.67\phi_c A_g F_{a(L)}] \tag{5.3a}$$

where f_a is the applied axial stress at service load conditions, and $F_{a(L)}$ is the allowable stress in ASD based on $K = 1$ (or $KL = L$). Substituting $\phi_c = 0.85$ and $A_g f_a = P_s$ on the left-hand side of this equation, and $1.67A_g F_{a(L)} = P_{n(L)}$ on the right-hand side, Eq. (5.3a) can be written as

---

[3] In ASD, the $f_a / F_a \le 0.75$ rule would be applied if the member stability interaction equation controls the design, whereas the $f_a / 0.6F_y \le 0.75$ rule would be applied if the beam-column interaction equation based solely on cross-section strength controls. The $f_a / F_a \le 0.75$ rule is more restrictive than $f_a / 0.6F_y \le 0.75$ within the context of LRFD, where there is no separate distinction between member stability and cross-section strength.

$$\frac{1.42}{(P_u / P_s)} P_u \leq 0.75 \, [\phi_c P_{n(L)}] \tag{5.3b}$$

in the context of AISC LRFD, where P_u is the factored axial load for strength design, P_s is the applied axial force at the corresponding service load level, and $P_{n(L)}$ is the nominal axial strength obtained from the LRFD column equations based on $K = 1$. If the maximum L/r limit of 35 (the second restriction in the above list) is considered along with the load factors for strength design in LRFD, and if A36 steel is assumed, the above equation translates to a limit on P_u ranging from $0.55P_y$ for $P_u / P_s = 1.3$ to $0.63P_y$ for $P_u / P_s = 1.5$ (1.3 to 1.5 representing the range of P_u / P_s for a large percentage of actual design cases in LRFD [Tide 1985]).

Equation (5.1) is applicable at nominal (i.e., unfactored) load levels in LRFD. However, this equation translates to a maximum allowable sidesway amplification at factored load levels[4],

$$B_2 = \frac{1}{1 - \dfrac{\sum_{all} P_u}{\sum_{non-leaner} P_L}} = \frac{1}{1 - \dfrac{\sum_{all} P_u}{\sum_{non-leaner} \dfrac{HL}{\Delta_{oh}}}} \tag{5.4}$$

that generally would be greater than 1.17, depending on the ratio $\sum_{all} P_u / \sum_{all} P_s$.

Although the above recommendations are cited in the AISC ASD Commentary [AISC 1989], the equivalent recommendations are not included in the LRFD Manual [AISC 1994]. Several other current limit states standards provide more restrictive provisions for when stability effects can be neglected. Since these provisions are in a limit states format, they can be discussed more succinctly. Eurocode 3 [CEN 1993] provides one of the most extensive discussions on this issue. It provides a limit on the ratio of the total factored gravity load to the elastic system or story buckling load at which the designer is allowed to neglect frame stability considerations in the calculation of the axial strength. Also, the engineer is allowed to base the design on first-order forces if this limit is not exceeded. This limit corresponds to the load level

---

[4] Although the developments in this chapter are tied directly to the story sidesway amplification factor as expressed by Eq. (5.4), it is recommended that the story amplification may be estimated by any means deemed acceptable by the engineer. The most common approximation for B_2 other than Eq. (5.4) is Eq. (C1-5) of the AISC LRFD Specification [AISC, 1993], which requires the calculation of nomograph effective length factors for the columns of the lateral-resisting system. Therefore, use of Eq. (C1-5) is not practical within the context of this chapter, which seeks to avoid the effective length calculations.

at which the total gravity load is *1/10* of the associated elastic buckling value, or at which the second-order amplification is roughly *11* percent (i.e., $B_2 = 1.11$). An equation that is identical to a maximum limit on B_2 of *1.11* is suggested as a practical way of checking this limit in building frame design. Frames meeting this limit are referred to as "non-sway". A similar limit is provided in the Australian AS4100 limit states design standard for assessment of when second order effects may be neglected in plastic design of building frames [SAA 1990].

The *1991 NEHRP Recommended Provisions for the Development of Seismic Regulations for New Buildings* [FEMA 1992] provide an equation for when P-Δ effects on the member forces and story drifts may be neglected in seismic design. One note of particular interest regarding the NEHRP provisions is they suggest that when calculating the vertical load for purposes of determining P-Δ forces for seismic design, the load factors need not exceed *1.0*. The NEHRP equations translate to a requirement that B_{2s} must be less than or equal to *1.11* for the P-Δ effects on the member forces and story drifts to be neglected.

5.3 Error in LRFD Beam-Column Interaction Equations by Neglecting Frame Stability

To the knowledge of the committee, with the exception of the work by Lu et al. and Liapunov discussed in [SSRC 1976], there are no published studies available that attempt to quantify the actual error associated with neglecting frame stability in the design of steel frames. Furthermore, the studies reviewed in [SSRC 1976] are not comprehensive. The frames considered involved an extensive range of practical frame designs, but none of the frames supported leaning columns, and none were representative of the most severe stability-critical cases such as the Kanchanalai frames [Kanchanalai 1977] utilized in the development of the AISC LRFD beam-column equations (see Chapter 1, Section 1.2). All the frames were highly redundant with substantial inelastic redistribution of forces prior to reaching their limits of maximum resistance. None of the studies conducted by Lu, et al. and by Liapunov considered problems in which some of the columns are subjected to weak-axis bending within the plane of the frame. This case is important to include in the assessment of when stability effects can be neglected in general. In this type of frame, the columns subjected to weak-axis bending tend to "lean" on the columns that are in strong-axis bending, even if the weak-axis columns are framed using FR type connections.

In the subsections below, the basic concepts for when frame stability effects can be neglected in calculating column axial strengths are discussed, and associated general equations are developed that quantify the actual error associated with neglecting frame stability in design by LRFD (i.e., the error caused by using $K = 1$ in determining the axial strength of a beam column member, without applying any notional horizontal loads to the frame). Section 5.3.5 summarizes the trends in the error and the behavior associated with these trends.

5.3.1 Basic Concepts

The key concepts behind the assessment of when $K = 1$ is acceptable for design are quite basic. Use of $K = 1$ tends to be acceptable when:

- KL/r is small, since the column strength varies little with large variations (or large error) in the effective length factor for small KL/r.

- the columns are heavily restrained at each end, and subjected to nearly full-reversed curvature bending under sidesway of the frame, provided the gravity loads are small in any framing that leans on the lateral-resisting system. Of course, if the ends of a column subjected to sidesway are prevented from rotation, and if the leaning column loads are zero, the exact solution for the effective length factor is $K = 1$.

- sufficient sidesway stiffness and strength is provided by some means in addition to the sidesway bending behavior of the frame members.

- the beam-column interaction check for the column members of the lateral resisting system is dominated by the moment term, with the contribution to the interaction equation from the axial strength ratio $P_u / \phi_c P_n$ being relatively small.

It is important to recognize that the only "real" error in the context of a design evaluation is the error in the value obtained from the beam-column interaction equations. For any combination of the above beneficial situations, the error in the calculated effective length factor may be tremendous, whereas the error in the evaluation of the beam-column strength may be small.

Although consideration of inelastic effective length, or use of an inelastic buckling analysis, leads to substantial economy in many practical situations [LeMessurier 1995], the influence of column inelasticity on the effective length becomes small in the limit that the columns are restrained by very heavy beam members, or in the unusual situation that the story buckling is elastic. As discussed in Chapter 1 and illustrated by the studies in Chapter 4, design checks based on inelastic buckling analysis or inelastic effective lengths are generally the most representative of the "true" physical strength. If elastic effective lengths or buckling models are employed, then usually a conservative value for P_n is obtained for cases where: (1) "small" to "intermediate" levels of rotational restraint are provided by the beams, and (2) significant distributed yielding due to axial force and compressive residual stresses is encountered prior to the columns reaching their maximum strength as concentrically loaded members. "Intermediate" levels of end rotational restraint and KL/r values such that the column axial strength involves significant yielding are considered to be the cases encountered most often in building design practice.

For reasons of generality and simplicity, the error relationships developed in this section are based on comparison of the beam-column interaction equations with $K = 1$ to the equivalent check in which a specific *elastic* story-buckling analysis or story-based effective length factor is employed. It is shown that these error relations give an upper-bound to the actual errors often encountered in practice. The "actual" errors, based on comparison to strengths obtained using "exact" inelastic buckling procedures, are illustrated for a representative suite of frame designs in Section 5.5. The largest errors expressed by the relations developed here are achieved only for relatively-flexible frames, in which the column being considered is very stocky (low L/r), with one end heavily restrained (low G values from Eqs. (2.10a) and (2.10b)) and one end effectively pinned, and subjected to high leaning-column effects and relatively low directly applied design axial loads. In these situations, only a small amount of distributed yielding occurs in the member at the buckling of the story, any distributed yielding that does occur has a negligible influence on the effective length because of the heavy end restraint, and the story depends almost entirely on this "critical" member for its lateral stability. Obviously, few frames are designed with such an arrangement.

5.3.2 Buckling Model Utilized for Error Calculations

The buckling model used in the subsequent sections is similar to the model associated with the K_{R_L} approach, outlined in Chapter 2. In the K_{R_L} approach, the resulting equations for the *elastic* effective length factor and the corresponding column buckling load are

$$K_{R_L} = \sqrt{\frac{1}{P_u}\frac{\pi^2 EI}{L^2}\frac{\sum_{all} P_u}{\sum_{non-leaner} P_L}\frac{1}{(0.85 + 0.15R_L)}} \tag{5.5a}$$

and

$$P_{e(R_L)} = \frac{\sum_{non-leaner} P_L}{\sum_{all} P_u}(0.85 + 0.15R_L)\, P_u \tag{5.5b}$$

where

$$\sum_{non-leaner} P_L = \frac{\sum_{non-leaner} HL}{\Delta_{oh}} = \sum_{non-leaner}\frac{\beta EI}{L^2} \tag{5.6}$$

and

$$R_L = \frac{\sum\limits_{leaner} P_u}{\sum\limits_{all} P_u} \tag{5.7}$$

The reader is referred to Chapter 2 and to Appendices B and C for discussions of these parameters and equations.

In [LeMessurier 1977], a more precise form of the above equations is presented that can be expressed as

$$K_{C_L} = \sqrt{\frac{1}{P_u}\frac{\pi^2 EI}{L^2}\frac{\sum\limits_{all} P_u + \sum\limits_{non-leaner} C_L P_u}{\sum\limits_{non-leaner} P_L}} = \sqrt{\frac{1}{P_u}\frac{\pi^2 EI}{L^2}\frac{\sum\limits_{all} P_u}{\sum\limits_{non-leaner} P_L}[1+(C_L)_{avg}]} \tag{5.8a}$$

and

$$P_{e(C_L)} = \frac{\sum\limits_{non-leaner} P_L}{\sum\limits_{all} P_u + \sum\limits_{non-leaner} C_L P_u} P_u = \frac{\sum\limits_{non-leaner} P_L}{\sum\limits_{all} P_u}\frac{1}{[1+(C_L)_{avg}]} P_u \tag{5.8b}$$

where C_L is the "clarification factor," which accounts for the influence of P-δ effects on the sidesway stability (see Section 2.4 and Appendix C), and

$$(C_L)_{avg} = \frac{\sum\limits_{non-leaner} C_L P_u}{\sum\limits_{all} P_u} \tag{5.9}$$

is the weighted average value of C_L over all the columns within the story (the C_L values for the leaning columns being equal to zero).

Equations (5.5) are equivalent to Eqs. (5.8), in that $(0.85 + 0.15 R_L)$ is simply a specific approximation for $\frac{1}{[1+(C_L)_{avg}]}$. In this chapter, the error relationships are developed in terms of the latter of these sets of equations. This permits discussion of the errors in general terms, considering the influence of the P-δ effects by selection of a value for $(C_L)_{avg}$. Equation (5.8a) may be written in an alternate form that highlights the parametric effects discussed in this chapter as:

$$K_{C_L} = \sqrt{\frac{1}{\frac{P_u}{P_y}\left(\frac{L}{r}\right)^2} \frac{\pi^2 E}{F_y} \frac{\sum_{all} P_u}{\sum_{non-leaner} P_L} [1 + (C_L)_{avg}]} \tag{5.10}$$

It is important to note that due to simplifying assumptions involved in its formulation, the K_{R_L} approach has important limits on its applicability. These limits have been elucidated in Sections 2.4.4.1 and Appendix C, and they may be expressed in terms of the parameters P_u and P_L as

$$\frac{P_u}{P_L} \leq 1.7 \frac{\sum_{all} P_u}{\sum_{non-leaner} P_L} \frac{1}{(0.85 + 0.15 R_L)} \tag{5.11a}$$

or in terms of the design vertical and shear forces as

$$\frac{P_u}{\sum_{all} P_u} \leq 1.7 \frac{H}{\sum_{non-leaner} H} \frac{1}{(0.85 + 0.15 R_L)} \tag{5.11b}$$

Equations (5.11), or the equivalent equations in Chapter 2, are required to avoid cases in which the elastic story-buckling capacity predicted by the K_{R_L} method is significantly in error due to:

(1) negative end restraint (i.e., negative G values from Eqs. (2.10a), (2.10b) and (2.11), or column end rotations in a first-order sidesway analysis that are in the opposite direction to the column chord rotations associated with the sidesway) coupled with high axial force, and/or

(2) inordinately large $L\sqrt{P/EI}$ values for a certain column or for several columns within the story (for $L\sqrt{P/EI} > \pi$, or $\frac{P}{\pi^2 EI/L^2} > 1$, the C_L effect can be larger than 0.216).

Equation (5.11b) indicates that the error in the estimate of the elastic story-buckling capacity, $\sum_{non-leaner} P_L \, (0.85 + 0.15 R_L)$, can become significant and unconservative when the fraction of the story vertical load supported by a column is substantially larger than the fraction of the story shear resisted by that member in a first-order lateral load analysis. This can happen, for example, when some of the columns are turned in weak-axis bending whereas others are oriented in strong-axis bending under the sidesway deflections of the frame.

Generally speaking, the buckling model adopted in this chapter does not require the above limitations. The large reduction in the story buckling capacity due to a weak heavily loaded column member or group of members is caused entirely by large P-δ effects. This extreme reduction may be represented by large C_L values in the weak heavily loaded member(s), or by a relatively large $(C_L)_{avg}$ value for the story. For severe cases such as those demonstrated in Section 2.4.4.1 and by LeMessurier [1993], individual column C_L values can be greater than 0.216 (the maximum value for C_L suggested in LeMessurier [1977]). The K_{R_L} method assumes a specific approximation for $(C_L)_{avg}$ that has a maximum value of $\frac{1}{0.85} - 1 = 0.176$. The limit expressed by Eqs. (5.11) disallows cases in which the unconservative error associated with the $(0.85 + 0.15R_L)$ approximation is non-negligible.

Unfortunately, no practical method exists at the present time for calculating $C_L > 0.216$. Furthermore, it is better to prevent the application of recommended procedures to highly irregular frames, rather than to make the procedures ultra-conservative for more ordinary design situations. Therefore, to develop recommendations for when frame stability effects may be neglected, a maximum $(C_L)_{avg}$ of 0.176 is assumed, and Eqs. (5.11) are required to avoid situations where the story buckling capacity is severely reduced due to large P-δ effects. Fortunately, potential violation of these equations can be determined by inspection in most practical cases.

It is important to recognize that to achieve $C_L \geq 0.176$ in an isolated column member that does not lean on any of the other story framing, the G value at one of the column ends must be less than approximately 0.2, and the G value at the opposite end of the column must be either less than 0.2 or approaching infinity (i.e., a pinned base condition). Therefore, the assumption of a maximum $(C_L)_{avg} = 0.176$, as in the K_{R_L} method of the AISC LRFD Specification Commentary, is expected to be conservative for most practical frame designs.

5.3.3 Relationship Between Error in Column Axial Strength and Error in Beam-Column Interaction Equations

The nominal strength of a member as a concentrically-loaded column based on $K = 1$ (i.e., $KL = L$) may be related to the column strength based on its "actual" effective length as

$$P_{n(L)} = (1 + e)P_n \tag{5.12a}$$

or vice-versa as

$$P_n = \frac{1}{(1+e)} P_{n(L)} \tag{5.12b}$$

where e is the error in the column axial strength associated with the use of $K = 1$, $P_{n(L)}$ is the nominal axial strength of the member as a column based on $K = 1$, and P_n is the actual value of the column strength based on the buckling model or effective length approach described in the previous section. It is desired to develop equations for the maximum error in the evaluation of the AISC LRFD beam-column interaction equations as a function of the error in the column axial strength, e. The maximum error in the LRFD beam-column interaction equations is expressed here by the symbol ε. If it is assumed that both $\dfrac{P_u}{\phi_c P_{n(L)}}$ and $\dfrac{P_u}{\phi_c P_n}$ are greater than or equal to 0.2, Eq. H1-1a of the LRFD Specification is the governing interaction equation both for design with $K = 1$ and with the actual effective length factor. Therefore, for a given allowable error ε, the "correct" interaction equation can be written as

$$\frac{P_u}{\phi_c P_n} + \frac{8}{9}\frac{M_u}{\phi_b M_n} = (1+\varepsilon) \tag{5.13}$$

at the maximum design limit. It should be emphasized that this equation is based on the "correct" column design strength $\phi_c P_n$. If the design is based on $K = 1$, and $\dfrac{P_u}{\phi_c P_{n(L)}}$ is greater than or equal to 0.2, the engineer would use the interaction equation

$$\frac{P_u}{\phi_c P_{n(L)}} + \frac{8}{9}\frac{M_u}{\phi_b M_n} \le 1 \tag{5.14a}$$

The column strengths $P_{n(L)}$ and P_n can be related by dividing Eq. (5.13) by $(1 + \varepsilon)$, and then substituting the resulting equation into the right hand side of Eq. (5.14a) to obtain

$$\frac{P_u}{\phi_c P_{n(L)}} + \frac{8}{9}\frac{M_u}{\phi_b M_n} \le \frac{1}{(1+\varepsilon)}\left(\frac{P_u}{\phi_c P_n} + \frac{8}{9}\frac{M_u}{\phi_b M_n}\right) \tag{5.14b}$$

By substituting Eq. (5.12b) for P_n in Eq. (5.14b) and solving this equation for ε, the following relationship between ε and e is obtained:

$$\varepsilon \leq \frac{eP_u}{\phi_c P_{n(L)}} \tag{5.15}$$

This equation shows that if the beam-column interaction equation is dominated by its moment term such that the axial strength ratio is small, substantial error (e) can be tolerated in the estimate of the column axial strength without exceeding a specified maximum error in the value of the beam-column interaction check (ε).

If both $\dfrac{P_u}{\phi_c P_{n(L)}}$ and $\dfrac{P_u}{\phi_c P_n}$ are less than 0.2, the equivalent relationship between e and ε is

$$\varepsilon \leq \frac{eP_u}{2\phi_c P_{n(L)}} \tag{5.16}$$

Furthermore, if $\dfrac{P_u}{\phi_c P_n}$ is greater than or equal to 0.2, but $\dfrac{P_u}{\phi_c P_{n(L)}}$ (which is less than $\dfrac{P_u}{\phi_c P_n}$, assuming that the actual effective length factor is greater than one, and thus that e is positive) is less than 0.2, the relationship between e and ε is

$$\varepsilon \leq \left[\frac{5}{9} + e\right]\frac{P_u}{\phi_c P_{n(L)}} - \frac{1}{9} \tag{5.17}$$

Based on Eqs. (5.15) through (5.17), it can be concluded that for a given error in the column strength, e, the maximum error in the beam-column interaction equations, ε, depends only on the ratio $\dfrac{P_u}{\phi_c P_{n(L)}}$. Furthermore, this ratio can be expressed in terms of the basic design parameters P_u / P_y, L / r, and F_y. Therefore, the relationship between e and ε depends solely on these three design parameters.

5.3.4 Calculation of Error in Column Axial Strength

The previous section developed the relationships between the error in the beam-column interaction equations, ε, and the error in the estimated column axial strength based on $K = 1$, e. In this section, the necessary expressions for e are obtained simply by considering the ratio of $P_{n(L)}$ to P_n. As stated by Eqs. (5.12), this ratio is equal to $(1 + e)$.

In general, each of the values $P_{n(L)}$ and P_n may be based either on the AISC LRFD inelastic or elastic column strength equations. If both values are greater than or equal to $0.39\, P_y$, then they are each governed by Eq. E2-2 of the AISC LRFD Specification [AISC 1993] and we may write

$$(1+e) = \frac{0.658^{\lambda^2_{c(L)}}}{0.658^{\frac{P_y}{P_u}\frac{\sum_{all} P_u}{\sum_{non-leaner} P_L}[1+(C_L)_{avg}]}} \tag{5.18}$$

where

$$\lambda^2_{c(L)} = \frac{1}{\pi^2}\left(\frac{L}{r}\right)^2 \frac{F_y}{E} \tag{5.19}$$

If both $P_{n(L)}$ and P_n are less than $0.39\, P_y$, then Eq. E2-3 of LRFD controls, and the ratio of these values becomes

$$(1+e) = \frac{\frac{P_y}{P_u}\frac{\sum_{all} P_u}{\sum_{non-leaner} P_L}[1+(C_L)_{avg}]}{\lambda^2_{c(L)}} \tag{5.20}$$

Otherwise, $P_{n(L)}$ is controlled by Eq. E2-2 and P_n is controlled by Eq. E2-3 (assuming positive e), and the ratio of the two column strengths is

$$(1+e) = \frac{0.658^{\lambda^2_{c(L)}}}{0.877}\frac{P_y}{P_u}\frac{\sum_{all} P_u}{\sum_{non-leaner} P_L}[1+(C_L)_{avg}] \tag{5.21}$$

By inspection of the above equations, it can be seen that the error e in the column axial strength $P_{n(L)}$ depends on P_u / P_y, L / r, F_y, $\frac{\sum_{all} P_u}{\sum_{non-leaner} P_L}$, and $(C_L)_{avg}$. This functional dependency also can be deduced from Eq. (5.10). Furthermore, the term $\frac{\sum_{all} P_u}{\sum_{non-leaner} P_L}$ is directly related to B_2 (Eq. (5.4)), and it may be written in terms of this parameter as

$$\frac{\sum_{all} P_u}{\sum_{non-leaner} P_L} = \frac{B_2 - 1}{B_2} \tag{5.22}$$

It is felt that the values obtained for the elastic sidesway amplification factor B_2 are in general more familiar than values of $\dfrac{\sum\limits_{all} P_u}{\sum\limits_{non-leaner} P_L}$. Therefore, Eq. (5.22) is used in this work to express the error e in terms of B_2.

5.3.5 Results and Observations

The equations that have been presented in Sections 5.3.3 and 5.3.4, show that the error in the beam-column interaction equations ε can be expressed generally as a function of B_2, $(C_L)_{avg}$, F_y, L/r, and P_u/P_y. The basic procedure for calculating this error is:

1. Determine e from Eqs. (5.18) through (5.22).

2. Compute ε from Eqs. (5.15) through (5.17).

The results for $B_2 = 1.11$, $(C_L)_{avg} = 0.176$, and $F_y = 36$ ksi are plotted for various L/r and P_u/P_y values in Fig. 5.1. The B_2 value of 1.11 is chosen because this is the limit proposed in Eurocode 3 [CEN 1993], and it is approximately the limit suggested in AS4100 [SAA 1990] for which frame stability effects may be neglected in design. The value $(C_L)_{avg} = 0.176$ represents an assumed upper-bound on the reduction in the story elastic buckling capacity due to P-δ effects, as per the discussion in Section 5.3.2.

The specific behavior underlying Fig. 5.1 is described in the section below. This is followed by a broader discussion of the factors that influence the maximum error in Section 5.3.5.2 and in Section 5.4, and by studies of specific designs in Section 5.5. These results are synthesized into simple design recommendations in Section 5.6.

5.3.5.1 Basic Attributes of the Error Relationships, and Influence of L/r and P_u/P_y on the Error ε

The following are key attributes associated with the plot in Fig. 5.1:

- Using this figure, the error ε may be determined for specified values of L/r and P_u/P_y. However, there are an infinite number of possible frames associated with any one of the points in the plot. This is a result of the fact that the buckling model (Eqs. (5.8) through (5.10)) does not require any values to be specified for the column end rotational restraint or for the leaning effects on the story being considered.

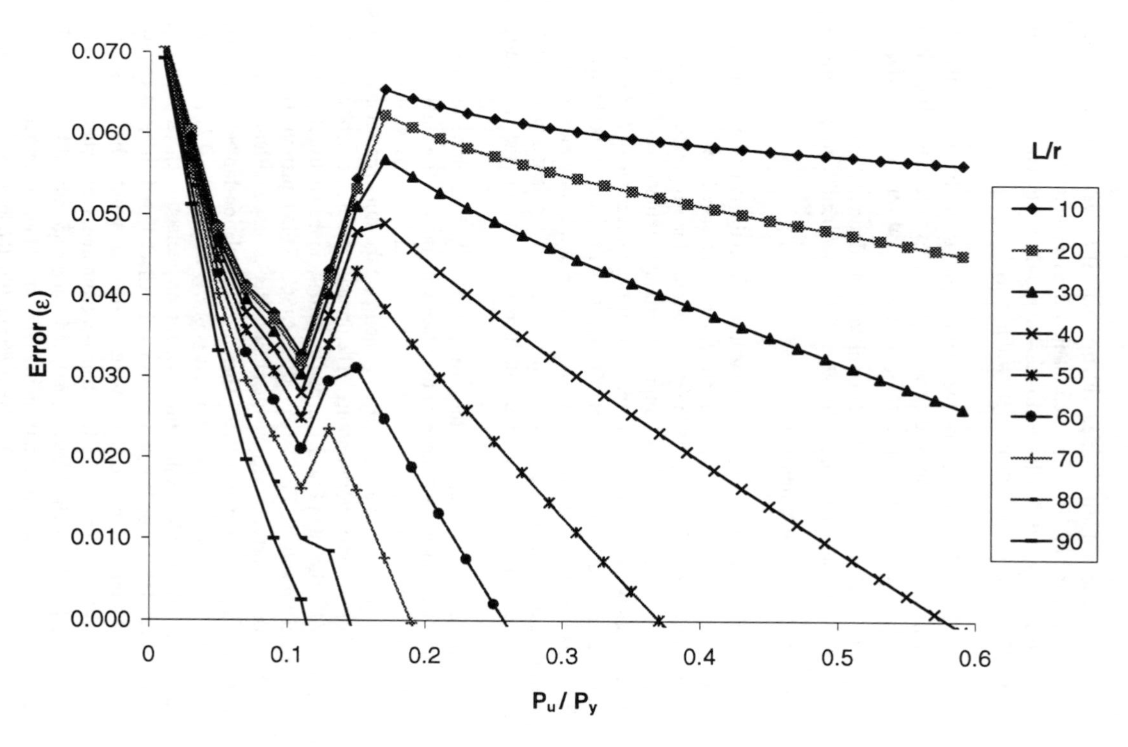

Figure 5.1. Error in beam-column interaction equations (ε) for $B_2 = 1.11$, $(C_L)_{avg} = 0.176$, and $F_y = 36$ ksi.

- Larger errors occur for the curves with the smallest L/r values. This is because, if B_2, $(C_L)_{avg}$, F_y and P_u / P_y are held constant, then as a column becomes stiffer by having its L/r reduced, it will tend to share to a greater extent in stabilizing the story. Thus its actual effective length will increase.

- For larger L/r and P_u / P_y values, the column tends to be "braced" by the other story framing. At the points where each of the curves intersect the horizontal axis in Fig. 5.1, the error is equal to zero because the actual value of K_{C_L} (Eq. (5.10)) is equal to one. For example, the error plots in the figure indicate that a column with $L/r = 90$ would have an actual value of $K_{C_L} = 1$, and therefore zero error, if P_u / P_y is equal to approximately *0.11*.

- The curves for each of the column L/r values exhibit the largest error when the yield load ratio P_u / P_y is near zero. However, the error reduces rapidly with increases in P_u / P_y for values of this parameter less than approximately *0.11*. Most of the curves reach a local minimum error approximately at $P_u / P_y = 0.11$, and then show a rapid increase in the error over a short range of P_u / P_y as the yield load ratio is increased further. The error for these curves peaks again at P_u / P_y between *0.11* and *0.17*, the exact location of this local maximum depending on which L/r curve is considered. For P_u / P_y larger than *0.17*, all the curves show a decrease in the error with increasing values of the yield load ratio.

- The reasons for the variation in the error with P_u / P_y, as discussed above, are as follows. The tendency for the error to reduce with increasing values of P_u / P_y has already been discussed, and is due to the fact that for larger values of P_u / P_y, the column participates less and less in stabilizing the story and actually may start to "lean" on the other story framing. The dramatic dip in the error curves for small P_u / P_y, shown in Fig. 5.1, is due to changes in the controlling beam-column interaction equations. For $P_u / P_y < 0.11$, AISC LRFD Equation H1-1b (Eq. (1.1b) in this report) governs the interaction checks based on both P_n and $P_{n(L)}$ for all the curves shown in the figure. Therefore, the relationship between e and ε is based on Eq. (5.16). However, for values of P_u / P_y between *0.11* and *0.17*, LRFD Eq. H1-1b controls in many cases for the interaction check based on $P_{n(L)}$, whereas Eq. H1-1a (Eq. 5.13) controls for the check based on P_n. As a result, Eq. (5.17) is utilized in calculating the error for these points. For $P_u / P_y \geq 0.17$, Eq. H1-1a always controls for both beam-column interaction checks (i.e., Eqs. (5.13) and (5.14a) are both applicable), and e and ε are related by Eq. (5.15). For the curves that show a local maximum in the error at an intermediate value of P_u / P_y (all the curves except for $L/r = 80$ and *90*), the local maximum point corresponds to the lowest value of P_u / P_y where Eq. H1-1a controls for both of the interaction checks.

- The AISC-LRFD inelastic column strength equation, E2-2 (Eq. 2.1 in this report), governs the calculation of both P_n and $P_{n(L)}$ for all the curves shown in Fig. 5.1

when P_u / P_y is greater than approximately *0.05*. Therefore, the elastic column strength equation (LRFD Equation E2-3) and the associated error equations (Eqs. (5.20) and (5.21)) have no influence on the errors for the majority of the points shown in the figure.

- In Section 5.5, it is shown that stability critical designs corresponding to $B_2 = 1.11$ and P_u / P_y less than about *0.05* are unusual cases, and that application of the desired recommendations for when frame stability may be neglected to these cases can be disallowed with a simple, easily calculated check. Basically, the types of frames giving the errors shown in Fig. 5.1 for $P_u / P_y < 0.05$ tend to have extremely large leaning column effects, with the lateral-resisting columns acting as a lateral spring that is propping up the leaning columns of the story. Also, as noted above, these frames tend to have their story vertical load strength governed by elastic story sidesway buckling. Therefore, the errors shown for P_u / P_y less than about *0.05* are not considered to be of any practical consequence.

Two important observations can be made based on Fig. 5.1 that have important implications with respect to practical restrictions on the use of $K = 1$ in design:

- Since for $P_u / P_y \geq 0.17$, the errors decrease with increasing P_u / P_y for all values of L/r, there appears to be no need to restrict the maximum value of P_u / P_y as suggested in [SSRC 1976] to limit the error associated with using $K = 1$.

- Similarly, the error plot demonstrates that there is no need to restrict the maximum L/r of the columns to control the error associated with using $K = 1$. By requiring a maximum limit on B_2, then either: (1) the slenderness values of at least some of the columns within the story must remain reasonably small, (2) the total gravity loads on the story must be small, or (3) an alternate mechanism (such as sidesway bracing) that gives adequate story lateral stiffness must be provided.

Based on Fig. 5.1, a "practical worst case" error can be estimated for any frames with $B_2 \leq 1.11$ and $F_y \geq 36$ ksi (the error reduces with larger F_y, as discussed in the next section). If $L/r = 10$ is assumed as a lower-bound value for the column slenderness, the maximum error for $P_u / P_y \geq 0.05$ is *0.065* or 6.5 percent (at $P_u / P_y = 0.17$). The actual value of K_{C_L} corresponding to this point is *7.38*. This type of design situation can only be achieved with very large leaning column effects, and/or very low rotational end restraint of the column (in which case, it is practically impossible to obtain $(C_L)_{avg} = 0.176$). The influence of reduced $(C_L)_{avg}$ values is demonstrated in the next section.

Furthermore, L/r values this low generally require the use of a beam type section. For example, if a *12* ft. story height and strong-axis bending is assumed, then for all the wide-flange sections with a nominal depth less than or equal to *16* inches listed in the AISC LRFD Manual [AISC 1994], the smallest L/r_x achievable is *17.5* (using a W14x808). Smaller values of L/r_x for this assumed story height are possible only with heavy W18 and W21 shapes, sections with nominal depth greater than or equal to *24* inches, or with built-up members. Furthermore, the L/d values for L/r_x less than *17.5* are often less than six, where d is the depth of the cross-section. For L/d values this low, the contribution of shear deformation of the column web to the sidesway deflections is typically non-negligible, and therefore, the column flexural buckling equations are not strictly valid. Therefore, it can be concluded that $L/r \cong 20$ is a better practical lower bound for assessment of the errors caused by using $K = 1$. If $L/r = 20$ is assumed as a lower-bound value for the column slenderness, the maximum error for $P_u/P_y \geq 0.05$ is reduced to *6.2* percent (at $P_u/P_y = 0.17$), with a corresponding value of $K_{C_L} = 3.69$.

5.3.5.2 Other Factors that Influence the Error ε

The previous section has studied in detail the attributes of the error associated with neglecting frame stability for a wide range of L/r and P_u/P_y values, and for constant parameters B_2, $(C_L)_{avg}$, and F_y. The influence of B_2, $(C_L)_{avg}$, F_y, and several other factors that do not show up in the error relations derived in Sections 5.3.3 and 5.3.4 is discussed in this section.

Sidesway Amplification Factor, B_2

Figure 5.2 is the same plot as in Fig. 5.1, but for $B_2 = 1.17$ instead of *1.11*. The value of *1.17* is selected since this is the minimum B_2 limit in LRFD based on the recommendations by Lu, et al. [1975], assuming that $\sum_{all} P_u / \sum_{all} P_s$ is greater than or equal to one (see Section 5.2).

Comparing Figs. 5.1 and 5.2, it can be seen that the error curves for different L/r values have the same shape in each of the plots, but the errors based on $B_2 = 1.17$ are somewhat larger. The maximum error for $B_2 = 1.17$ is *10.4* percent (again for $L/r = 10$ and $P_u/P_y = 0.17$). This error might be judged to be excessive.

Fig. 5.3 is a plot of the errors associated with an extreme value of $B_2 = 1.6$. For values of B_2 larger than approximately *1.4*, the dip in the error plots for $P_u/P_y <$ *0.13* to *0.17* disappears such that ε actually decreases monotonically with increasing P_u/P_y for all values of L/r. For these plots, it has been found that if design is restricted to cases for which the story vertical load capacity is controlled by inelastic sidesway buckling, the maximum errors associated with the smallest P_u/P_y values can be avoided. This restriction can be checked by inspection in most cases, or in general by using the simple, approximate equation

$$\left[S_L = \frac{\sum_{non-leaner} P_y}{\sum_{non-leaner} P_L} N \right] \leq 2.25 \tag{5.23}$$

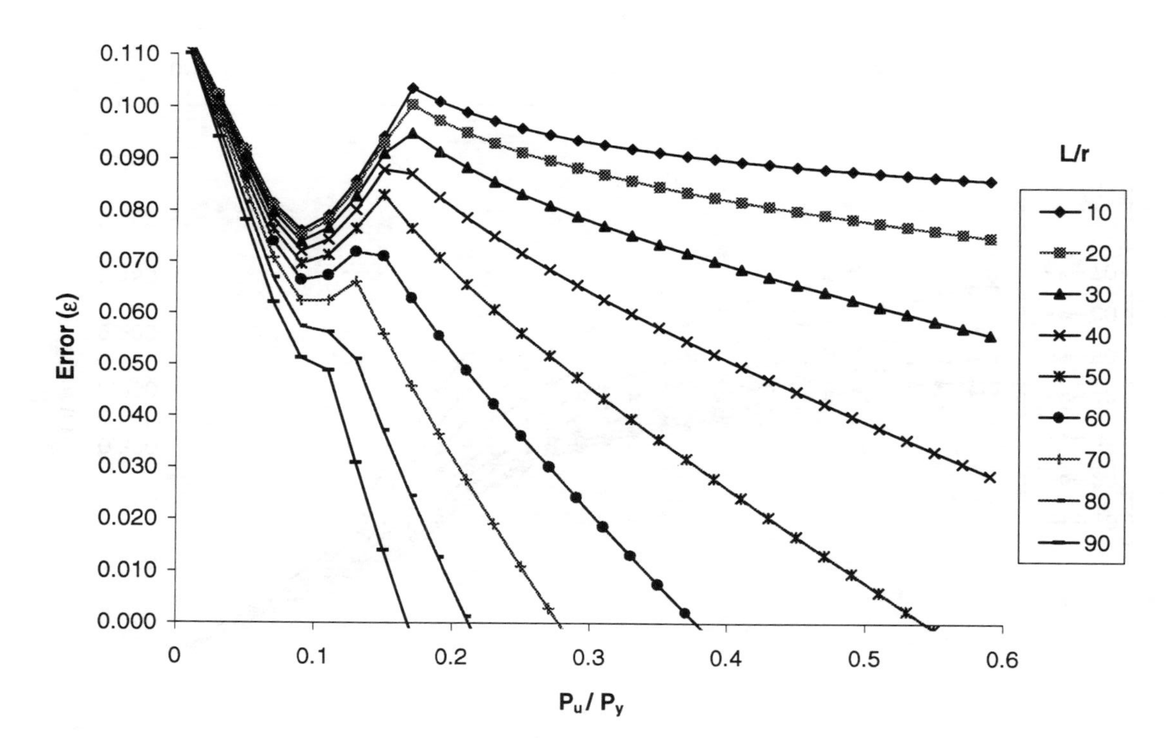

Figure 5.2. Error in beam-column interaction equations (ε) for $B_2 = 1.17$, $(C_L)_{avg} = 0.176$, and $F_y = 36$ ksi.

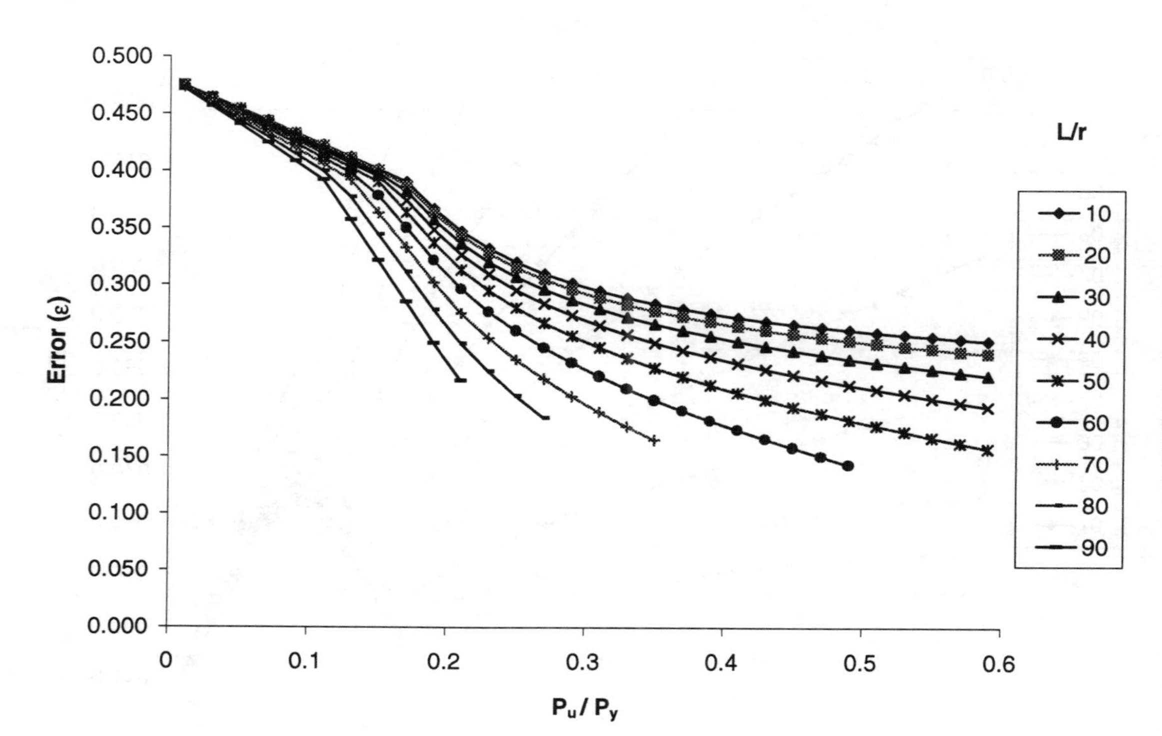

Figure 5.3. Error in beam-column interaction equations (ε) for $B_2 = 1.6$, $(C_L)_{avg} = 0.176$, and $F_y = 36$ ksi.

where

$$N = \frac{\sum\limits_{all} P_u}{\sum\limits_{non-leaner} P_u} = \frac{1}{1 - R_L} \tag{5.24}$$

is the ratio of the total vertical load on the story to the vertical load applied directly to the lateral resisting columns. The parameter S_L is an estimate of the sum of the yield loads for all the columns of the story to the elastic story buckling capacity, and is used by LeMessurier [1995] in a similar way to limit the application of simplified equations for column effective length. In Figs. 5.1 and 5.2, use of this limit requires $P_u / P_y \geq 0.05$; in Fig. 5.3, this limit restricts P_u / P_y to values greater than approximately 0.19.

Figure 5.4 is a graph of the maximum errors (ε_{max}) from a wide range of plots similar to Figs. 5.1 through 5.3 for $S_L \leq 2.25$, $(C_L)_{avg} \leq 0.176$, $L/r \geq 10$, and $F_y \geq 36$ ksi. Also shown on this graph is an approximate equation for this error in terms of B_2 :

$$\varepsilon_{max} = 0.5 B_2 (B_2 - 1) \tag{5.25}$$

This equation slightly underestimates the upper-bound error for $B_2 \leq 1.25$ (the largest underestimate is 9.9 versus 10.4 percent at $B_2 = 1.17$). It increasingly overestimates the error for increasing values of $B_2 > 1.25$ (at $B_2 = 1.4$, Eq. (25) gives $\varepsilon_{max} = 0.28$ whereas the error equations predict a maximum error of 0.26; at $B_2 = 1.6$, this equation predicts an error of 48 percent whereas the "actual" maximum ε is only 37 percent). It has been explained that it is practically impossible for $(C_L)_{avg}$ to exceed the above limit for the combinations of parameters that produce the maximum errors. The limit on S_L prevents the application of $K = 1$ to any types of frames for which the error is larger than given by Eq. (25). The above limits on L/r and F_y are simply minimum values that would be expected in practice.

From Fig. 5.4, it can be concluded that by limiting B_2, the maximum error associated with neglecting frame stability can be controlled, although the actual errors vary substantially based on the other parameters involved. Also, as noted in the previous section, the "practical worst case" error will be slightly less than these values.

Clarification Factor for P-δ Effects, $(C_L)_{avg}$

Figure 5.5 is identical to Fig. 5.1, except that $(C_L)_{avg}$ is set to 0.056 instead of 0.176. The value 0.056 is representative of maximum $(C_L)_{avg}$ values that would be obtained in actual designs based on the maximum error parameters (i.e., $L/r = 10$ or 20, and $P_u / P_y = 0.17$). This is the largest value of $(C_L)_{avg}$ obtained for the designs with $P_u / P_y = 0.17$ to be discussed in Section 5.5. These plots show that the maximum errors based on the elastic story-buckling model outlined in Section 5.3.2 are 5.8 and

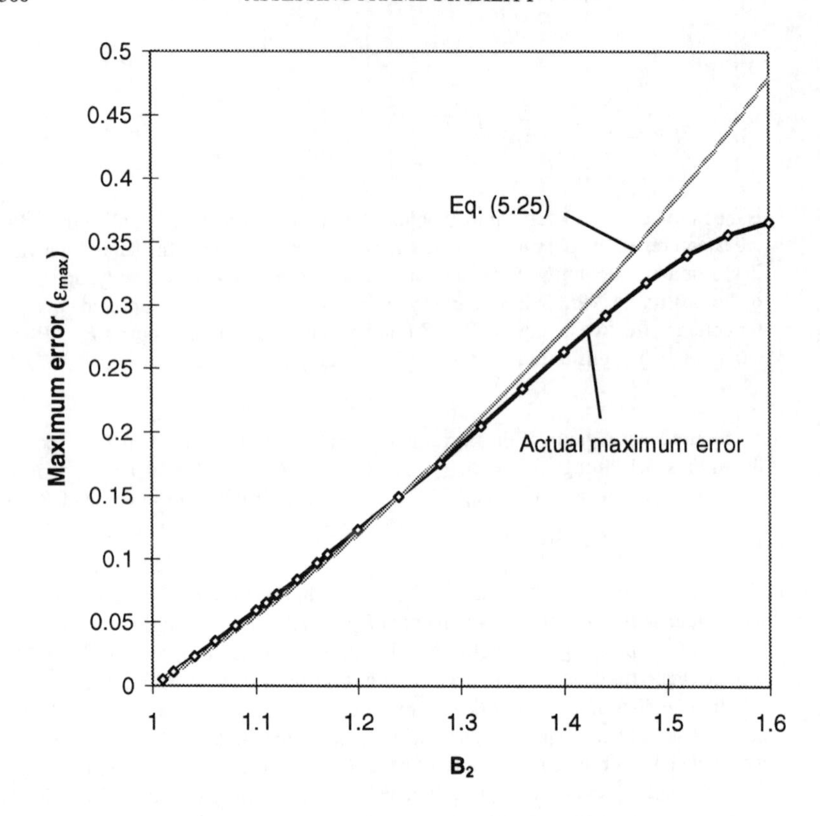

Figure 5.4. Upper-bound of maximum errors versus B_2 for $S_L \leq 2.25$, $(C_L)_{avg} \leq 0.176$, $L/r \geq 10$, and $F_y \geq 36$ ksi.

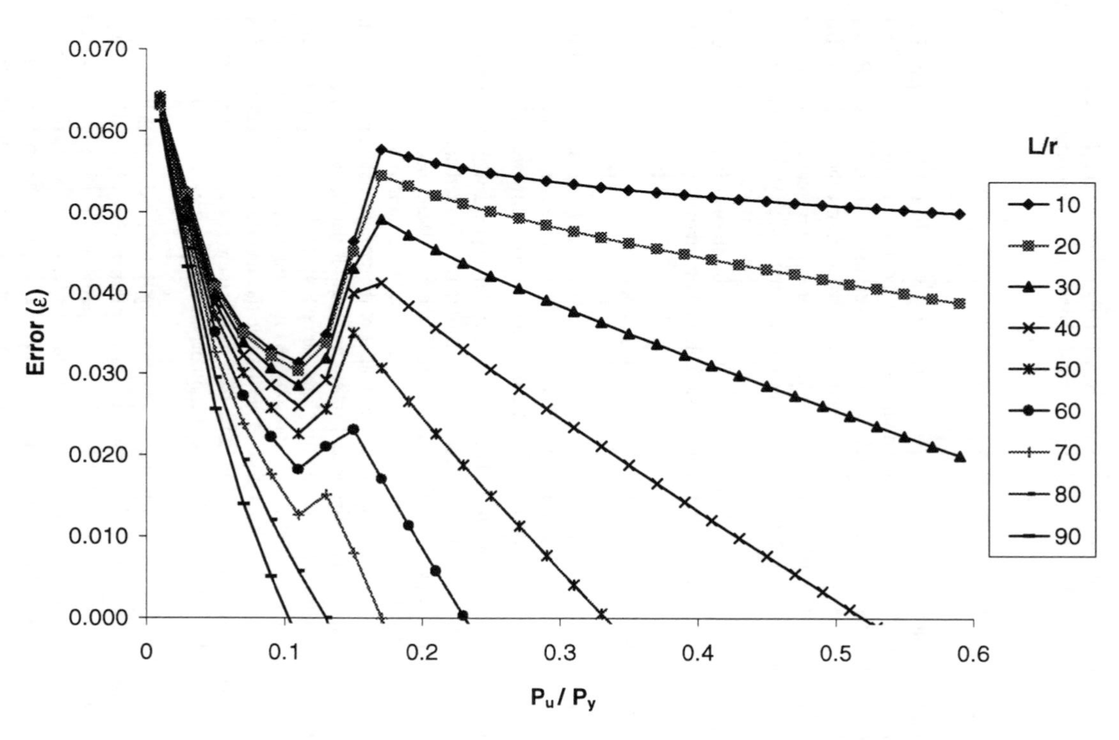

Figure 5.5. Error in beam-column interaction equations (ε) for $B_2 = 1.11$, $(C_L)_{avg} = 0.056$, and $F_y = 36$ ksi.

5.5 percent for $L/r = 10$ and *20* respectively, when B_2 is limited to *1.11*. Even if a highly conservative $(C_L)_{avg}$ of *0.216* is assumed, the maximum errors are only *6.8* and *6.5* percent for $L/r = 10$ and *20*.

Yield Stress, F_y

Figure 5.6 shows the errors for $B_2 = 1.11$, $(C_L)_{avg} = 0.176$, and $F_y = 50$ ksi. For this value of the yield stress, the maximum error is still *6.5* percent for $L/r = 10$. It is reduced slightly to *6.1* percent for $L/r = 20$. However, for the larger L/r and P_u/P_y values, shown in the plot, the reduction in the error by increasing the yield stress from *36* to *50* ksi is substantial. The reduction in the error ε with increases in F_y is primarily due to decreases in the error in the column strength, as specified by Eqs. (5.18) through (5.21). Considering Eq. (5.18), the numerator decreases with increases in F_y, whereas the denominator is unaffected by changes in the yield stress (P_u/P_y being handled as an independent variable in these studies).

Beam-Columns Controlled by Out-of-Plane Strength

For all the cases that have been considered thus far, it has been assumed that the design is based on $K = 1$, and the error in the beam-column interaction equations compared to the design based on K_{C_L} has been determined. However, in many cases, a building frame may be subjected to sidesway in its plane, with its members oriented in strong-axis bending in this plane, and effectively braced at each of the floor levels in the out-of-plane direction. If this is the case, then since LRFD uses the same interaction equation format for checking stability for both in-plane and out-of-plane limit states (with $\phi_c P_n$ based on the smaller of the strong-axis column strength in the plane of the frame, or the weak-axis strength out-of-plane), the beam-column design is often controlled by the out-of-plane interaction check. This is the case if K_yL_y is greater than $K_xL_x/(r_x/r_y)$. A reasonable lower-bound for the r_x/r_y of rolled wide-flange shapes is *1.6*. Therefore, if we assume that an effective length $K_y = 1$ is used for the out-of-plane check, and that $L_x = L_y = L$, the weak-axis column strength would control the design of wide-flange columns unless K_x is greater than *1.6*.

Figure 5.7 shows the error plot comparable to Fig. 5.1 if the design is based on $K = 1.6$ rather than $K = 1$. The error equations for this plot are identical to those presented in Sections 5.3.3 and 5.3.4 with the exception that Eq. (5.19) is multiplied by $K^2 = 1.6^2 = 2.56$. In this plot, the maximum error for $L/r=10$ is *6.4* percent at $P_u/P_y = 0.17$ (a change of only *0.1* percent from Fig. 5.1), and the corresponding effective length factor is *4.61*. Essentially, for this low value of column slenderness, the leaning column effects are so high, and/or the column end rotational restraint is so small, that designing the column with the larger effective length has negligible effect on the error. However, for $L/r = 20$, the maximum error is *5.5* percent (reduced from *6.2* percent in Fig. 5.1). In this plot, the errors reduce more sharply as P_u/P_y and L/r are increased than in any of the other similar plots that have been presented.

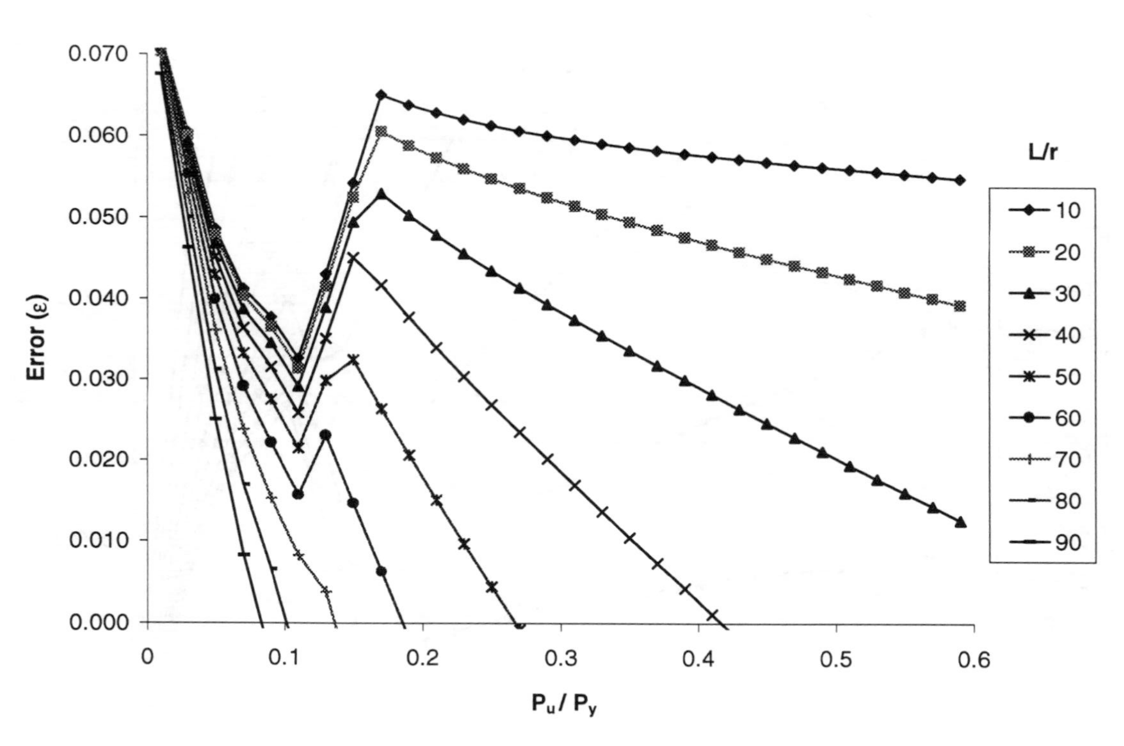

Figure 5.6. Error in beam-column interaction equations (ε) for $B_2 = 1.11$, $(C_L)_{avg} = 0.176$, and $F_y = 50$ ksi.

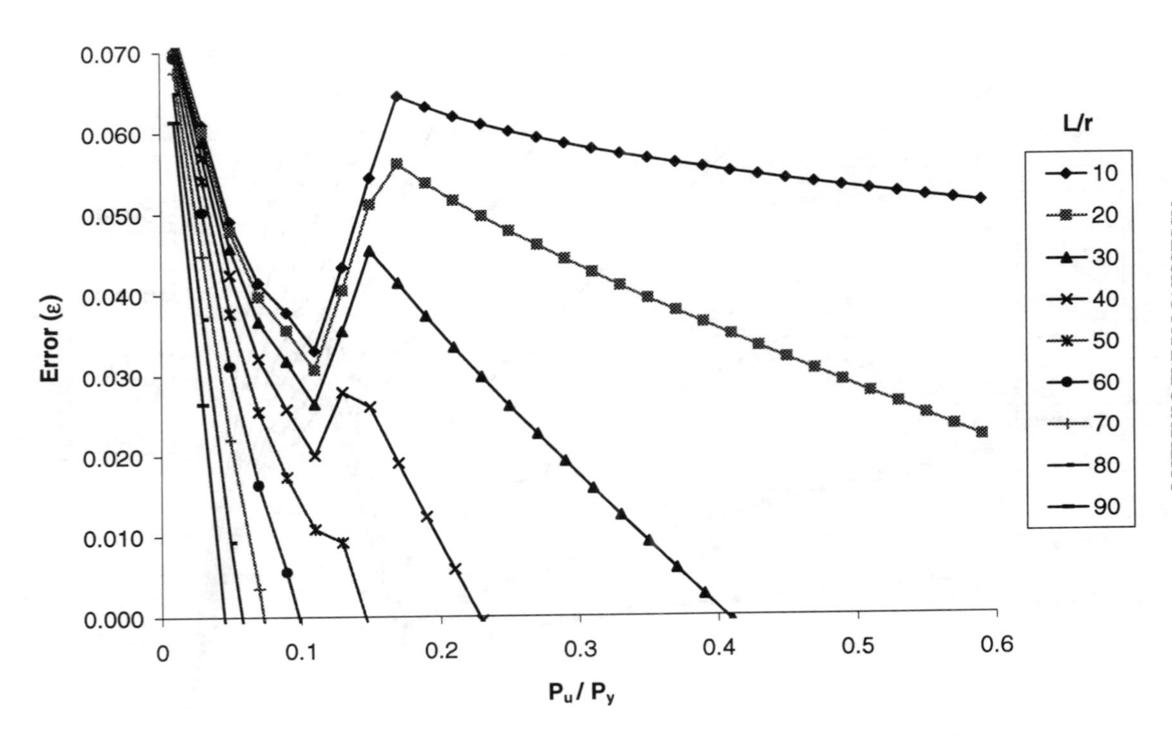

Figure 5.7. Error ε based on use of $K = 1.6$; values plotted for $B_2 = 1.11$, $(C_L)_{avg} = 0.176$, and $F_y = 36$ ksi.

It is interesting to consider the errors obtained if the beneficial effects discussed thus far in this section are combined. If the design is based on $B_2 = 1.11$, $F_y = 36$ ksi and $K = 1.6$, and if $(C_L)_{avg}$ is taken as 0.056 as discussed previously in this section, then the maximum errors become 5.6 percent and 4.8 percent for $L/r = 10$ and 20 respectively. If $B_2 = 1.11$, $(C_L)_{avg} = 0.056$, and $F_y = 50$ ksi, and if $K = 1.6$ is used for design, then the maximum errors are 5.5 and 4.3 percent for $L/r = 10$ and 20.

Interaction Equation Value Less than One

In the development of the error relationships, it is assumed that the value of the beam-column interaction equation based on $K = 1$ is equal to one (see Eq. (5.14a)). This assumes that the design is controlled by strength, whereas if the design is controlled by drift or some other serviceability criterion, the interaction equation will have a value less than one. If this is the case, the error is reduced rapidly as the value of the interaction equation reduces below one. For example, if the beam-column interaction check evaluates to less than approximately 0.94 and B_2 is less than 1.11, it can be stated that the "practical worst case" unconservative error based on all the above considerations will be negligible. In other words, if the engineer wishes to ensure that his or her design is conservative, the design can still be based on $K = 1$ if B_2 is less than 1.11 and the beam-column interaction values are limited to 0.94. Alternatively, based on Fig. 5.2, the design can be based conservatively on $K = 1$ if B_2 is less than 1.17 and the beam-column interaction values are limited to 0.90. In fact, a general limit on the interaction equation value can be set as $\dfrac{1}{(1+\varepsilon_{max})}$ based on Eq. (5.25).

A question can arise when considering how to apply the above approach in the context of beam-column design where the out-of-plane flexural buckling check may control the value for the axial strength P_n. The most precise way of handling the beam-column interaction checks with the above approach is to make two checks: one for lateral-torsional strength of the member with P_n based on an out-of-plane flexural buckling model and a maximum allowed value of 1.0, and one for in-plane strength based on the in-plane flexural buckling model with a maximum allowed value less than one. At the expense of some extra conservatism in checking the lateral-torsional strength, a single interaction equation value may be computed and limited to

$$\frac{1}{(1+\varepsilon_{max})}.$$

Redundancy and Inelastic Reserve Strength

In building frame designs, there is usually some redundancy and an associated amount of inelastic reserve strength within the structural system. This attribute of the behavior was relied upon heavily in the prior research outlined in [SSRC, 1976] and discussed in Section 5.2, where a B_2 limit of 1.17 for use of $K = 1$ was in effect recommended. Although inelastic force redistribution increases the actual strength of framing systems above that estimated by elastic design procedures, generally it cannot be counted upon to allow the use of $K = 1$ unless a more detailed second-order inelastic analysis check is utilized such as discussed in [SSRC, 1993].

Column Inelastic Stiffness Reduction and Reversed Curvature Bending

As previously discussed, the effective length is substantially reduced in many practical designs due to distributed plasticity and the resulting inelastic stiffness reduction within the columns. The use of the elastic buckling model outlined in Section 5.3.2 allows development of the direct relationships between the error ε and the *elastic* sidesway amplification factor B_2 that have been presented. Since the determination the "exact" inelastic buckling capacity requires in general an iterative analysis, development of relations between B_2 and the beam-column interaction error based on an inelastic buckling model is much more involved.

Also, as noted in Section 5.3.1, it is generally expected that when the columns of a frame are subjected to nearly full-reversed curvature bending under sidesway, the possibility of basing the design on $K = 1$ is increased. This aspect of the behavior also is not reflected directly in the error relationships that have been developed in Sections 5.3.3 and 5.3.4. The influence of reversed-curvature bending on the error can only be considered by studying the severe error cases with specific frame designs in which practical values for the end rotational restraints are introduced. For instance, as will be shown in Section 5.5, the maximum error cases in Figs. 5.1 through 5.7 require practically zero rotational end restraints or extraordinary leaning column effects if the column being considered is nearly in full-reversed curvature bending.

The implications that these behavioral effects have on the actual error associated with using $K = 1$ is investigated in Section 5.4 by considering the development and use of approximate inelastic effective length equations recently presented by LeMessurier [1995], and by study of a suite of frame designs in Section 5.5

5.4 LeMessurier's Inelastic Buckling Studies, Approximate Effective Length Equations, and Implications for Use of $K = 1$

Before illustrating the "actual" error trends with example designs in Section 5.5, it is useful to consider an alternate "drift-controlled design" approach recently presented by LeMessurier [1995]. LeMessurier performed an extensive number of "exact" inelastic buckling studies using the frame shown in Fig. 5.8. Although this is a specific framing arrangement, and thus the associated "exact" buckling analysis is not as general as the buckling model utilized in the previous section, the parameters of this frame may be varied to consider a wide range of the design characteristics achievable in practice. Based on curve fitting to and generalization from the results of these studies, LeMessurier developed approximate equations for the inelastic effective length factor which produce values for P_n that are never more than three percent in

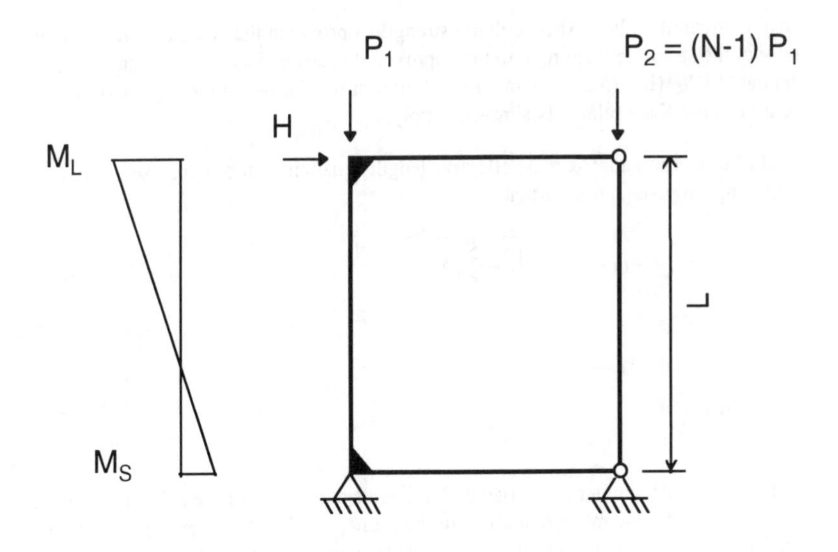

Figure 5.8. LeMessurier's study frame.

error compared to the "exact" column strengths - provided that a suitable limit on the ratio of the story yield strength to the approximate elastic story buckling capacity, denoted by S_L (Eq. (5.23)), is satisfied. Furthermore, LeMessurier's approximate equations for K are relatively simple to apply.

LeMessurier's approximate effective length equations can be expressed in the following single-equation format:

$$K_A = [1 + (1 - \alpha)^4]\sqrt{1 + \frac{5}{6}(N - 1)} \tag{5.26}$$

where

$$\alpha = \frac{M_S}{M_L} \tag{5.27}$$

is the ratio of the smaller and larger end moments at the ends of the column, obtained from a first-order lateral load analysis of the frame, and N is given by Eq. (5.24). It is useful to note that the ratio of the gravity loads supported by leaning columns to the vertical load supported directly by the lateral resisting system, $\dfrac{\sum_{leaner} P_u}{\sum_{non-leaner} P_u}$, is equal to N - 1.

LeMessurier [1995] has shown that the unconservative error in P_n based on Eq. (5.26) is less than three percent for all parametric variations of the frame shown in Fig. 5.8 as long as S_L is less than 0.45. This parameter is essentially an approximation of the ratio of the yield strength to the elastic story buckling capacity of the lateral resisting system. Also, LeMessurier [1995] demonstrates that Eq. (5.26) can be applied to predict with reasonable accuracy the story vertical load strength for both regular and irregular frames (i.e., frames in which P_u / P_y and/or $L\sqrt{P/EI}$ varies significantly among the columns of the lateral resisting system).

It is useful to compare and contrast the results of LeMessurier's procedure with those based on a B_2 limit of 1.11 and use of $K = 1$, as suggested in the previous section. The following are some of the key items that should be considered:

- At the expense of having to compute α, LeMessurier's approximate effective length equations have a broader range of applicability in terms of B_2 limits. For a flexible frame subjected to relatively large gravity load and small lateral load, Eq. (5.26) will allow a simple and fast design check of the frame (although not as simple and fast as using $K = 1$), whereas the $B_2 = 1.11$ limit for use of $K = 1$ may be violated. LeMessurier [1995] points out that if the total vertical load on the story is $0.75 \sum_{non-leaner} P_y N$ for example, the sidesway amplifier B_2 may be written as

$$B_2 = \frac{1}{1 - 0.75 S_L} \tag{5.28}$$

Therefore, if a frame has an S_L of 0.45, LeMessurier's equations are still accurate up to $B_2 = 1.5$. The primary reason for the larger range in B_2 values for this approach is that LeMessurier attempts to approximate the actual inelastic effective length, whereas the use of $K = 1$ completely neglects the sidesway stability problem. For instance, if a column has one end pinned, α is equal to zero and the minimum possible value of K_A is 2.0 (for $N = 0$). If any columns are leaned on the lateral resisting system, K_A is always greater than one. The approach discussed in the previous sections allows the use $K = 1$ for each of these cases, and thus requires much more restrictive limits to maintain the error to an acceptable level.

- LeMessurier's equations are derived by limiting the error in the axial strength P_n to less than three percent, regardless of the relative amount of axial load and bending moment in the members. They do not consider the fact that larger error can be tolerated in P_n if the interaction equations are dominated by the moment term (assuming that the second-order elastic moment is calculated accurately). The S_L limits for use of LeMessurier's equations can be violated by practical frames subjected to relatively small gravity loads, particularly if the columns are subjected to nearly full-reversed curvature bending under sidesway. In this type of situation, the $B_2 = 1.11$ limit still ensures that $K = 1$ can be used with an upper-bound error in the beam-column interaction check less than six percent.

- For cases in which the columns are subjected to near reversed curvature bending and in which the leaning column effects are zero, Eq. (5.26) indicates that $K = 1$ can be used with negligible error in the calculation of P_n. In fact, LeMessurier explains in his paper that for frames with S_L less than 0.45, $P_{n(L)}$ is within three percent of the "exact" axial strength as long as α is greater than or equal to 0.75 and the leaning column effects are negligible. The condition $\alpha \geq 0.75$ corresponds to an inflection point being located within the middle $1/7$ of the column length. This proof, combined with the observation that any unconservatism of Eq. (5.26) in predicting overall story vertical load strength is actually reduced for irregular frames [LeMessurier 1995], provides a significant range of cases for which the use of $K = 1$ is acceptable. Combined with the conclusion of the previous section that frame stability effects may be neglected in determining the axial strength when $B_2 \leq 1.11$, an even wider number of cases is covered for which the design can be based on $K = 1$. Alternatively, if the value of the interaction equation is restricted to $\frac{1}{(1+\varepsilon_{max})}$ as discussed in the previous section, a wide range of cases can be designed using $K = 1$.

5.5 Example Frames

In this section, LeMessurier's study frame (shown in Fig. 5.8) is selected to illustrate the actual errors associated with the use of $K = 1$ (compared to the design checks based on "exact" inelastic effective lengths) for a wide range of designs associated with the error plots in Figs. 5.1 and 5.5. Since $L/r = 20$ is considered to be a practical lower bound for the column slenderness in many building frames, and since the errors in Figs. 5.1 and 5.5 are largest for smaller values of L/r, all the specific designs considered here are based on a story height of *11* ft. and a W14x211 in strong-axis bending for the left-hand lateral-resisting column. This produces a column slenderness of $L/r_x = 11(12)/6.55 = 20.15$. It should be noted that the W14x211 section is the heaviest Group 3 W shape available [AISC 1994].

The length of the girders is selected as *11* ft., and five different girder sizes are considered giving elastic G values (see Eqs. (2.10a), (2.10b), and (2.11)) equal to *0.243* (W33x354), *0.507* (W36x170), *1.002* (W21x201), *1.993* (W21x111), and *4.000* (W18x76). Both pinned base conditions ($G = \infty$) and restrained base conditions (finite G) are considered. For the cases with restrained base conditions, the girder at the base is set such that α (Eq. (5.27)) is approximately equal to *0.75*.

The vertical load on the lateral-resisting column, P_1, is set at three different levels: the smaller of $0.6P_y$ or the story vertical load corresponding to $B_2 = 1.11$, $0.17P_y$ (the axial load corresponding to the largest error in the previous error plots), and $0.05P_y$ (representing a design with extremely small load on the columns of the lateral resisting system). In all of the cases, the load on the right-hand leaning column, P_2, is set such that B_2 is equal to *1.11*.

Some of the key design attributes for the frames, with P_n based on an "exact" inelastic buckling analysis, and M_u based on an "exact" second-order elastic analysis (but neglecting overturning effects on the column axial forces), are summarized in Tables 5.1 and 5.2. Table 5.1 shows the results for pinned base conditions, whereas the results in Table 5.2 are for both ends of the left-hand column restrained but with a smaller girder on the bottom, specified such that $\alpha \approx 0.75$ (Eq. (5.27)). The reader is referred to [LeMessurier 1995] for a summary of the equations associated with the "exact" inelastic buckling analysis of this frame. For each design, the tables list:

- the fraction of the yield load on the lateral-resisting column, P_1 / P_y,
- the ratio of the gravity load on the leaning and lateral resisting columns, $N - 1 = P_2 / P_1$,

Table 5.1. Study frames with pinned-base conditions ($\alpha = 0$).

Girder	G	N-1 = 0	$P_1 / P_y = 0.17$	$P_1 / P_y = 0.05$
W33x354	0.243	$P_1 / P_y = 0.526$ $H = 35.0$ k $\Delta_{oh} / L = 1/338$ $S_L = 0.188$ $(C_L)_{avg} = 0.172$ $K_{CL} = 2.08$ $e_{el} = 7.36$ % $e_{inel} = 6.79$ % $\varepsilon_{el} = 4.66$ % $\varepsilon_{inel} = 4.29$ %	$N\text{-}1 = 2.09$ $H = 76.7$ k $\Delta_{oh} / L = 1/154$ $S_L = 0.583$ $(C_L)_{avg} = 0.056$ $K_{CL} = 3.47$ $e_{el} = 26.6$ % $e_{inel} = 25.1$ % $\varepsilon_{el} = 5.44$ % $\varepsilon_{inel} = 5.14$ %	$N\text{-}1 = 9.52$ $H = 83.5$ k $\Delta_{oh} / L = 1/142$ $S_L = 1.98$ $(C_L)_{avg} = 0.016$ $K_{CL} = 6.28$ $e_{el} = 127$ % $e_{inel} = 127$ % $\varepsilon_{el} = 3.83$ % $\varepsilon_{inel} = 3.82$ %
W36x170	0.507	$P_1 / P_y = 0.471$ $H = 41.5$ k $\Delta_{oh} / L = 1/255$ $S_L = 0.211$ $(C_L)_{avg} = 0.138$ $K_{CL} = 2.17$ $e_{el} = 8.21$ % $e_{inel} = 6.95$ % $\varepsilon_{el} = 4.64$ % $\varepsilon_{inel} = 3.93$ %	$N\text{-}1 = 1.77$ $H = 76.8$ k $\Delta_{oh} / L = 1/138$ $S_L = 0.583$ $(C_L)_{avg} = 0.050$ $K_{CL} = 3.46$ $e_{el} = 26.5$ % $e_{inel} = 23.6$ % $\varepsilon_{el} = 5.41$ % $\varepsilon_{inel} = 4.81$ %	$N\text{-}1 = 8.41$ $H = 83.5$ k $\Delta_{oh} / L = 1/127$ $S_L = 1.98$ $(C_L)_{avg} = 0.015.$ $K_{CL} = 6.27$ $e_{el} = 127$ % $e_{inel} = 127$ % $\varepsilon_{el} = 3.82$ % $\varepsilon_{inel} = 3.81$ %
W21x201	1.002	$P_1 / P_y = 0.393$ $H = 50.7$ k $\Delta_{oh} / L = 1/175$ $S_L = 0.252$ $(C_L)_{avg} = 0.096$ $K_{CL} = 2.33$ $e_{el} = 9.89$ % $e_{inel} = 7.33$ % $\varepsilon_{el} = 4.67$ % $\varepsilon_{inel} = 3.46$ %	$N\text{-}1 = 1.31$ $H = 76.8$ k $\Delta_{oh} / L = 1/115$ $S_L = 0.583$ $(C_L)_{avg} = 0.041$ $K_{CL} = 3.45$ $e_{el} = 26.2$ % $e_{inel} = 21.1$ % $\varepsilon_{el} = 5.35$ % $\varepsilon_{inel} = 4.32$ %	$N\text{-}1 = 6.86$ $H = 83.5$ k $\Delta_{oh} / L = 1/106$ $S_L = 1.98$ $(C_L)_{avg} = 0.012$ $K_{CL} = 6.27$ $e_{el} = 127$ % $e_{inel} = 126$ % $\varepsilon_{el} = 3.81$ % $\varepsilon_{inel} = 3.78$ %
W21x111	1.993	$P_1 / P_y = 0.295$ $H = 62.2$ k $\Delta_{oh} / L = 1/107$ $S_L = 0.335$ $K_{CL} = 2.63$ $(C_L)_{avg} = 0.054$ $e_{el} = 13.5$ % $e_{inel} = 8.16$ % $\varepsilon_{el} = 4.79$ % $\varepsilon_{inel} = 2.90$ %		

Table 5.2. Study frames for which the lateral-resisting column is nearly in full-reversed curvature bending ($\alpha \approx 0.75$).

Girders	G values	$P_1 / P_y = 0.6$	$P_1 / P_y = 0.17$	$P_1 / P_y = 0.05$
W33x354	$G_S = 0.243$	N-1 = 1.48	N-1 = 7.77	N-1 = 28.8
W21x201	$G_L = 1.002$	H = 26.9 k	H = 77.1 k	H = 83.6 k
		$\Delta_{oh} / L = 1/1250$	$\Delta_{oh} / L = 1/435$	$\Delta_{oh} / L = 1/401$
		$S_L = 0.165$	$S_L = 0.583$	$S_L = 1.98$
		$(C_L)_{avg} = 0.047$	$(C_L)_{avg} = 0.013$	$(C_L)_{avg} = 0.004$
		$K_{CL} = 1.84$	$K_{CL} = 3.40$	$K_{CL} = 6.24$
		$e_{el} = 5.23$ %	$e_{el} = 25.3$ %	$e_{el} = 125$ %
		$e_{inel} = 3.00$ %	$e_{inel} = 19.4$ %	$e_{inel} = 124$ %
		$\varepsilon_{el} = 3.77$ %	$\varepsilon_{el} = 5.18$ %	$\varepsilon_{el} = 3.76$ %
		$\varepsilon_{inel} = 2.17$ %	$\varepsilon_{inel} = 3.97$ %	$\varepsilon_{inel} = 3.73$ %
W21x201	$G_S = 1.002$	N-1 = 0.610	N-1 = 4.68	N-1 = 18.3
W21x111	$G_L = 1.993$	H = 27.0 k	H = 77.1 k	H = 83.6 k
		$\Delta_{oh} / L = 1/807$	$\Delta_{oh} / L = 1/282$	$\Delta_{oh} / L = 1/260$
		$S_L = 0.165$	$S_L = 0.583$	$S_L = 1.98$
		$(C_L)_{avg} = 0.031$	$(C_L)_{avg} = 0.009$	$(C_L)_{avg} = 0.003$
		$K_{CL} = 1.83$	$K_{CL} = 3.39$	$K_{CL} = 6.24$
		$e_{el} = 5.12$ %	$e_{el} = 25.2$ %	$e_{el} = 125$ %
		$e_{inel} = 1.45$ %	$e_{inel} = 14.4$ %	$e_{inel} = 123$ %
		$\varepsilon_{el} = 3.69$ %	$\varepsilon_{el} = 5.15$ %	$\varepsilon_{el} = 3.75$ %
		$\varepsilon_{inel} = 1.04$ %	$\varepsilon_{inel} = 2.95$ %	$\varepsilon_{inel} = 3.70$ %
W21x111	$G_S = 1.993$	N-1 = 0.027	N-1 = 2.63	N-1 = 11.3
W18x76	$G_L = 4.000$	H = 27.0 k	H = 77.1 k	H = 83.6 k
		$\Delta_{oh} / L = 1/514$	$\Delta_{oh} / L = 1/180$	$\Delta_{oh} / L = 1/166$
		$S_L = 0.165$	$S_L = 0.583$	$S_L = 1.98$
		$(C_L)_{avg} = 0.023$	$(C_L)_{avg} = 0.007$	$(C_L)_{avg} = 0.002$
		$K_{CL} = 1.82$	$K_{CL} = 3.39$	$K_{CL} = 6.24$
		$e_{el} = 5.06$ %	$e_{el} = 25.1$ %	$e_{el} = 125$ %
		$e_{inel} = 0.36$ %	$e_{inel} = 10.4$ %	$e_{inel} = 123$ %
		$\varepsilon_{el} = 3.65$ %	$\varepsilon_{el} = 5.14$ %	$\varepsilon_{el} = 3.75$ %
		$\varepsilon_{inel} = 0.26$ %	$\varepsilon_{inel} = 2.13$ %	$\varepsilon_{inel} = 3.69$ %

- the factored horizontal load, H, required such that the interaction equation based on $K = 1$ (Eq. 5.14a) is equal to one,

- the first-order drift caused by the above lateral load, Δ_{oh} / L,

- the approximate ratio of the yield load to the elastic buckling capacity of the story, S_L (Eq. (5.23)),

- the weighted average value of the clarification factor, $(C_L)_{avg}$ (Eq. (5.9)),

- the effective length factor K_{C_L} (Eq. 5.10),

- the errors associated with $P_{n(L)}$, based on comparison to P_n obtained with the exact elastic and the exact inelastic effective length factors, e_{el} and e_{inel} respectively, and

- the errors in the evaluation of the beam-column interaction equation based on comparison to the interaction equation values obtained with the "exact" P_n values, ε_{el} and ε_{inel}.

Some of the important characteristics of these designs, which illustrate the behavior discussed in the previous two sections, are as follows:

- The S_L values range from 0.165, exhibited by the frames in the first column of Table 5.2, to 1.98 for the frames listed in the third column of each of the tables. If frames with P_1 / P_y less than 0.05 are considered, the value of S_L rapidly increases beyond a value of 2.25. For these frames, this corresponds approximately to a transition from inelastic to elastic sidesway buckling as the mode of failure associated with the story's total vertical load strength. Furthermore, the value of $N - 1$ approaches infinity as P_1 / P_y approaches zero (the leaning column load is set such that $B_2 = 1.11$). As noted previously in Section 5.3.5.1, the error in using $K = 1$ for the design of the lateral-resisting column is larger in this type of situation than any other cases. However, it can be argued that these types of frames are very unusual, and if the engineer does wish to design such a frame, he or she probably should consider the frame stability effects. Consideration of frame stability in the design of these types of frames can be enforced by restricting the use of $K = 1$ to cases in which S_L is less than 2.25.

- The error ε_{inel} is largest for the cases in the top row of the $P_1 / P_y = 0.17$ column for each of the tables. These cases have the smallest values of G (i.e., the heaviest girders) and an axial force on the lateral resisting column equal to that associated with the maximum error in Figs. 5.1 and 5.5. Due to the heavy end restraint and the relatively low P_1 / P_y in these frames, the column inelastic stiffness reduction has little effect on the errors. Relatively large leaning column loads are required to achieve $B_2 = 1.11$ in each of these cases. For the design with pinned base conditions (Table 5.1), ε_{inel} is *5.14* percent, and for the design with both ends of the lateral-resisting column restrained such that this member is subjected nearly to full-reversed curvature bending, ε_{inel} is *3.97* percent. These values are believed to be the most representative of actual upper-bound errors that would be encountered in practical frame design.

- The error ε_{inel} is smallest for the cases in the bottom-left corner of the tables. These cases have the largest values of G (i.e., the lightest girders) and the largest axial force on the lateral resisting column. As a result, they obtain the most benefit from column inelastic stiffness reduction. For the case with the lightest girders and $\alpha \approx 0.75$, the effect of column inelasticity reduces the error ε_{inel} to only *0.26* percent (see the lower-left corner of Table 5.2).

- For comparable cases (i.e., for the columns of each of the tables corresponding to $P_l / P_y = 0.17$), the frames with the pinned base conditions (Table 5.1) exhibit larger errors than the frames with end restraint at both ends of the lateral resisting column (Table 5.2). This is caused by the fact that much larger leaned column loads are required to achieve $B_2 = 1.11$ for the frames with the restrained bases, thus resulting in smaller $(C_L)_{avg}$ values for these frames.

- For the cases involving very small axial force on the lateral resisting column and pinned base conditions ($P_l / P_y = 0.05$ in Table 5.1), the lateral drift under the design load H is excessive. If factors on the lateral load of *1.3* and *0.75* are assumed for strength and serviceability design respectively, and if it is assumed that the second-order drift at *service load* levels is to be restricted to *1/200* (a very liberal criterion), then any of the frames with Δ_{oh} / L at the strength design load levels shown in the tables greater than *1/104* would be considered unserviceable. It can be seen that many of the frames with pinned-base conditions (Table 5.1) are very flexible and only narrowly meet this serviceability requirement. The frames with nearly full-reversed curvature bending in the lateral-resisting column (Table 5.2) would not have any problem meeting the above or a more restrictive serviceability limit.

Based on these observations and the generalization discussed by LeMessurier [1995] that the errors in predicting story vertical load capacity actually decrease for stories with irregular P_u / P_y and/or $L\sqrt{P/EI}$ values, it may be concluded that six percent is a reasonably good upper bound on the error ε_{inel} caused by using $K = 1$ in any practical frame design in which story-based effective length calculations are valid and B_2 is limited to a maximum value of *1.11*. In fact, based on the largest values of ε_{inel} exhibited in Tables 5.1 and 5.2 and other similar studies, it can be concluded that the maximum "practical worst case" error is closer to four percent for columns subjected to nearly full-reversed curvature bending ($\alpha \geq 0.75$) in sidesway, and five percent for columns with pinned base conditions.

5.6 Design Recommendations

Based on the studies presented in this chapter, the following limits are recommended for assessing when a design can be based on $K = 1$ without consideration of equivalent geometric imperfection or notional load effects:

- If $B_2 \leq 1.11$ and $S_L \leq 2.25$, use $K = 1$ for calculation of column in-plane axial strengths.

  An upper bound on the maximum error is approximately six percent, with the actual maximum error being on the order of five percent for pinned base conditions and four percent for cases involving nearly full-reversed curvature bending ($\alpha \geq 0.75$) for combinations of practical design parameters.

  This recommendation is subject to the same restrictions as the story-based effective length equations of Chapter 2. That is, if the story has any columns that are relatively slender and are loaded heavily compared to the other columns, then Eq. (5.11b) should be checked to ensure that the story buckling strength is not substantially reduced by the weak column(s). Also, in general, the design problem must be one in which the use of story-buckling models, such as the ones outlined in Chapter 2 and applied in this chapter, are valid. For instance, the axial forces in the girders must be negligible, or the girder stiffnesses must be reduced to account for these effects. Also, it is assumed that the reduction in stiffness due to cantilever bending action of slender frames does not significantly reduce the system buckling strength. It is assumed in general that the buckling behavior is a story-by-story phenomenon, with limited interstory buckling interactions. Finally, as discussed in Chapter 2, due to the greater inelastic stiffness reduction in columns with larger P_u/P_y values, the story-based inelastic effective lengths in columns that have smaller P_u/P_y can be larger than the corresponding elastic KL values. In certain cases, the unconservative error in the beam-column interaction equations can be significant for such members. For cases in which the lateral resisting columns have similar P_u/P_y, and/or the P_u/P_y values on all the lateral resisting columns are significantly less than one at the limit of the structure's resistance, and/or the structure has substantial redundancy, these unconservative errors are likely to be negligible. However, for frames that (1) contain columns with substantially different levels of P_u/P_y and (2) have little redundancy, the column strengths may need to be based on an inelastic buckling analysis (or the corresponding inelastic effective lengths) for adequate assessment of the inelastic stability effects

- If the engineer wishes to ensure that a design based on use of $K = 1$ (without notional load) is conservative, the above procedure may be applied and the beam-column interaction values should be limited to 0.94.

- This procedure can be extended to frames that do not satisfy a B_2 limit of *1.11* by limiting the maximum beam-column interaction check to the value $\frac{1}{(1+\varepsilon_{max})}$, where ε_{max} is given by Eq. (5.25).

If none of the above limits are met, but the assumptions regarding basic story-by-story flexural buckling behavior are valid, the K_{R_L} approach without iteration on the inelastic stiffness reduction, as outlined in Chapter 2, and LeMessurier's approximate equations for the inelastic effective length factors are two of the simplest approaches to obtain valid effective lengths. Also, the notional load approach provides a simple solution that accounts for the frame stability effects. LeMessurier's equations are sufficiently accurate for many cases in which the beam-column behavior is dominated by the moment term of the interaction equation, even if S_L is greater than *0.45*. However, limits on the use of LeMessurier's equations for these types of cases have not been developed at the present time.

References

AISC [1989], *Specifications for Structural Steel Buildings: Allowable Stress Design and Plastic Design*, 9th Ed., American Institute of Steel Construction, Chicago, IL.

AISC [1993], *Load and Resistance Factor Design Specification for Structural Steel Buildings*, 2nd Ed., American Institute of Steel Construction, Chicago, IL.

AISC [1994], *Manual of Steel Construction, Load and Resistance Factor Design*, 2nd Ed., American Institute of Steel Construction, Chicago, IL.

CEN [1993], *ENV 1993-1-1 Eurocode 3: Design of Steel Structures, Part 1.1 -- General Rules and Rules for Buildings*, European Committee for Standardization, Brussels.

FEMA [1992], "NEHRP Recommended Provisions for the Development of Seismic Regulations for New Buildings, 1991 Edition," Earthquake Hazards Reduction Series 16, Federal Emergency Management Agency.

Kanchanalai, T. [1977], "The Design and Behavior of Beam-Columns in Unbraced Steel Frames," *AISI Project No. 189, Report No. 2*, Civil Engineering/Structures Research Lab., Univ. of Texas, Austin, TX, 300 pp.

LeMessurier, W.J. [1995], "Simplified K Factors for Stiffness Controlled Designs," *Restructuring: America and Beyond*, Proceedings of Structures Congress XIII, ASCE, Boston, MS, pp. 1797-1812.

LeMessurier, W.J. [1993], "Discussion of the Proposed LRFD Commentary to Chapter C of the Second Edition of the AISC Specification," presentation to the ASCE Technical Committee on Load and Resistance Factor Design, April 18, 1993, Irvine, CA.

LeMessurier, W.J. [1977], "A Practical Method for Second Order Analysis. Part 2: Rigid Frames," *Engineering Journal*, AISC, 2nd quarter, pp. 49-67.

Liapunov, S. [1973], "Ultimate Load Studies of Plane Multistory Steel Rigid Frames," Ph.D. dissertation, New York University, New York, NY.

Liapunov, S. [1974]. "Ultimate Strength of Plane Multistory Steel Rigid Frames," *Journal of the Structural Division*, ASCE, 100(ST8), 1643-1655.

Lu, L.W., Ozer, E., Daniels, J.H., Okten, O.S., and Morino, S. [1975], "Effective Column Length and Frame Stability, Frame Stability and Design of Columns in Unbraced Multistory Steel Frames," Fritz Engineering Laboratory Report No. 375.2, Lehigh Univ., Bethlehem, PA.

Salmon, C.G. and Johnson, J.E. [1996], *Steel Structures: Design and Behavior*, 4th ed., Harper Collins, New York, NY, 1024 pp.

SAA [1990], *AS4100-1990, Steel Structures*, Standards Association of Australia, Sydney, Australia.

SSRC [1976], *Guide to Stability Design Criteria for Metal Structures*, 3rd Ed., B.G. Johnston, (ed.), Structural Stability Research Council, Wiley, 616 pp.

Tide, R.H.R. [1985], "Reasonable Column Design Equations," *Proceedings, 1985 Annual Technical Session*, Structural Stability Research Council, 47-55.

CHAPTER 6

EXAMPLES

This chapter presents a number of examples to elucidate the behavior and procedures for use of the techniques for accounting for structural stability presented in this document. The examples include two-dimensional and three-dimensional frames having both fully-restrained and partially-restrained connections.

Chapters 2 through 4 each include several illustrative examples specific to the approaches to account for stability described in those chapters. The examples presented in this chapter, in turn, compare and contrast several of the techniques presented, thus illustrating the behavior of these approaches and elucidating the limitations and assumptions embedded in the procedures. This chapter also shows, step-by-step, how to apply these effective length and notional load procedures to complex structures.

The first study, in Section 6.1, investigates the behavior of a two-dimensional, one-story frame which includes leaning columns. The second example is an unsymmetric, two-dimensional, two-bay, two-story frame which illustrates some of the limitations of using the AISC nomograph directly for the calculation of axial strength. The third example is a three-dimensional building in which the computations are outlined for a beam-column subjected to biaxial bending plus axial force. The fourth example compares and contrasts several related approaches for assessing stability of frames which incorporate partially-restrained connections.

6.1 Two-Dimensional, Two-Bay, One-Story, Unbraced Frame With Leaning Columns

To further demonstrate the approaches outlined in Chapters 2 and 4, the frame in Figure 6.1 is analyzed in this example. This frame was first evaluated by LeMessurier [1977] using AISC ASD provisions. The center column of the frame has been modified for this example, and it is analyzed using the provisions of the LRFD Specification [AISC 1993].

In the design of the center column (a W14x43) the following assumptions are made:

- All members are assumed to be compact and therefore not susceptible to local buckling.
- Strong axis bending of all members is assumed.
- Columns are braced at the base supports, at their connections to the beams, and, out-of-plane, at their midpoints. $K = 1$ for weak axis buckling.

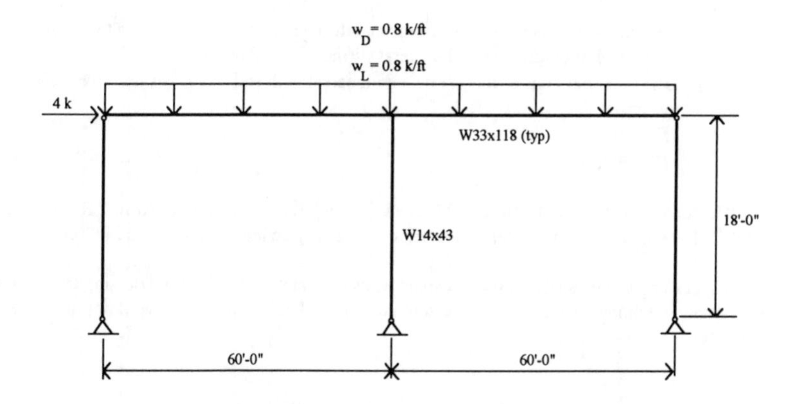

Figure 6.1: Two-dimensional, One-Story, One-Bay Frame with Leaning Columns

- The connection between the girders and the center column is assumed to be fully restrained, providing all of the lateral stiffness for the frame.
- The outer columns of this frame are pinned at both ends ($K = 1$ is used for their design).
- $F_y = 36$ ksi.
- $E = 29000$ ksi.

The load combination, $1.2Dead + 0.5Live + 1.3Wind$ (corresponding to AISC LRFD Eq. 4-4 [AISC, 1993]), governs the design of this frame. The service loads are given in Fig. 6.1.

This example shows how to use the approaches of Chapters 2 and 4 for assessing the stability of frames containing leaning columns, and of the potential effects of leaning columns on frame stability.

6.1.1 Computation of Strength of the Center Column in the Frame

The strength of the columns in this examples are determined using the following approaches:

- Nomograph-Based Effective Length Factor, K_{K_n} (Section 2.4.1)
- Practical Story-Based Effective Length Factor, K_{R_L} (Section 2.4.2)
- Notional Load (Chapter 4)

Comprehensive second-order elastic frame analysis [White and Hajjar, 1991], which accounts for both P-Δ and P-δ effects, was used to obtain the required forces for design. The combined axial force in the two leaning columns is 62.6 kips. The required axial force and bending moment at the top of the center column are:

$$P_u = 100.6 \text{ kips}$$
$$M_{ux} = 1470 \text{ kip-inches}$$

The nominal strong axis bending moment capacity of the center column is computed according to the AISC LRFD Specification [AISC, 1993]:

$$M_{nx} = M_{px} = 2506 \text{ kip-inches}$$

The above values, as well as the value of P_n computed in the following sections, are used in the LRFD beam-column interaction equation (Eq. (1.1)).

To investigate inelastic behavior, the inelastic stiffness reduction factor $\hat{\tau}$ is calculated as per Sections 2.1 and 2.2. For this example, Eq. (2.7) is evaluated conservatively by using P_u / ϕ_c in place of P_n. For the center column:

$$P_y = A_g F_y = (12.6)(36) = 453.6 \text{ kips}$$

$$\frac{P_u/\phi_c}{P_y} = \frac{100.6/0.85}{453.6} = 0.261 \le 0.39$$

Therefore, $\hat{\tau} = 1$ and no inelastic stiffness reduction is used.

Weak axis buckling of the center column is given by:

$$L_y = \frac{18'}{2} = 9' = 108''$$

$$K_y = 1.0$$

$$\lambda_{cy} = \frac{K_y L_y}{r_y \pi}\sqrt{\frac{F_y}{E}} = \frac{(1.0)(108)}{(1.89)(\pi)}\sqrt{\frac{36}{29000}} = 0.641$$

6.1.1.1 Nomograph-Based Effective Length Factor K_{K_n} (Section 2.4.1)

This method assumes the story has a capacity equal to the sum of the Euler buckling loads of each column, computed using the effective length factor based upon the sidesway uninhibited nomograph. Based upon this procedure, and noting that the columns in this frame, both leaning and non-leaning, have the same length, a buckling load, P_{eK_n} (Eq. (2.25a)), may be computed for the non-leaning column in the story, subject to the limit of Eq. (2.73):

$$P_{e\hat{\tau}(K_n)} = \frac{P_u}{\sum_{all} P_u} \sum_{non-leaner} \frac{\pi^2 \hat{\tau} EI}{(K_n L)^2} \le 1.6 \frac{\pi^2 \hat{\tau} EI}{(K_n L)^2} \qquad (2.25a)$$

This buckling load would then be used directly in Eq. (2.2) to compute λ_c.

However, an effective length factor may also be computed using this approach. This is also used here to permit direct comparison with the computations based on the direct use of a nomograph effective length factor. Therefore, Eq. (2.25b) may be used to calculate a K factor for the center column, subject to the limit of Eq. (2.74):

$$K_{K_n} = \sqrt{\frac{\hat{\tau}I}{P_u} \frac{\sum_{all} P_u}{\sum_{non-leaner} \frac{\hat{\tau}I}{K_n^2}}} \geq \sqrt{\frac{5}{8}} K_n \tag{2.25b}$$

The summation of the axial force in the center column and the two leaning columns is:

$$\sum_{all} P_u = 100.6 + 62.6 = 163.2 \text{ kips}$$

In this example the far ends of the girders framing into the center column are pinned rather than rigid. Therefore, as discussed in LRFD Commentary Chapter C [AISC, 1993] and Section 2.3.1, the girder stiffnesses are modified by adjusting their lengths:

$$L_g' = L_g \left[2 - \frac{M_F}{M_N} \right] \tag{2.11a}$$

where $M_F = 0$ for both girders. Therefore, $L_g' = 2 \cdot L_g = 120'$

$$L_x = L_c = 18 \text{ feet}$$

$$G_A = \frac{\sum \frac{I_c}{L_c}}{\sum \frac{I_g}{L_g'}} = \frac{\frac{428}{18}}{2\frac{5900}{120}} = 0.242$$

$$G_B = \infty \text{ (pinned base)}$$

From the nomograph, $K_n = 2.08$ for strong axis buckling. Therefore, with $I_x = 428 \text{ in}^4$:

$$\sum_{non-leaner} \frac{\hat{\tau}I}{K_n^2} = \frac{(1.0)(428)}{\left(2.08^2\right)} = 98.9$$

Limits: $\sqrt{\frac{5}{8}} K_n = \sqrt{\frac{5}{8}}(2.08) = 1.64$ or $1.6P_e = 1.6\frac{\pi^2 \hat{\tau}EI}{(K_n L)^2} = 606.9 \text{ kips}$

Therefore, the story-based effective length factor equals:

$$K_{K_n} = \sqrt{\frac{(1.0)(428)}{(100.6)}\frac{(163.2)}{(98.9)}} = 2.65 > 1.64$$

or, alternately,

$$P_{e\hat{\tau}(K_n)} = \frac{P_u}{\sum_{all} P_u} \sum_{non-leaner} \frac{\pi^2 \hat{\tau} EI}{(K_n L_x)^2} = \frac{100.6}{163.2} \frac{\pi^2 (1.0)(29000)(428)}{[(2.08)(216)]^2} = 374.1 < 606.9$$

$$\lambda_{cx} = \frac{K_{K_n} L_x}{r_x \pi}\sqrt{\frac{Fy}{E}} = \frac{(2.65)(216)}{(5.82)(\pi)}\sqrt{\frac{36}{29000}} = \sqrt{\frac{P_y}{P_{e(K_n)}}} = \sqrt{\frac{453.6}{374.1}} = 1.10$$

$$\lambda_{cy} = 0.641$$

Strong axis buckling governs:

$$P_n = 0.658^{\lambda_{cx}^2} P_y = (0.658^{1.10^2})(453.6) = 272.6 \text{ kips}$$

This axial strength is 50 kips lower than that predicted (but not shown here) by the nomograph K factor value of 2.08. The interaction equation yields:

$$\frac{P_u}{\phi_c P_n} + \frac{8}{9}\frac{M_{ux}}{\phi_b M_{nx}} = \frac{100.6}{(0.85)(272.6)} + \frac{8}{9}\frac{1470}{(0.90)(2506)} = 0.434 + 0.579 = 1.01 \qquad \text{OK}$$

This story-based method captures the destabilizing effects of the leaned columns, showing the center column to be just adequate. Using the nomograph K factor (2.08) yields an interaction equation value of 0.938, since it does not account for the destabilizing effects of the loading on the two leaning columns.

6.1.1.2 Practical Story-Based Effective Length Factor K_{R_L} (Section 2.4.2)

This method assumes the story has a buckling load equal to its first order sidesway resistance multiplied by a reduction factor which accounts for second-order effects. Each column's buckling load is then given by:

$$P_{e\hat{\tau}(R_L)} = \frac{P_u}{\sum_{all} P_u} \frac{\sum_{non-leaner} HL}{\Delta_{oh}} (0.85 + 0.15 R_L) \qquad (2.28a)$$

where the column's buckling load is limited to the value derived in Section 2.4.4.1:

$$P_{eR_L} \leq 1.7 \frac{HL}{\Delta_{oh}} \tag{2.67}$$

Alternately, an effective length factor K_{R_L} of the center column may be calculated according to Eq. (2.28b), subject to the limiting value of Eq. (2.68):

$$K_{R_L} = \sqrt{\frac{\hat{\tau} I_x}{P_u} \frac{\pi^2 E}{L^2} \frac{\sum\limits_{all} P_u}{\frac{\sum\limits_{non-leaner} HL}{\Delta_{oh}}(0.85 + 0.15 R_L)}} \geq \sqrt{\frac{\pi^2}{1.7} \frac{\hat{\tau} EI}{L^2} \frac{\Delta_{oh}}{HL}} \tag{2.28b}$$

The following values are used in this example:

$$\sum_{leaner} P_u = 62.6 \text{ kips}$$

$$\sum_{all} P_u = 100.6 + 62.6 = 163.2 \text{ kips}$$

$$R_L = \frac{\sum\limits_{leaner} P_u}{\sum\limits_{all} P_u} = \frac{62.6}{163.2} = 0.384$$

As discussed in Sections 2.4.2 and 2.4.5, the term $\sum\limits_{non-leaning} HL \Big/ \Delta_{oh}$ is the first order sidesway resistance of the story, which may be obtained from a first-order elastic analysis in which the frame is subjected only to lateral forces. As described in Sections 2.3.1 and 2.4.5, for this analysis the total gravity load at each story is applied as the lateral load. Because the analysis is first order linear elastic, the magnitude of the lateral loads is arbitrary; rather, the ratio of column shears to story deflection is required. The results of this analysis are given below:

Shear in Center Column = 163.2 kips
Average Lateral Deflection =50.3 inches

Therefore:

$$\frac{\sum\limits_{non-leaner} HL}{\Delta_{oh}} = \frac{(163.2)(216)}{50.3} = 700.8 \text{ kips}$$

Limits: $K_{R_L} \geq \sqrt{\frac{\pi^2}{1.7}\frac{\hat{\tau} EI_x}{L^2}\frac{\Delta_{oh}}{HL}} = \sqrt{\frac{\pi^2}{1.7}\frac{(1.0)(29000)(428)}{216^2}\frac{1}{700.8}} = 1.48$ or

$$P_{e\hat{\tau}(R_L)} < 1.7\frac{HL}{\Delta_{oh}} = 1.7\frac{163.2(216)}{50.3} = 1191 \text{ kips}$$

Therefore:

$$K_{R_L} = \sqrt{\frac{(1.0)(428)}{(100.6)}\frac{\pi^2(29000)}{(216)^2}\frac{(163.2)}{(700.8)[0.85+(0.15)(0.384)]}} = 2.59 > 1.48$$

$$P_{e\hat{\tau}(R_L)} = \frac{Pu}{\sum\limits_{all} Pu}\frac{\sum\limits_{non-leaner} HL}{\Delta_{oh}}(0.85+0.15R_L) \text{ kips}$$

$$= \frac{100.6}{163.2}(700.8)[0.85+(0.15)(0.384)] = 392.1 < 1191 \text{ kips}$$

$$\lambda_{cx} = \frac{K_{R_L} L_x}{r_x \pi}\sqrt{\frac{Fy}{E}} = \frac{(2.59)(216)}{(5.82)(\pi)}\sqrt{\frac{36}{29000}} = \sqrt{\frac{P_y}{P_{e\hat{\tau}(R_L)}}} = \sqrt{\frac{453.6}{392.1}} = 1.08$$

$$\lambda_{cy} = 0.641$$

Strong axis buckling governs. Therefore:

$$P_n = 0.658^{\lambda_{cx}^2} P_y = (0.658^{1.08^2})(453.6) = 278.3 \text{ kips}$$

The axial strength is comparable to that computed using K_{K_n}. The interaction equation yields:

$$\frac{P_u}{\phi_c P_n} + \frac{8}{9}\frac{M_{ux}}{\phi_b M_{nx}} = \frac{100.6}{(0.85)(278.3)} + \frac{8}{9}\frac{1470}{(0.90)(2506)} = 0.425 + 0.579 = 1.00 \quad \text{No good}$$

The approach, which here accounts for the P-δ conservatively, gives nearly identical results to using K_{K_n}, thus indicating this column is subjected to a relatively high P-δ effect.

6.1.1.3 Notional Load (Chapter 4)

In the notional load approach, the sum across each story of the axial forces in the columns due to gravity loads on the frame are multiplied by the notional load coefficient, ξ, and the product is applied as a lateral load at the floor level, acting in the same direction as any primary applied lateral load, for all load combinations:

$$\xi = 0.005$$

$$H_{not} = \xi \sum_{all} P_u = (0.005)(163.2) = 0.816 \text{ kips}$$

For this example, a second-order elastic analysis [White and Hajjar, 1991], including application of the notional load applied, gives the following forces in the center column:

Factored Axial Load = $P_n = 100.6$ kips (unchanged due to notional load)
Maximum Factored Strong Axis Moment = $M_{ux} = 1706$ kip-inches

Here, the notional load has increased the maximum bending moment in the center column by 16%, while the axial load remains unchanged (as compared to forces listed in Section 6.1.1). In this approach, an effective length factor of one may be used. Therefore:

$$\lambda_{cx} = \frac{(1)L_x}{r_x \pi}\sqrt{\frac{F_y}{E}} = \frac{216}{(5.82)(\pi)}\sqrt{\frac{36}{29000}} = 0.416$$

$$\lambda_{cy} = 0.641$$

$$P_{nx} = 0.658^{\lambda_{cx}^2} P_y = (0.658^{0.416^2})(453.6) = 421.9 \text{ kips}$$

$$P_{ny} = 0.658^{\lambda_{cx}^2} P_y = (0.658^{0.641^2})(453.6) = 381.9 \text{ kips}$$

Since $P_{nx} > P_{ny}$, both an in-plane and an out-of-plane check must be performed (see Chapter 4). For the in-plane check, the required bending moment and axial force include the effects of the notional load:

$$\frac{P_u}{\phi_c P_{nx}} + \frac{8}{9}\frac{M_{ux}}{\phi_b M_{nx}} = \frac{100.6}{(0.85)(421.9)} + \frac{8}{9}\frac{1706}{(0.90)(2506)} = 0.281 + 0.672 = 0.953 \quad \text{OK}$$

For the out-of-plane check, the required bending moment and axial force need not include the notional load (see Chapter 4):

$$\frac{P_u}{\phi_c P_{nx}} + \frac{8}{9}\frac{M_{ux}}{\phi_b M_{nx}} = \frac{100.6}{(0.85)(381.9)} + \frac{8}{9}\frac{1470}{(0.90)(2506)} = 0.310 + 0.579 = 0.889$$

As with the previous two approaches, strong axis buckling is seen to control. However, here, the column is clearly shown to be acceptable. This is because the notional load approach is calibrated to capture the effects of inelastic flexural buckling. The two effective length procedures, using an iterative process to compute $\hat{\tau}$, would have yielded lower interaction equation values (i.e., on the order of that computed using the notional load approach). Using the non-iterative values for $\hat{\tau}$ yields a more conservative design using the effective length procedures in the previous two sections.

6.1.1.4 Summary of Results

The interaction equation values for the center column are summarized in the following table:

Table 6.1 Comparison of Interaction Equation Values

Center Column	**Interaction Equation Value using K_{K_n}**	**Interaction Equation Value using K_{R_L}**	**Interaction Equation Value using Notional Load**
W14x43	1.01	1.00	0.953

To investigate the use of an effective length factor equal to one and no notional load, the limit on the B_2 factor proposed in Chapter 5 is investigated. The B_2 factor for this frame, according to formula C2-4 of AISC LRFD [1993] is:

$$B_2 = \frac{1}{1 - \dfrac{\sum_{all} P_u}{\sum_{non-leaner} HL/\Delta_{oh}}} = \frac{1}{1 - \dfrac{163.2}{(163.2)(216)/50.32}} = 1.30$$

This breaches the limit of 1.11 proposed in Chapter 5, indicating that the stability of the non-leaning column in this frame must be assessed during design.

6.2 Unsymmetrical Two-Dimensional, Two-Bay, Two-Story Unbraced Frame

This example analyzes the two-dimensional unsymmetrical frame of Figure 6.2 using the methods described in Chapters 2 and 4. This frame is a modified version of one investigated by Ziemian [1990]. The following assumptions are made in the analysis of this frame:

- All members are assumed to be compact and therefore not susceptible to local buckling.
- Strong axis bending of all members is assumed.
- Columns are braced at the base supports and at their connections to beams; the first story columns are also braced out-of-plane at their midpoints. $K = 1$ for weak axis buckling.
- The frames are spaced 25' on center, with the uniform load on the girders varying across the bays and stories as shown in Figure 6.2.
- All beam-to-column connections are fully restrained.
- For economy, the same column size is used for the full height of the structure.
- $Fy = 36$ ksi.
- $E = 29000$ ksi.

The gravity load combination, $1.2Dead + 1.6Live$ (corresponding to AISC LRFD Eq. 4-2 [AISC, 1993]), governs the design of this frame. The service loads are given in Fig. 6.2.

This example serves two purposes. First, it shows how to use the approaches of Chapter 2 and 4 for assessing stability for multi-story frames. Second, this practical, unsymmetric plane frame offers an excellent example of the strengths and weaknesses of the methods discussed in this report.

6.2.1 Computation of Strength of the Columns in the Frame

The strengths of the columns in this example are determined using the following approaches:

- Nomograph-Based Effective Length Factor, K_{K_n} (Section 2.4.1)
- Practical Story-Based Effective Length Factor, K_{R_L} (Section 2.4.2)
- Notional Load (Chapter 4)

For each effective length procedure, both the elastic and inelastic K factors are calculated. Inelasticity is assumed to be confined to the columns.

Comprehensive second-order elastic frame analysis [White and Hajjar, 1991], which accounts for both P-Δ and P-δ effects, was used to obtain the required forces for design. The required forces, as well as the nominal moment capacity according to the LRFD Specification [AISC, 1993] are summarized below for each of the six columns in the frame:

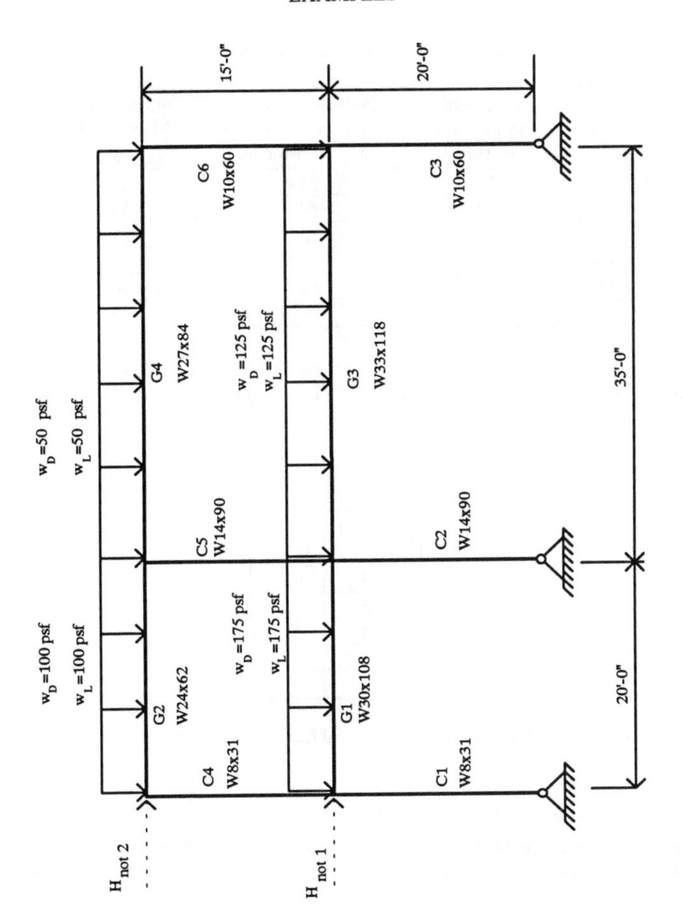

Figure 6.2 Two-Dimensional, Two-Story, Two-Bay Frame

Table 6.2 Results of Second-Order Elastic Analysis for Two-Story Frame

Column	Required Axial Force P_u (kips)	Maximum Required Bending Moment M_{ux} (kip-inches)	Nominal Bending Moment M_{nx} (kip-inches)
C1	139	51.4	1094
C2	491	659	5652
C3	184	689	2686
C4	54.7	285	986.7
C5	155	1470	5652
C6	52.7	1610	2573

The inelastic stiffness reduction factor, $\hat{\tau}$, is computed for each column from Eq. (2.7), using a value of P_u/ϕ_c to permit this computation to be non-iterative. The value of P_u is obtained from the table above. These values of $\hat{\tau}$ are:

Table 6.3 Approximate Inelastic Stiffness Reduction Factor $\hat{\tau}$

Column	$\hat{\tau}$
C1	0.946
C2	0.828
C3	1.0
C4	1.0
C5	1.0
C6	1.0

Only columns C1 and C2 have a compressive axial force sufficient to cause a reduction in stiffness.

For comparison, and for use in the computation of $P_{e\hat{\tau}(K_n)}$ and K_{K_n}, the strong axis elastic and inelastic sidesway uninhibited nomograph effective length factors for the six columns are presented below, along with the value of the interaction equation for each column resulting from the use of these effective length factors to compute P_n. For the calculation of these K factors, the girder lengths are adjusted to account for the location of the girder inflection point, as discussed in LRFD Commentary Chapter C [AISC, 1993] and Section 2.3.1. The bending moments used in this modification are obtained from a first-order linear elastic lateral load analysis of the frame, for which the total gravity load at each story is applied as the lateral load. These bending moments should *not* be obtained from the second-order elastic analysis used to determine required member forces. In this example, because of this modification, some of the resulting corrected G factors are negative (see Section 2.3.1 for a discussion of negative G factors). In these cases, Eq. (2.9) is used to calculate the effective length, since the nomograph does not address negative G values. For column C4, this results in a sidesway uninhibited nomograph effective length factor that is less than 1.0. A G factor of infinity is used at the base supports. These results are tabulated below:

Table 6.4 Nomograph K Factors and Column Interaction Values

a. Elastic Case

Column	K_{nx}	Interaction Equation Value
C1	1.975	1.379
C2	2.144	0.991
C3	2.054	0.916
C4	0.993	0.583
C5	1.157	0.474
C6	1.067	0.759

b. Inelastic Case

Column	K_{nx}	Interaction Equation Value
C1	1.976	1.380
C2	2.134	0.988
C3	2.054	0.916
C4	0.993	0.583
C5	1.152	0.474
C6	1.067	0.759

It may be observed that the calculated K factors and the interaction equations do not vary significantly between the elastic and inelastic cases. This is due in part to the conservative manner with which the inelastic stiffness reduction factor is calculated. While calculating $\hat{\tau}$ in this fashion is much simpler, it fails to capture the full extent of inelastic behavior in the columns.

The value of nominal axial strength, P_n, for use in the LRFD interaction equations (Eq. (1.1)) is computed in the following sections. The calculations for columns C1 and C4 are shown as representative of those of the entire frame.

6.2.1.1 Nomograph-Based Effective Length Factor, K_{K_n} (Section 2.4.1) - Elastic Case

This method assumes the story has a capacity equal to the sum of the Euler buckling loads of each column, computed using the effective length factor based upon the sidesway uninhibited nomograph. Based upon this procedure and noting that all columns in each story have the same length, a buckling load, $P_{e\hat{\tau}(K_n)}$ (Eq. (2.25a)), may be computed for each non-leaning column in the story. This buckling load could then be used directly in Eq. (2.2) to compute λ_c. However, an effective length factor may also be computed using this approach. This is done here to permit direct comparison with the computations of the other approaches, including the direct use of a nomograph effective length factor.

Equation (2.25b) may be used to calculate a story-based effective length factor for each column, subject to the limit of Eq. (2.74):

$$K_{K_n} = \sqrt{\frac{\hat{\tau}I}{P_u} \frac{\sum_{all} P_u}{\sum_{non-leaner} \frac{\hat{\tau}I}{K_n^2}}} \geq \sqrt{\frac{5}{8}} K_n \quad (2.25b)$$

For the case of computing elastic effective length factors, $\hat{\tau}$ is taken as 1.0 for all columns. Also, in this example, all columns contribute to the lateral resistance of the frame. Therefore, both summation terms in the above equation apply to all columns in the story considered.

The elastic nomograph K factors from Table 6.4 and the *in-plane* moments of inertia of the columns are listed below:

Table 6.5 Elastic Nomograph Effective Length Factors and Moments of Inertia of Two-Story Frame

Column	K_n	I (in^4)
C1	1.975	110
C2	2.144	999
C3	2.054	341
C4	0.993	110
C5	1.157	999
C6	1.067	341

The computations then proceed as follows:

Column C1

$$\sum_{all} P_u = 139 + 491 + 184 = 814 \text{ kips}$$

$$\sum_{non-leaner} \frac{\hat{\tau}I}{K_n^2} = \frac{(1.0)110}{1.975^2} + \frac{(1.0)999}{2.144^2} + \frac{(1.0)341}{2.054^2} = 326.4 \text{ in}^4$$

Limit: $\sqrt{\frac{5}{8}}K_n = \sqrt{\frac{5}{8}}1.975 = 1.561$

$$K_{K_n x} = \sqrt{\frac{\hat{\tau}I}{P_u}\frac{\sum_{all} P_u}{\sum_{non-leaner}\frac{\hat{\tau}I}{K_n^2}}} = \sqrt{\frac{(1.0)110}{139}\frac{814}{326.4}} = 1.404 \quad < \quad 1.561$$

Therefore, the limiting value governs: $K_{K_n x} = 1.561$, and:

$$\lambda_{cx} = \frac{K_{K_n x} L_x}{r_x \pi}\sqrt{\frac{F_y}{E}} = \frac{(1.561)(240)}{(3.47)\pi}\sqrt{\frac{36}{29000}} = 1.211$$

$$\lambda_{cy} = \frac{K_y L_y}{r_y \pi}\sqrt{\frac{F_y}{E}} = \frac{(1.0)(240/2)}{2.02\pi}\sqrt{\frac{36}{29000}} = 0.666$$

In-plane behavior controls flexural buckling. Note that using the elastic, corrected, sidesway uninhibited nomograph effective length factor (1.975) shows this column to buckle elastically ($\lambda_c > 1.5$), whereas this story-based approach indicates inelastic buckling behavior.

$$P_n = 0.658^{\lambda_{cx}^2} P_y = 0.658^{(1.211)^2}(9.13)(36) = 177.9 \text{ kips}$$

$$\frac{P_u}{\phi_c P_n} + \frac{8}{9}\frac{M_{ux}}{\phi_b M_{nx}} = \frac{139}{(0.85)(177.9)} + \frac{8}{9}\frac{51.4}{(0.90)(1094)} = 0.966 \qquad \text{OK}$$

While the nomograph methods show this column to be more than 30% overstressed, this method shows that the column is acceptable. This is due to the story-based approaches accounting for the stronger columns in the story providing support to this weak column.

Column C4

$$\sum_{all} P_u = 54.7 + 155 + 52.7 = 262 \text{ kips}$$

$$\sum_{non-leaner}\frac{\hat{\tau}I}{K_n^2} = \frac{(1.0)110}{0.993^2} + \frac{(1.0)999}{1.157^2} + \frac{(1.0)341}{1.067^2} = 1157 \text{ in}^4$$

Limit: $\sqrt{\frac{5}{8}}K_n = \sqrt{\frac{5}{8}}0.993 = 0.785$

$$K_{K_n x} = \sqrt{\frac{\hat{\tau} I}{P_u} \frac{\sum\limits_{all} P_u}{\sum\limits_{non-leaner} \frac{\hat{\tau} I}{K_n^2}}} = \sqrt{\frac{(1.0)110}{54.7}\frac{262}{1157}} = 0.675 \; < \; 0.785$$

The limit governs; $K_{K_n x} = 0.785$

$$\lambda_{cx} = \frac{K_{K_n x} L_x}{r_x \pi} \sqrt{\frac{F_y}{E}} = 0.457$$

$$\lambda_{cy} = \frac{K_y L_y}{r_y \pi} \sqrt{\frac{F_y}{E}} = 0.999$$

Out-of-plane behavior controls:

$$P_n = 0.658^{\lambda_{cy}^2} P_y = 0.658^{(0.999)^2}(9.13)(36) = 216.4 \text{ kips}$$

$$\frac{P_u}{\phi_c P_n} + \frac{8}{9} \frac{M_{ux}}{\phi_b M_{nx}} = \frac{54.7}{(0.85)(216.4)} + \frac{8}{9} \frac{285}{(0.90)(986.7)} = 0.583 \qquad \text{OK}$$

The results for the other four columns are summarized below. Note that using the nomograph effective length factor directly (Table 6.4a) shows column C2 to be understressed, while the story-based approach shows it to be slightly overstressed. This is because this stronger column must support the weaker columns in the story prior to incipient buckling. The nomograph K factor cannot account fully for this effect.

Table 6.6 Summary of Results using K_{K_n} -- Elastic Case

Column	K_{K_n}	Interaction Equation Value
C1	1.561	0.966
C2	2.253	1.026
C3	2.151	0.960
C4	0.785	0.583
C5	1.209	0.474
C6	1.211	0.759

6.2.2.2 Nomograph-Based Effective Length Factor, K_{K_n} (Section 2.4.1) - Inelastic Case

To compute inelastic values of K_{K_n} using Eq. (2.25), the story buckling capacity is based upon the use of the inelastic nomograph effective length factor (Table 6.4a), and an inelastic stiffness reduction factor, $\hat{\tau}$ (from Table 6.3).

<u>Column C1</u>

$$\sum_{all} P_u = 139 + 491 + 184 = 814 \text{ kips}$$

$$\sum_{non-leaner} \frac{\hat{\tau} I}{K_n^2} = \frac{(0.946)(110)}{1.976^2} + \frac{(0.828)(999)}{2.134^2} + \frac{(1.0)(341)}{2.054^2} = 289 \text{ in}^4$$

Limit: $\sqrt{\frac{5}{8}} K_n = \sqrt{\frac{5}{8}} 1.976 = 1.562$

$$K_{K_n x} = \sqrt{\frac{\hat{\tau} I}{P_u} \frac{\sum_{all} P_u}{\sum_{non-leaner} \frac{\hat{\tau} I}{K_n^2}}} = \sqrt{\frac{(0.946)(110)}{139} \frac{814}{289}} = 1.451 \quad < \quad 1.562$$

Therefore, the limiting value governs: $K_{K_n x} = 1.562$

$$\lambda_{cx} = \frac{K_{K_n x} L_x}{r_x \pi} \sqrt{\frac{F_y}{E}} = 1.212$$

$$\lambda_{cy} = \frac{K_y L_y}{r_y \pi} \sqrt{\frac{F_y}{E}} = 0.666$$

In-plane behavior controls.

$$P_n = 0.658^{\lambda_{cx}^2} P_y = 0.658^{(1.214)^2} (9.13)(36) = 177.8 \text{ kips}$$

$$\frac{P_u}{\phi_c P_n} + \frac{8}{9} \frac{M_{ux}}{\phi_b M_{nx}} = \frac{139}{(0.85)(177.8)} + \frac{8}{9} \frac{51.4}{(0.90)(1094)} = 0.967 \qquad \text{OK}$$

Because the limiting value of K_{K_n} governs for this column, the inclusion of the inelastic stiffness reduction in the computation has little effect as compared to the elastic case.

Column C4

$$\sum_{all} P_u = 54.7 + 155 + 52.7 = 262 \text{ kips}$$

$$\sum_{non-leaner} \frac{\hat{\tau} I}{K_n^2} = \frac{(1.0)(110)}{0.993^2} + \frac{(1.0)(999)}{1.152^2} + \frac{(1.0)(341)}{1.067^2} = 1164 \text{ in}^4$$

$$\text{Limit: } \sqrt{\frac{5}{8}} K_n = \sqrt{\frac{5}{8}} 0.993 = 0.785$$

$$K_{K_n x} = \sqrt{\frac{\hat{\tau} I}{P_u} \frac{\sum_{all} P_u}{\sum_{non-leaner} \frac{\hat{\tau} I}{K_n^2}}} = \sqrt{\frac{(1.0)(1101)}{54.7} \frac{263}{1164}} = 0.673 < 0.785$$

The limit governs: $K_{K_n x} = 0.785$

$$\lambda_{cx} = \frac{K_{K_n x} L_x}{r_x \pi} \sqrt{\frac{F_y}{E}} = 0.457$$

$$\lambda_{cy} = \frac{K_y L_y}{r_y \pi} \sqrt{\frac{F_y}{E}} = 0.999$$

Out-of-plane behavior controls. The nominal axial strength and interaction equation for this column is:

$$P_n = 0.658^{\lambda_{cy}^2} P_y = 0.658^{(0.999)^2}(9.13)(36) = 216.4 \text{ kips}$$

$$\frac{P_u}{\phi_c P_n} + \frac{8}{9} \frac{M_{ux}}{\phi_b M_{nx}} = \frac{54.7}{(0.85)(216.4)} + \frac{8}{9} \frac{285}{(0.90)(986.7)} = 0.583 \qquad \text{OK}$$

The results for the other four columns are summarized below:

Table 6.7 Summary of Results using K_{K_n} -- Inelastic Case

Column	K_{K_n}	Interaction Equation Value
C1	1.562	0.967
C2	2.178	1.002
C3	2.286	1.030
C4	0.785	0.583
C5	1.205	0.474
C6	1.208	0.759

The effects of inelastic stiffness are most apparent here for column C3, which has an interaction equation value less than one when elastic effective length factors are used, but is slightly overstressed when inelastic effective length factors are used. Since the other columns in the story are weakened due to inelastic stiffness reduction, column C3 becomes stronger relative to columns C1 and C2, and its effective length factor increases. This effect is discussed in Section 2.2.

6.2.2.3 Practical Story-Based Effective Length Factor K_{R_L} (Section 2.4.2) - Elastic Case

This method assumes the story has a capacity equal to its first order sidesway resistance multiplied by a reduction factor which accounts for second-order effects. Based upon this procedure, a buckling load, $P_{e\hat{\tau}(R_L)}$ (Eq. (2.28a)), may be computed for each non-leaning column in the story. This buckling load may then be used directly in Eq. (2.2) to compute λ_c. However, an effective length factor may also be computed using this approach. As with the calculations presented for K_{K_n}, the effective length factors are computed here so as to permit direct comparison with the results of the other approaches, including the direct use of a nomograph effective length factor.

The columns within each of this frame's stories have the same length. Therefore, Eq. (2.28b) may be used to calculate a K factor for each column, subject to the limit of Eq. (2.68):

$$K_{R_L} = \sqrt{\frac{\hat{\tau}I}{P_u}\frac{\pi^2 E}{L^2}\frac{\sum\limits_{all} P_u}{\dfrac{\sum\limits_{non-leaner} HL}{\Delta_{oh}}(0.85+0.15R_L)}} \geq \sqrt{\frac{\pi^2}{1.7}\frac{\hat{\tau}EI}{L^2}\frac{\Delta_{oh}}{HL}} \qquad (2.28b)$$

where

$$R_L = \frac{\sum_{leaner} P_u}{\sum_{all} P_u} = 0$$

since there are no leaning columns in this frame. The results of the analysis to compute $\sum_{non-leaning} HL \Big/ \Delta_{oh}$ follow:

First Story[1]:

Column Shears: $H_1 = 0.676$ kips
$H_2 = 5.48$ kips
$H_3 = 1.98$ kips

Average Lateral Deflection: $\Delta_{oh} = 0.960$ inches

$$\frac{\sum_{non-leaning} HL}{\Delta_{oh}} = \frac{240(0.676 + 5.48 + 1.98)}{0.960} = 2033 \text{ kips}$$

Note that in this story, all shears act in the same direction due to the loading (as would be customary), so the sign of all of the shears is positive.

Second Story:

Column Shears: $H_4 = 0.437$ kips
$H_5 = 1.30$ kips
$H_6 = 0.88$ kips

Average Lateral Deflection: $\Delta_{oh} = 0.063$ inches

$$\frac{\sum_{non-leaner} HL}{\Delta_{oh}} = \frac{180(0.437 + 1.30 + 0.88)}{0.063} = 7450 \text{ kips}$$

With little loss of accuracy, in this example the interstory drift of each column is taken as equal to the average story lateral deflection.

For the elastic case, $\hat{\tau}$ is taken as 1.0 for all columns.

---

[1]The subscripts 1, 2, and 3 refer to columns C1, C2, and C3 respectively.

Column C1

$$\sum_{all} P_u = 139 + 491 + 184 = 814 \text{ kips}$$

$$\text{Limit: } \sqrt{\frac{\pi^2}{1.7} \frac{(1.0)(29000)(110)}{(240)^2} \frac{0.96}{(0.676)(240)}} = 1.380$$

$$K_{R_L x} = \sqrt{\frac{(1.0)110}{139} \frac{\pi^2 (29000)}{(240)^2} \frac{(814)}{(2033)(0.85)}} = 1.362 \quad < \quad 1.380$$

Therefore, the limiting value governs: $K_{R_L x} = 1.380$

$$\lambda_{cx} = \frac{K_{R_L x} L_x}{r_x \pi} \sqrt{\frac{F_y}{E}} = 1.070$$

$$\lambda_{cy} = \frac{K_y L_y}{r_y \pi} \sqrt{\frac{F_y}{E}} = 0.666$$

In-plane behavior controls.

$$P_n = 0.658^{\lambda_c^2} P_y = 0.658^{(1.070)^2} (9.13)(36) = 203.5 \text{ kips}$$

$$\frac{P_u}{\phi_c P_n} + \frac{8}{9} \frac{M_{ux}}{\phi_b M_{nx}} = \frac{139}{(0.85)(203.5)} + \frac{8}{9} \frac{51.4}{(0.90)(1094)} = 0.851 \qquad \text{OK}$$

As was also found using K_{K_n}, this story-based approach shows that the column is acceptable, while the nomograph methods show this column to be more than 30% overstressed. The interaction equation value for this column is actually lower than that obtained by using K_{K_n}. This is because the limiting values of these effective length factors are invoked in the calculations for column C1. While K_{R_L} is generally more conservative than K_{K_n} (see Section 2.4.2), its limiting value is actually less conservative than that of K_{K_n} (see Section 2.4.4.2).

Column C4

$$\sum_{all} P_u = 54.7 + 155 + 52.7 = 262 \text{ kips}$$

$$\text{Limit: } \sqrt{\frac{\pi^2}{1.7}\frac{(1.0)(29000)(110)}{(180)^2}\frac{0.063}{(0.437)(180)}} = 0.679$$

$$K_{R_L x} = \sqrt{\frac{(1.0)110}{54.7}\frac{\pi^2(29000)}{(180)^2}\frac{(262)}{(7450)(0.85)}} = 0.858 > 0.679$$

Therefore $K_{R_L x} = 0.858$

$$\lambda_{cx} = \frac{K_{R_L x} L_x}{r_x \pi}\sqrt{\frac{F_y}{E}} = 0.499$$

$$\lambda_{cy} = \frac{K_y L_y}{r_y \pi}\sqrt{\frac{F_y}{E}} = 0.999$$

Out-of-plane behavior controls:

$$P_n = 0.658^{\lambda_{cy}^2} P_y = 0.658^{(0.999)^2}(9.13)(36) = 216.4 \text{ kips}$$

$$\frac{P_u}{\phi_c P_n} + \frac{8}{9}\frac{M_{ux}}{\phi_b M_{nx}} = \frac{54.7}{(0.85)(216.4)} + \frac{8}{9}\frac{285}{(0.90)(986.7)} = 0.583 \qquad \text{OK}$$

The results for the other four columns are summarized below:

Table 6.8 Summary of Results using K_{R_L} -- Elastic Case

Column	K_{R_L}	Interaction Equation Value
C1	1.380	0.851
C2	2.182	1.003
C3	2.084	0.929
C4	0.858	0.583
C5	1.536	0.474
C6	1.539	0.759

6.2.2.4 Practical Story-Based Effective Length Factor K_{R_L} (Section 2.4.2) - Inelastic Case

To incorporate the effects of inelasticity into this approach, a new first-order sidesway analysis is performed based upon the in-plane moments of inertia of the columns being reduced by their respective inelastic stiffness reduction factors, $\hat{\tau}$. These factors are tabulated in Table 6.3 for every column in the frame. Results of this new analysis are given below:

First Story:

Column Shears:

$$H_1 = 0.717 \text{ kips}$$
$$H_2 = 5.20 \text{ kips}$$
$$H_3 = 2.22 \text{ kips}$$

Average Lateral Deflection: $\Delta_{oh}^{inelastic} = 1.08$ inches

$$\frac{\sum\limits_{non-leaner} HL}{\Delta_{oh}^{inelastic}} = \frac{240(0.717 + 5.20 + 2.22)}{1.08} = 1806 \text{ kips}$$

Second Story:

Column Shears:

$$H_4 = 0.425 \text{ kips}$$
$$H_5 = 1.39 \text{ kips}$$
$$H_6 = 0.807 \text{ kips}$$

Average Lateral Deflection: $\Delta_{oh}^{inelastic} = 0.063$ inches

$$\frac{\sum\limits_{non-leaner} HL}{\Delta_{oh}^{inelastic}} = \frac{180(0.425 + 1.39 + 0.807)}{0.063} = 7512 \text{ kips}$$

The inelastic effective length factor is then given by:

$$K_{R_L} = \sqrt{\frac{\hat{\tau}I}{P_u} \frac{\pi^2 E}{L^2} \frac{\sum\limits_{all} P_u}{\frac{\sum\limits_{non-leaner} HL}{\Delta_{oh}^{inelastic}}(0.85 + 0.15R_L)}} \geq \sqrt{\frac{\pi^2}{1.7} \frac{\hat{\tau}EI}{L^2} \frac{\Delta_{oh}^{inelastic}}{HL}} \qquad (2.28b)$$

Column C1

$$\sum_{all} P_u = 139 + 491 + 184 = 814 \text{ kips}$$

$$\text{Limit: } \sqrt{\frac{\pi^2}{1.7}\frac{(0.946)(29000)(110)}{(240)^2}\frac{1.08}{(0.717)(240)}} = 1.383$$

$$K_{R_L x} = \sqrt{\frac{(0.946)(110)}{139}\frac{\pi^2(29000)}{(240)^2}\frac{(814)}{(1806)(0.85)}} = 1.404 \quad > \quad 1.383$$

Therefore $K_{R_L x} = 1.404$, and :

$$\lambda_{cx} = \frac{K_{R_L x} L_x}{r_x \pi}\sqrt{\frac{F_y}{E}} = 1.089$$

$$\lambda_{cy} = \frac{K_y L_y}{r_y \pi}\sqrt{\frac{F_y}{E}} = 0.666$$

In-plane behavior controls.

$$P_n = 0.658^{\lambda_c^2} P_y = 0.658^{(1.102)^2}(9.13)(36) = 200.1 \text{ kips}$$

$$\frac{P_u}{\phi_c P_n} + \frac{8}{9}\frac{M_{ux}}{\phi_b M_{nx}} = \frac{139}{(0.85)(200.1)} + \frac{8}{9}\frac{51.4}{(0.90)(1094)} = 0.864 \qquad \text{OK}$$

The weakening of the story as a whole due to the inelasticity causes a rise in the interaction equation value relative to the case of using $\hat{\tau} = 1$.

Column C4

$$\sum_{all} P_u = 54.7 + 155 + 52.7 = 262 \text{ kips}$$

$$\text{Limit: } \sqrt{\frac{\pi^2}{1.7}\frac{(29000)(1.0)(110)}{(180)^2}\frac{0.063}{(0.425)(180)}} = 0.686$$

$$K_{R_L x} = \sqrt{\frac{(1.0)(110)}{54.7}\frac{\pi^2(29000)}{(180)^2}\frac{(262)}{(7512)(0.85)}} = 0.855 \quad > \quad 0.686$$

Therefore $K_{R_L x} = 0.855$ and:

$$\lambda_{cx} = \frac{K_{R_L x} L_x}{r_x \pi} \sqrt{\frac{F_y}{E}} = 0.497$$

$$\lambda_{cy} = \frac{K_y L_y}{r_y \pi} \sqrt{\frac{F_y}{E}} = 0.999$$

Out-of-plane behavior controls:

$$P_n = 0.658^{\lambda_{cy}^2} P_y = 0.658^{(0.999)^2}(9.13)(36) = 216.4 \text{ kips}$$

$$\frac{P_u}{\phi_c P_n} + \frac{8}{9}\frac{M_{ux}}{\phi_b M_{nx}} = \frac{54.7}{(0.85)(216.4)} + \frac{8}{9}\frac{285}{(0.90)(986.7)} = 0.583 \qquad \text{OK}$$

The results for the other four columns are summarized below:

Table 6.9 Summary of Results using K_{R_L} -- Inelastic Case

Column	K_{R_L}	Interaction Equation Value
C1	1.404	0.864
C2	2.107	0.980
C3	2.211	0.990
C4	0.855	0.583
C5	1.529	0.474
C6	1.533	0.759

6.2.2.5 Notional Load (Chapter 4)

As described in Chapter 4, in this approach, a notional lateral load is applied to each floor level and is included in every load combination that is used to check a member's strength. The notional load at any given story is equal to the sum of the applied gravity loads at that story times the notional load coefficient, ξ:

$$H_{not} = \xi \sum_{all} P_u^{applied} \qquad (\xi = 0.005)$$

See Chapter 4 and Section 6.3 for an example of how to compute a more refined value for the notional load for frames consisting of more than one column; here ξ is taken as 0.005 to illustrate the most conservative approach for utilizing a notional load.

For the frame in this example:

$$\sum_{all} P_u^{applied} = 551 \text{ kips (first floor)}$$

$$H_{not\_1} = (0.005)(551) = 2.76 \text{ kips}$$

$$\sum_{all} P_u^{applied} = 262 \text{ kips (second floor)}$$

$$H_{not\_2} = (0.005)(262) = 1.31 \text{ kips}$$

A new, comprehensive second-order elastic analysis of the frame is run with these notional loads applied simultaneously with the governing factored gravity loads. As this structure is unsymmetric, the notional load must be applied separately in each lateral direction. The results of the pair of analyses are given below:

Table 6.10 Results of Second-Order Elastic Analysis with Notional Load

a. Notional Load Acting to the Right

Column	**Axial Force P_u (kips)**	**Maximum Required Bending Moment M_{ux} (kip-inches)**
C1	136	79.1
C2	491	475
C3	186	1095
C4	54.5	261
C5	155	1463
C6	52.4	1655

b. Notional Load Acting to the Left

Column	Axial Force P_u (kips)	Maximum Required Bending Moment M_{ux} (kip-inches)
C1	142	181
C2	490	1794
C3	181	282
C4	55.0	310
C5	155	1482
C6	52.4	1569

One can see large variations in the moments between the two analyses. This is because the moments due to the notional loads tend to offset the effects of the moments caused by the gravity loads on the beams.

Column C1

$$\lambda_{cx} = \frac{L_x}{r_x \pi}\sqrt{\frac{F_y}{E}} = 0.776$$

$$\lambda_{cy} = \frac{L_y}{r_y \pi}\sqrt{\frac{F_y}{E}} = 0.666$$

In-plane behavior controls, even with the use of an effective length factor equal to one, because of the out-of-plane brace at mid-height of the column. Therefore, only an interaction equation for in-plane effects need be checked (see Chapter 4):

$$P_n = 0.658^{\lambda_{cx}^2} P_y = 0.658^{(0.776)^2}(9.13)(36) = 255.5 \text{ kips}$$

Notional load applied to the right:

$$\frac{P_u}{\phi_c P_n} + \frac{8}{9}\frac{M_{ux}}{\phi_b M_{nx}} = \frac{136}{(0.85)(255.5)} + \frac{8}{9}\frac{79.1}{(0.90)(1094)} = 0.698 \qquad \text{OK}$$

Notional load applied to the left:

$$\frac{P_u}{\phi_c P_n} + \frac{8}{9}\frac{M_{ux}}{\phi_b M_{nx}} = \frac{142}{(0.85)(255.5)} + \frac{8}{9}\frac{181}{(0.90)(1094)} = 0.817 \qquad \text{OK}$$

Column C4

$$\lambda_{cx} = \frac{L_x}{r_x\pi}\sqrt{\frac{F_y}{E}} = 0.582$$

$$\lambda_{cy} = \frac{L_y}{r_y\pi}\sqrt{\frac{F_y}{E}} = 0.999$$

Out-of-plane behavior controls. Therefore, two interaction equations must be checked (see Chapter 4), one for in-plane action, and one for out-of-plane action:

In-plane check:

$$P_{nx} = 0.658^{\lambda_{cx}^2} P_y = 0.658^{(0.582)^2}(9.13)(36) = 285.3 \text{ kips}$$

Notional load applied to the right:

$$\frac{P_u}{\phi_c P_{nx}} + \frac{8}{9}\frac{M_{ux}}{\phi_b M_{nx}} = \frac{54.4}{(0.85)(285.3)} + \frac{8}{9}\frac{261}{(0.90)(986.7)} = 0.486 \qquad \text{OK}$$

Notional load applied to the left:

$$\frac{P_u}{\phi_c P_{nx}} + \frac{8}{9}\frac{M_{ux}}{\phi_b M_{nx}} = \frac{55.0}{(0.85)(285.3)} + \frac{8}{9}\frac{310}{(0.90)(986.7)} = 0.537 \qquad \text{OK}$$

As outlined in Chapter 4, for the out-of-plane check, the required bending moment from an analysis which does not include notional load may be used (e.g., obtained from Table 6.2):

$$P_{ny} = 0.658^{\lambda_{cy}^2} P_y = 0.658^{(0.999)^2}(9.13)(36) = 216.4 \text{ kips}$$

$$\frac{P_u}{\phi_c P_{ny}} + \frac{8}{9}\frac{M_{ux}}{\phi_b M_{nx}} = \frac{54.7}{(0.85)(216.4)} + \frac{8}{9}\frac{285}{(0.90)(986.7)} = 0.583 \qquad \text{OK}$$

The out-of-plane interaction controls, but the column remains satisfactory. The results for the other four columns follow.

Table 6.11 Summary of Results using Notional Load

Column	Governing Interaction Equation Value	Governing Interaction Equation Direction	Governing Direction of Notional Load Application
C1	0.817	In-plane	To the left
C2	0.968	In-plane	To the left
C3	0.807	In-plane	To the right
C4	0.583	Out-of-plane	Both are the same
C5	0.474	Out-of-plane	Both are the same
C6	0.768	In-plane	To the right

6.2.2.6 Summary of Results

The following table provides a comparison of the values yielded by the interaction equations for each of the methods outlined in this example.

Table 6.12 Comparison of Interaction Equation Values

Column Designation	Interaction Eq. Value using K_{K_n} - Elastic	Interaction Eq. Value using K_{K_n} - Inelastic	Interaction Eq. Value using K_{R_L} - Elastic	Interaction Eq. Value using K_{R_L} - Inelastic	Interaction Eq. Value using Notional Load
C1	0.966	0.967	0.851	0.864	0.817
C2	1.026	1.002	1.003	0.980	0.968
C3	0.960	1.030	0.929	0.990	0.807
C4	0.583	0.583	0.583	0.583	0.583
C5	0.474	0.474	0.474	0.474	0.474
C6	0.759	0.759	0.759	0.759	0.768

Note that columns C4 and C5 are controlled by out-of-plane flexural buckling; thus, the values of the interaction equation for these columns do not change between methods. Column C6 is controlled by in-plane flexural buckling, but it feels little to no effect from inelastic stiffness reduction.

The notional load approach yields results for column C3 which are lower than the results from the two effective length procedures for that column. This is because non-iterative values for $\hat{\tau}$ have been used in this example. Column C3 is dominated by bending moment, since it frames into a long beam span. Consequently, the column size is large relative to the applied axial force, and using P_n/P_y to compute $\hat{\tau}$ iteratively in Eq. (2.7) would yield an inelastic stiffness reduction that is less than one, thus potentially lowering the resulting value of the interaction equation produced by the effective length procedures, i.e., results closer to that obtained by using the notional load procedure. Recall from Chapter 4 that the notional load approach is calibrated to results from inelastic advanced analyses, which requires the use of a $\hat{\tau}$ factor computed iteratively to provide comparable results for the effective length procedures.

Alternately, the fact that columns C1 and C2, which do experience some inelastic stiffness reduction in the computations shown above, yield more comparable values for the interaction equation using the notional load approach attests to the implicit conservative nature of the notional load calibration.

As a final note, the B_2 factor for the first story of this frame equals:

$$B_2 = \frac{1}{1 - \dfrac{\sum\limits_{all} P_u}{\sum\limits_{non-leaner} \dfrac{HL}{\Delta_{oh}}}} = \frac{1}{1 - \dfrac{814}{2033}} = 1.67$$

The limit on the B_2 factor of 1.11, established in Chapter 5, is clearly breached by this first story, thus confirming the need to assess stability for the columns in this story. Alternately, the second story has a B_2 factor of 1.04 and a value of the story stiffness coefficient (defined in Chapter 5) of:

$$S_L = \frac{N \sum\limits_{non-leaner} P_y}{\sum\limits_{non-leaner} \dfrac{HL}{\Delta_{oh}}} = \frac{(1)(9.13 + 26.5 + 17.6)(36)}{7450} = 0.26$$

where $N = 1$ for this frame (see Chapter 5). The limit on S_L established in Chapter 5 is 2.25. For the second story, both B_2 and S_L are well below their limit. Consequently, even though the effective length factors for columns C5 and C6 are on the order of 1.5, this story is sufficiently stiff, and is so lightly loaded with gravity load, that the columns of the second story could have been designed using effective length factors of 1.0 (without the use of a notional load) without introducing unacceptable errors into the computations.

6.3 Three-Dimensional Unbraced Frame

This example demonstrates how to apply the story-based effective length approaches and the notional load approach to a structure that is supported laterally by rigid frames in both principal, orthogonal directions. The details of the three-story building are shown in Figs. 6.3 to 6.5. The service dead load is 85 psf, the service live load is 100 psf (unreduced), the service live roof load is 80 psf, and the service wind load is 30 psf. In the analysis, the full, factored wind load is applied in one direction, and 30% of the factored wind load is applied in the orthogonal direction. Because the building is unbraced in both principal directions, and because wind is applied from two orthogonal directions, the corner columns in the structure are subjected to biaxial bending plus axial force. The design computations for two of the corner beam-columns will be shown in this section. Preliminary design yields the following member sizes for the structure's beam-columns:

Table 6.13 Column Sizes for 3D Building

Column Designation	Size
A1, A4, D1, D4	W14x61
B1, B4, C1, C4	W14x82
B2, B3, C2, C3	W12x53

Other assumptions made in the design of the frame are listed below:

- All members are assumed to be compact and therefore not susceptible to local buckling.
- Columns are braced at the base supports and at their connections to the beams.
- All connections are either fully rigid or perfectly pinned.
- To simplify construction, all columns extend the full height of the structure.
- $K = 1$ is used for design of the leaning columns.
- Live load reduction and wind uplift are ignored.
- $F_y = 50$ ksi
- $E = 29000$ ksi.

In this example, only column lines 1 and 4 contribute to lateral stiffness in the north-south direction. Column lines 2 and 3 in the north-south direction lean on the unbraced frames of column lines 1 and 4. In the east-west direction, the unbraced end frames along column lines A and D provide the lateral stiffness, while the columns of column lines B and C lean on the unbraced frames. The corner columns contribute to the lateral stability in both directions.

This example outlines the calculations to determine the strength of one of the corner columns of the bottom and middle stories of the structure. Preliminary analysis shows that these corner columns are governed by wind acting in the north-south direction. The governing load combination is $1.2Dead + 1.3Wind + 0.5Live + 0.5Live\_Roof$.

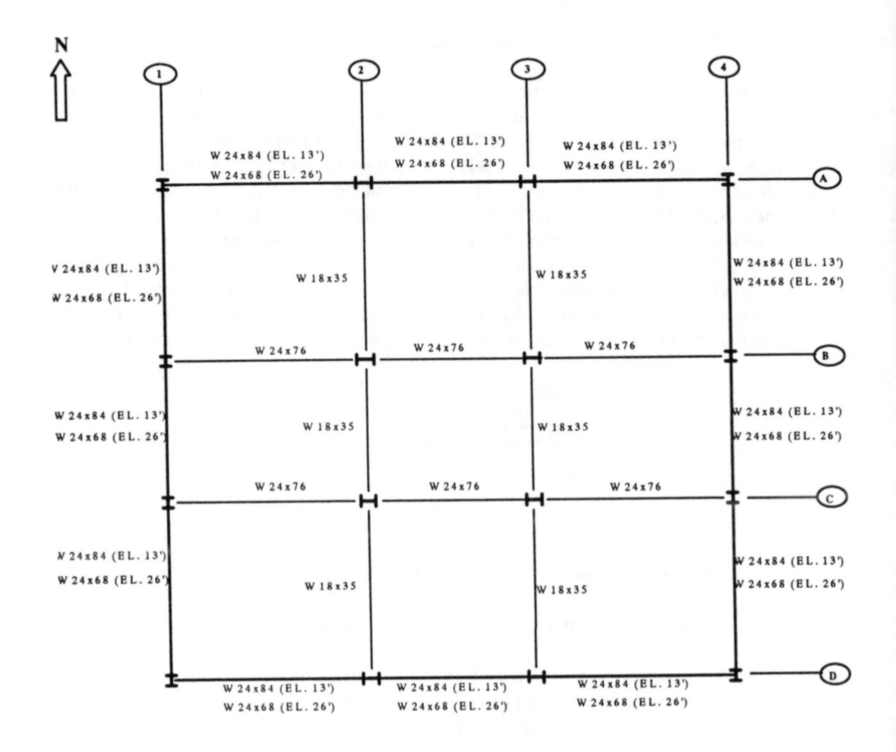

Figure 6.3 Typical Floor Framing Plan for Three-Dimensional Building

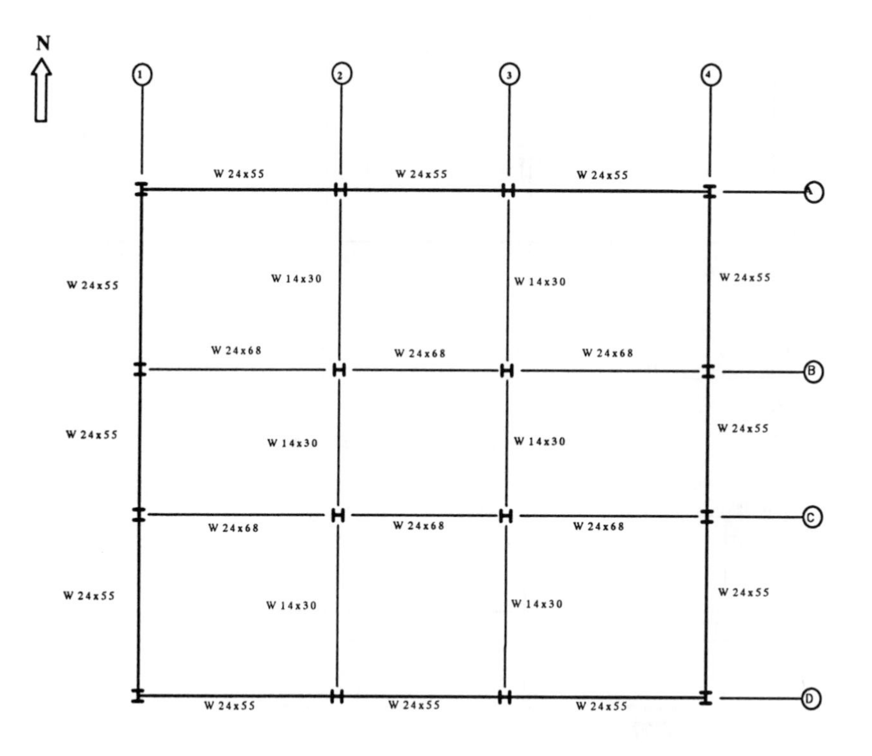

Figure 6.4 Roof Framing Plan for Three-Dimensional Building

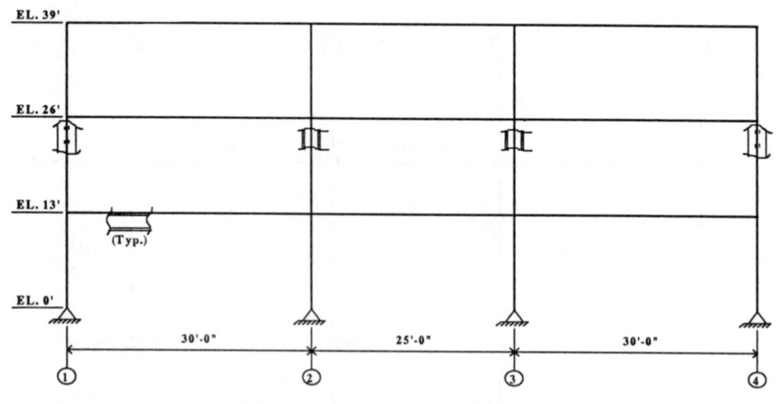

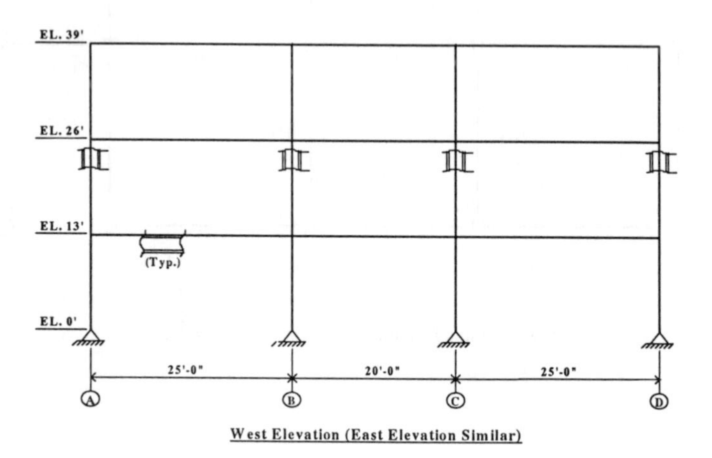

Figure 6.5 Lateral-Resistance System Elevations for Three-Dimensional Building

Therefore, referring to Figure 6.3, by applying 100% of the wind load from the south and 30% of the factored wind load from the west, overturning effects make column A4 the critical corner column.

Comprehensive second-order elastic frame analysis [White and Hajjar, 1991] is used to obtain the required forces for design. Table 6.14 gives results from this analysis.

Table 6.14 Results of Second-Order Elastic Analysis for 3D Building

a. Bottom Story Column Loads

Column	**Axial Force** P_u **(kips)**	**Strong Axis Moment** M_{ux} **(in-kips)**	**Weak Axis Moment** M_{uy} **(in-kips)**
A1	90.7	2410	69.0
A2	160	1400	259
A3	172	945	258
A4	98.1	2410	293
B1	157	3600	83.0
B2	276	237	160
B3	276	237	160
B4	157	3600	83.0
C1	145	3660	83.0
C2	276	236	160
C3	276	237	160
C4	145	3660	84.0
D1	42.4	2250	68.0
D2	160	1390	259
D3	172	949	259
D4	49.9	2250	299

b. Middle Story Column Loads

Column	Axial Force P_u (kips)	Strong Axis Moment M_{ux} (in-kips)	Weak Axis Moment M_{uy} (in-kips)
A1	52.4	902	123
A2	108	872	259
A3	111	193	258
A4	55.0	905	354
B1	102	1410	83.0
B2	182	237	161
B3	182	236	161
B4	102	1420	83.0
C1	96.8	1490	83.0
C2	182	236	161
C3	182	237	161
C4	96.8	1490	83.0
D1	38.2	630	128
D2	108	867	258
D3	111	198	258
D4	40.9	627	359

6.3.1 Computation of Strength of Column A4

The buckling loads of the columns in this example are determined using the following approaches:

- Nomograph-Based Effective Length Factor, K_{K_n} (Section 2.4.1)
- Practical Story-Based Effective Length Factor, K_{R_L} (Section 2.4.2)
- Notional Load (Chapter 4)

Based upon the use of P_u (as per Section 2.2) from Table 6.14, it is found that $\hat{\tau} = 1$ for each column. No inelastic stiffness reductions are taken in this example.

Column A4 has a nominal moment capacity, computed per the AISC LRFD Specification [AISC, 1993], of $\phi_b M_{nx} = 4590$ kip-inches and $\phi_b M_{ny} = 1476$ in-kips.

6.3.1.1 Nomograph-Based Effective Length Factor, K_{K_n} (Section 2.4.1)

In each principal direction, Eq. (2.25b) is used to calculate a K factor, subject to the limit of Eq. (2.74):

$$K_{K_n} = \sqrt{\frac{\hat{\tau} I_x}{P_u} \frac{\sum_{all} P_u}{\sum_{non-leaner} \frac{\hat{\tau} I}{K_n^2}}} \geq \sqrt{\frac{5}{8}} K_{K_n} \tag{2.25b}$$

Alternately, the elastic critical load is given by the following equation, but will not be calculated here for brevity (see Section 6.1 for sample computations using this equation):

$$P_{e\hat{\tau}(K_n)} = \frac{P_u}{\sum_{all} P_u} \sum_{non-leaner} \frac{\pi^2 \hat{\tau} EI}{(K_n L)^2} < 1.6 P_e = 1.6 \frac{\pi^2 \hat{\tau} EI}{(K_n L)^2} \tag{2.25a}$$

The nomograph effective length factor, K_n, is computed using corrected G factors (Eq. (2.11a)). The far and near girder moments (relative to the column under consideration) are based on a first order analysis in which the gravity loads on each story are removed and applied laterally (see Section 2.3). Because stability of the corner columns must be considered in both directions, two separate first-order analyses were performed, and the resulting girder moments are provided in Table 6.15:

Table 6.15 Analysis Results for Determining K_n (Column A4)

Story	**Direction**	**Column End**	M_F **(in-kips)**	M_N **(in-kips)**	L_g' **(ft)**	**Corrected *G* Factor**
Bottom	North-South	Bottom	-----	-----	-----	10
		Top	40480	47130	28.52	0.9854
	East-West	Bottom	-----	-----	-----	10
		Top	37950	20080	3.305	0.0191
Middle	North-South	Bottom	40480	47130	28.52	0.9854
		Top	17110	18600	27.00	1.453
	East-West	Bottom	37950	20080	3.305	0.0191
		Top	18820	11940	12.71	0.1143

Modified girder stiffnesses for the other columns in each story are determined similarly, using end moments obtained from the same analyses. Once these stiffnesses are found, the nomograph is used to find the *K* factor for each column in the story. Table 6.16 provides these effective length factors, as well as the calculations necessary for the determination of the term $\sum_{non-leaner} I/K_n^2$:

Table 6.16 Calculation of $\sum_{non-leaner} I/K_n^2$ for each Column in the Story

Story	Direction	Column Designation	K_n	I	$\frac{I}{K_n^2}$
Bottom	North-South	A1, A4, D1, D4	1.87	640	183.0
		B1, B4, C1, C4	1.78	882	278.4
	East-West	A1, A4, D1, D4	1.66	107	38.83
		A2, A3, D2, D3	1.82	882	266.3
Middle	North-South	A1, A4, D1, D4	1.38	640	336.1
		B1, B4, C1, C4	1.20	882	612.5
	East-West	A1, A4, D1, D4	1.03	107	100.9
		A2, A3, D2, D3	1.30	882	521.9

Note that the moments of inertia for columns A1, A4, D1, and D4 in the east-west direction are for the weak axis for each column. From this table:

Bottom Story, North-South Direction:

$$\sum_{non-leaner} \frac{\hat{\tau} I}{K_n^2} = 4(1.0)(183.0) + 4(1.0)(278.4) = 1846$$

Bottom Story, East-West Direction:

$$\sum_{non-leaner} \frac{\hat{\tau} I}{K_n^2} = 4(1.0)(38.83) + 4(1.0)(266.3) = 1221$$

Middle Story, North-South Direction:

$$\sum_{non-leaner} \frac{\hat{\tau} I}{K_n^2} = 4(1.0)(336.1) + 4(1.0)(612.5) = 3794$$

Middle Story, East-West Direction:

$$\sum_{non-leaner} \frac{\hat{\tau} I}{K_n^2} = 4(1.0)(100.9) + 4(1.0)(521.9) = 2491$$

Next, from the second-order analysis results presented in Table 6.14, $\sum_{all} P_u$ may be computed:

Bottom Story: $\sum_{all} P_u = 2654$ kips

Middle Story: $\sum_{all} P_u = 1759$ kips

The effective length factor, K_{K_n} may now be calculated, and the axial strength may be determined for column A4 (noting that $I_x = 640 \text{ in}^4$ and $I_x = 107 \text{ in}^4$ for this W14x61):

Bottom Story, North-South Direction:

$$K_{K_n x} = \sqrt{\frac{\hat{\tau} I_x}{P_u} \frac{\sum_{all} P_u}{\sum_{non-leaner} \frac{\hat{\tau} I}{K_n^2}}} = \sqrt{\frac{(1.0)640}{98.1}\frac{2654}{1846}} = 3.06$$

Bottom Story, East-West Direction:

$$K_{K_n y} = \sqrt{\frac{\hat{\tau} I_y}{P_u} \frac{\sum_{all} P_u}{\sum_{non-leaner} \frac{\hat{\tau} I}{K_n^2}}} = \sqrt{\frac{(1.0)107}{98.1}\frac{2654}{1221}} = 1.54$$

Middle Story, North-South Direction:

$$K_{K_n x} = \sqrt{\frac{\hat{\tau} I_x}{P_u} \frac{\sum_{all} P_u}{\sum_{non-leaner} \frac{\hat{\tau} I}{K_n^2}}} = \sqrt{\frac{(1.0)640}{55.0}\frac{1759}{3794}} = 2.32$$

Middle Story, East-West Direction:

$$K_{K_n y} = \sqrt{\frac{\hat{\tau} I_y}{P_u} \frac{\sum_{all} P_u}{\sum_{non-leaner} \frac{\hat{\tau} I}{K_n^2}}} = \sqrt{\frac{(1.0)107}{55.0}\frac{1759}{2491}} = 1.17$$

All of the above effective length factors are well above their limiting value of $\sqrt{5/8}K_n$, where K_n is given in Table 6.16.

The interaction equation may now be evaluated for column A4:

Bottom Story, North-South Direction:

$$\lambda_{cx} = \frac{K_{K_nx} L}{r_x \pi}\sqrt{\frac{F_y}{E}} = \frac{(3.06)(13)(12)}{(5.98)\pi}\sqrt{\frac{50}{29000}} = 1.06$$

Bottom Story, East-West Direction:

$$\lambda_{cy} = \frac{K_{K_ny} L}{r_y \pi}\sqrt{\frac{F_y}{E}} = \frac{(1.54)(13)(12)}{(2.45)\pi}\sqrt{\frac{50}{29000}} = 1.29$$

Therefore, weak axis buckling controls the axial strength of the bottom story corner column. The interaction equation is:

$$P_y = A_g F_y = 17.9(50) = 895 \text{ kips}$$

$$P_n = 0.658^{\lambda_c^2} P_y = 0.658^{(1.29^2)}(895) = 446 \text{ kips}$$

$$\frac{P_u}{\phi_c P_n} + \frac{8}{9}\left[\frac{M_{ux}}{\phi_b M_{nx}} + \frac{M_{uy}}{\phi_b M_{ny}}\right] = \frac{98.1}{(0.85)(446)} + \frac{8}{9}\left[\frac{2410}{4590} + \frac{293}{1476}\right] = 0.902 \quad \text{OK}$$

The beam-column is adequate.

Middle Story, North-South Direction:

$$\lambda_{cx} = \frac{K_{K_nx} L}{r_x \pi}\sqrt{\frac{F_y}{E}} = \frac{(2.32)(13)(12)}{(5.98)\pi}\sqrt{\frac{50}{29000}} = 0.800$$

Middle Story, East-West Direction:

$$\lambda_{cy} = \frac{K_{K_ny} L}{r_y \pi}\sqrt{\frac{F_y}{E}} = \frac{(1.17)(13)(12)}{(2.45)\pi}\sqrt{\frac{50}{29000}} = 0.985$$

Therefore, weak axis buckling controls the axial capacity of the middle story corner column:

$$P_n = 0.658^{\lambda_c^2} P_y = 0.658^{(0.985^2)}(895) = 596 \text{ kips}$$

$$\frac{P_u}{2\phi_c P_n} + \frac{M_{ux}}{\phi_b M_{nx}} + \frac{M_{uy}}{\phi_b M_{ny}} = \frac{55.0}{2(0.85)(596)} + \frac{905}{4590} + \frac{354}{1476} = 0.491 \qquad \text{OK}$$

Because the middle story corner column is the same size as the bottom story corner column, but more lightly loaded due to gravity, it is understressed.

6.3.1.2 Practical Story-Based Effective Length Factor K_{R_L} (Section 2.4.2)

In this section, an effective length factor K_{R_L} of the center column is calculated according to Eq. (2.28b), subject to the limiting value of Eq. (2.68):

$$K_{R_L} = \sqrt{\frac{\hat{\tau} I_x}{P_u} \frac{\pi^2 E}{L^2} \frac{\sum\limits_{all} P_u}{\frac{\sum\limits_{non-leaner} HL}{\Delta_{oh}}(0.85 + 0.15 R_L)}} \geq \sqrt{\frac{\pi^2}{1.7} \frac{\hat{\tau} E I_x}{L^2} \frac{\Delta_{oh}}{HL}} \qquad (2.28b)$$

Alternately, the buckling load is given by the following equation, but will not be calculated here for brevity (see Section 6.1 for sample computations using this equation):

$$P_{e\hat{\tau}(R_L)} = \frac{P_u}{\sum\limits_{all} P_u} \frac{\sum\limits_{non-leaner} HL}{\Delta_{oh}}(0.85 + 0.15 R_L) > 1.7 \frac{HL}{\Delta_{oh}} \qquad (2.28a)$$

where

$$R_L = \sum_{leaner} P_u \Big/ \sum_{all} P_u$$

In these equations, the term $\sum\limits_{non-leaner} HL \Big/ \Delta_{oh}$ is the first-order sidesway stiffness of the story, obtained from a first-order elastic analysis in which applied gravity loads on a story are removed and applied laterally at the floor levels (see Section 2.3). In this example, it is necessary to perform two such analyses, one in each of the building's primary directions. The results of these analyses are given in Table 6.17.

Table 6.17 Results of Analysis for Computation of $\sum_{non-leaner} HL \Big/ \Delta_{oh}$

Story	Column Designation	Story Deflection Δ_{oh} (inches) (N-S dir.)	Column Shear H_i (kips) (N-S dir.)	Story Deflection Δ_{oh} (inches) (E-W dir.)	Column Shear H_i (kips) (E-W dir.)
Bottom	A1, A4, D1, D4	24.9	258	33.8	92
	A2, A3, D2, D3		--		572
	B1, B4, C1, C4		404		--
Middle	A1, A4, D1, D4	11.6	167	12.6	129
	A2, A3, D2, D3		--		534
	B1, B4, C1, C4		358		--

Therefore:

Bottom Story, North-South Direction:

$$\sum_{non-leaner} HL \Big/ \Delta_{oh} = \frac{(4(258)+4(404))(13)(12)}{24.9} = 16590 \text{ kips}$$

Bottom Story, East-West Direction:

$$\sum_{non-leaner} HL \Big/ \Delta_{oh} = \frac{(4(92)+4(572))(13)(12)}{33.8} = 12260 \text{ kips}$$

Middle Story, North-South Direction:

$$\sum_{non-leaner} HL \Big/ \Delta_{oh} = \frac{(4(167)+4(358))(13)(12)}{11.6} = 28240 \text{ kips}$$

Middle Story, East-West Direction:

$$\sum_{non-leaner} HL \Big/ \Delta_{oh} = \frac{(4(129)+4(534))(13)(12)}{12.7} = 32580 \text{ kips}$$

Next, from the second-order analysis results presented in Table 6.14, $\sum_{all} P_u$ and $\sum_{leaner} P_u$ may be computed for the calculation of R_L. Note that the perimeter columns which are designated as "leaning" differ depending on the direction of lateral load.

Bottom Story, North-South Direction:

$$\sum_{all} P_u = 2654 \text{ kips}$$

$$\sum_{leaner} P_u = 2(160) + 2(172) + 4(276) = 1768 \text{ kips}$$

$$R_L = \frac{\sum_{leaner} P_u}{\sum_{all} P_u} = \frac{1768}{2654} = 0.666$$

Bottom Story, East-West Direction:

$$\sum_{leaner} P_u = 2(145) + 2(157) + 4(276) = 1708 \text{ kips}$$

$$R_L = \frac{\sum_{leaner} P_u}{\sum_{all} P_u} = \frac{1708}{2654} = 0.644$$

Middle Story, North-South Direction:

$$\sum_{all} P_u = 1759 \text{ kips}$$

$$\sum_{leaner} P_u = 2(108) + 2(111) + 4(182) = 1166 \text{ kips}$$

$$R_L = \frac{\sum_{leaner} P_u}{\sum_{all} P_u} = \frac{1166}{1759} = 0.663$$

Middle Story, East-West Direction:

$$\sum_{leaner} P_u = 2(96.8) + 2(102) + 4(182) = 1125 \text{ kips}$$

$$R_L = \frac{\sum\limits_{leaner} P_u}{\sum\limits_{all} P_u} = \frac{1125}{1759} = 0.639$$

The effective length factor may now be computed for the corner columns on the bottom and middle stories:

Bottom Story, North-South Direction:

$$K_{R_L x} = \sqrt{\frac{(1.0)640}{98.1} \frac{\pi^2 (29000)}{[(13)(12)]^2} \frac{2654}{(16590)[0.85 + (0.15)(0.666)]}} = 3.59$$

which is greater than the limiting value

$$\sqrt{\frac{\pi^2}{1.7} \frac{\hat{\tau} E I_x}{L^2} \frac{\Delta_{oh}}{HL}} = \sqrt{\frac{\pi^2}{1.7} \frac{(1.0)(29000)(640)}{[(13)(12)]^2} \frac{24.9}{(258)(13)(12)}} = 1.66$$

Bottom Story, East-West Direction:

$$K_{R_L y} = \sqrt{\frac{(1.0)107}{98.1} \frac{\pi^2 (29000)}{[(13)(12)]^2} \frac{2654}{(12260)[0.85 + (0.15)(0.644)]}} = 1.71$$

which is greater than the limiting value

$$\sqrt{\frac{\pi^2}{1.7} \frac{\hat{\tau} E I_y}{L^2} \frac{\Delta_{oh}}{HL}} = \sqrt{\frac{\pi^2}{1.7} \frac{(1.0)(29000)(107)}{[(13)(12)]^2} \frac{33.8}{(92)(13)(12)}} = 1.32$$

Middle Story, North-South Direction:

$$K_{R_L x} = \sqrt{\frac{(1.0)640}{55.0} \frac{\pi^2 (29000)}{[(13)(12)]^2} \frac{1759}{(28240)[0.85 + (0.15)(0.663)]}} = 2.99$$

which is greater than the limiting value

$$\sqrt{\frac{\pi^2}{1.7} \frac{\hat{\tau} E I_x}{L^2} \frac{\Delta_{oh}}{HL}} = \sqrt{\frac{\pi^2}{1.7} \frac{(1.0)(29000)(640)}{[(13)(12)]^2} \frac{11.6}{(167)(13)(12)}} = 1.40$$

Middle Story, East-West Direction:

$$K_{R_L y} = \sqrt{\frac{(1.0)107}{55.0}\frac{\pi^2(29000)}{[(13)(12)]^2}\frac{1759}{(32580)[0.85+(0.15)(0.639)]}} = 1.14$$

which is greater than the limiting value

$$\sqrt{\frac{\pi^2}{1.7}\frac{\hat{\tau}EI}{L^2}\frac{\Delta_{oh}}{HL}} = \sqrt{\frac{\pi^2}{1.7}\frac{(1.0)(29000)(107)}{[(13)(12)]^2}\frac{12.6}{(129)(13)(12)}} = 0.681$$

The slenderness ratios for these columns may now be computed:

Bottom Story, North-South Direction:

$$\lambda_{cx} = \frac{K_{R_L x}L}{r_x \pi}\sqrt{\frac{F_y}{E}} = \frac{(3.59)(13)(12)}{(5.98)\pi}\sqrt{\frac{50}{29000}} = 1.23$$

Bottom Story, East-West Direction:

$$\lambda_{cy} = \frac{K_{R_L y}L}{r_y \pi}\sqrt{\frac{F_y}{E}} = \frac{(1.71)(13)(12)}{(2.45)\pi}\sqrt{\frac{50}{29000}} = 1.43$$

Middle Story, North-South Direction:

$$\lambda_{cx} = \frac{K_{R_L x}L}{r_x \pi}\sqrt{\frac{F_y}{E}} = \frac{(2.99)(13)(12)}{(5.98)\pi}\sqrt{\frac{50}{29000}} = 1.03$$

Middle Story, East-West Direction:

$$\lambda_{cy} = \frac{K_{R_L y}L}{r_y \pi}\sqrt{\frac{F_y}{E}} = \frac{(1.14)(13)(12)}{(2.45)\pi}\sqrt{\frac{50}{29000}} = 0.959$$

Weak axis buckling controls the capacity of the bottom story, while strong axis buckling controls buckling of the middle story. This is because column A4 in the middle story has girders framing in at both its top and bottom, and thus experiences significant restraint about its weak axis and, correspondingly, a significant P-δ effect in that direction. Therefore, the assumed value of the P-δ effect inherent in the formulation of K_{R_L} is not as conservative for buckling of the middle story column about its weak axis as it is for buckling of the bottom story column about its strong axis.

The computation of the interaction equation values follows:

Bottom Story:

$$P_n = 0.658^{\lambda_c^2} P_y = 0.658^{(1.43^2)}(895) = 380.2 \text{ kips}$$

$$\frac{P_u}{\phi_c P_n} + \frac{8}{9}\left[\frac{M_{ux}}{\phi_b M_{nx}} + \frac{M_{uy}}{\phi_b M_{ny}}\right] = \frac{98.1}{(0.85)(380.2)} + \frac{8}{9}\left(\frac{2405}{4590} + \frac{293}{1476}\right) = 0.946 \text{ OK}$$

The bottom story beam-column is adequate.

Middle Story:

$$P_n = 0.658^{\lambda_c^2} P_y = 0.658^{(1.03^2)}(895) = 574.1 \text{ kips}$$

$$\frac{P_u}{2\phi_c P_n} + \frac{M_{ux}}{\phi_b M_{nx}} + \frac{M_{uy}}{\phi_b M_{ny}} = \frac{55.0}{(2)(0.85)(574.1)} + \frac{905}{4590} + \frac{354}{1476} = 0.493 \qquad \text{OK}$$

The middle story column is adequate.

Table 6.18 summarizes the calculated effective length factors and governing interaction equation values of the columns in the bottom and middle stories of this structure for the load combination in which the north-south direction of wind is dominant. It is seen that the values of K_{R_L} are consistently higher that the values of K_{K_n}. This is due to the inherent, generally conservative assumption of the magnitude of the P-δ effect used in the formulation of K_{R_L}. However, while the use of K_{R_L} might conservatively underestimate the capacity of a column, it is arguably simpler to calculate than K_{K_n}.

Also, overturning effects from the applied wind load lessen the axial compression forces in the columns along lines 1 and D. Because these columns have lower axial forces, they provide a larger share of lateral stability to the structure than the columns in compression due to the wind load. Therefore, one may see that these columns have larger values of their effective length factors. Columns which do not contribute to lateral stability in a given direction have their effective length factor set equal to 1.0 for design.

Table 6.18 Comparison of K_{K_n} and K_{R_L} Interaction Values

Story	Column Designation	North-South K_{K_n}	North-South K_{R_L}	East-West K_{K_n}	East-West K_{R_L}	Interaction Value K_{K_n}	Interaction Value K_{R_L}
Bottom	A1	3.161	3.660	1.638	1.927	0.809	0.902
	A2	1.0	1.0	3.541	4.166	0.613	0.689
	A3	1.0	1.0	3.415	4.018	0.553	0.625
	A4	3.085	3.572	1.575	1.853	0.902	0.946
	B1	2.817	3.262	1.0	1.0	0.807	0.839
	B2	1.0	1.0	1.0	1.0	0.729	0.729
	B3	1.0	1.0	1.0	1.0	0.729	0.729
	B4	2.858	3.310	1.0	1.0	0.809	0.842
	C1	2.937	3.401	1.0	1.0	0.805	0.838
	C2	1.0	1.0	1.0	1.0	0.729	0.729
	C3	1.0	1.0	1.0	1.0	0.729	0.729
	C4	2.980	3.451	1.0	1.0	0.808	0.843
	D1	4.624	5.354	2.288	2.683	0.746	0.834
	D2	1.0	1.0	3.384	3.969	0.597	0.660
	D3	1.0	1.0	3.264	3.828	0.539	0.599
	D4	4.325	5.009	2.109	2.473	0.886	0.974
Middle	A1	2.364	3.046	1.211	1.234	0.348	0.350
	A2	1.0	1.0	2.427	2.474	0.347	0.347
	A3	1.0	1.0	2.389	2.434	0.233	0.233
	A4	2.324	2.994	1.182	1.204	0.491	0.493
	B1	1.987	2.559	1.0	1.0	0.348	0.350
	B2	1.0	1.0	1.0	1.0	0.540	0.540
	B3	1.0	1.0	1.0	1.0	0.540	0.540
	B4	2.000	2.576	1.0	1.0	0.348	0.351
	C1	2.042	2.630	1.0	1.0	0.358	0.361
	C2	1.0	1.0	1.0	1.0	0.540	0.540
	C3	1.0	1.0	1.0	1.0	0.540	0.540
	C4	2.057	2.649	1.0	1.0	0.358	0.361
	D1	2.769	3.567	1.386	1.411	0.279	0.282
	D2	1.0	1.0	2.374	2.416	0.345	0.345
	D3	1.0	1.0	2.337	2.379	0.234	0.234
	D4	2.695	3.472	1.340	1.364	0.436	0.439

6.3.1.3 Notional Load (Chapter 4)

As discussed in Chapter 4, a notional load is applied at every story in the structure, with a magnitude of

$$H_{not} = \xi \sum_{all} P_u^{applied}$$

where $\sum_{all} P_u^{applied}$ is the sum of the applied axial loads for all of the columns in the story and

$$\xi = \xi_0 k_s k_L k_y k_c$$

where

$$\xi_o = 0.005$$

and k_s and k_L are correction factors that are taken as 1.0 in this example (see Chapter 4). The correction factor for yield stress, k_y equals:

$$k_y = 22\sqrt{\frac{F_y}{E}} = 22\sqrt{\frac{50}{29000}} = 0.914$$

while k_c is a statistical correction factor accounting for the number of columns in each unbraced frame n_c:

$$k_c = \sqrt{0.5 + \frac{1}{n_c}} \leq 1.0$$

For this example, $n_c = 4$ and

$$k_c = \sqrt{0.5 + \frac{1}{4}} = 0.866 \leq 1.0$$

Applying all of the above correction factors, the coefficient for the notional load becomes

$$\xi = \xi_0 k_s k_L k_y k_c = (0.005)(0.914)(0.866) = 0.0040$$

For this example of a structure that is unbraced in both primary directions, the notional load H_{not} is conservatively applied in both primary directions simultaneously:

Bottom story: $H_{not} = 0.004(904.4) = 3.62$ kips
Middle story: $H_{not} = 0.004(904.4) = 3.62$ kips

where $\sum_{all} P_u^{applied}$ is based on the applied gravity loads on a story resulting from the load combination $1.2D + 1.3W + 0.5L + 0.5L_r$ (Table 6.14). The calculated notional loads H_{not} are applied horizontally in the same direction as the applied wind.

The forces in column A4 resulting from a second-order elastic analysis (including notional load and wind acting in both directions) yields the following internal forces:

Table 6.19 Internal Forces in Column A4 Including Notional Load

Story	P_u **(kips)**	M_{ux} **(in-kips)**	M_{uy} **(in-kips)**
Bottom	101	2579	349
Middle	56.2	966	392

The axial strength of these beam-columns may be computed using an effective length factor equal to 1.0:

$$\lambda_{cx} = \frac{L}{r_x \pi}\sqrt{\frac{F_y}{E}} = \frac{(13)(12)}{(5.98)\pi}\sqrt{\frac{50}{29000}} = 0.345$$

$$\lambda_{cy} = \frac{L}{r_y \pi}\sqrt{\frac{F_y}{E}} = \frac{(13)(12)}{(2.45)\pi}\sqrt{\frac{50}{29000}} = 0.842$$

Weak axis flexural buckling governs. Therefore:

Bottom Story:

$$P_n = 0.658^{\lambda_c^2} P_y = 0.658^{(0.842^2)}(895) = 665.2 \text{ kips}$$

$$\frac{P_u}{2\phi_c P_n} + \frac{M_{ux}}{\phi_b M_{nx}} + \frac{M_{uy}}{\phi_b M_{ny}} = \frac{101}{2(0.85)(665.2)} + \frac{2579}{4590} + \frac{349}{1474} = 0.888$$

Middle Story:

$$P_n = 0.658^{\lambda_c^2} P_y = 0.658^{(0.842^2)}(895) = 665.2 \text{ kips}$$

$$\frac{P_u}{2\phi_c P_n} + \frac{M_{ux}}{\phi_b M_{nx}} + \frac{M_{uy}}{\phi_b M_{ny}} = \frac{56.2}{(2)(0.85)(665.2)} + \frac{966}{4590} + \frac{392}{1476} = 0.526 \qquad \text{OK}$$

For the bottom story column, the notional load approach results in a slightly less conservative prediction of member capacity than the two effective length procedures. This is most likely due to the fact that, had an iteratively computed τ factor been used, this column would have exhibited inelastic stiffness reduction, thus lowering its interaction equation values resulting from using an effective length approach. Alternately, the lightly loaded middle story column exhibits a slightly more conservative interaction value than when the two effective length procedures are used. However, in either case, the simplicity of the technique is readily apparent in this example.

6.3.1.4 Summary of Results

As a final note, the B_2 value for the first story ranges from 1.19 in the North-South direction to 1.27 in the East-West direction, clearly indicating, as per Chapter 5, that stability must be assessed when designing the columns of this story. Alternately, the B_2 factors for the two directions of the lightly loaded middle story are well below the limiting value of 1.11 established in Chapter 5, and the S_L value for the two directions of this story are well below the limiting value of 2.25 established in Chapter 5. Consequently, effective length factors of 1.0 could be used in both directions for all columns in this story without incurring unacceptable errors.

The following table provides a comparison of the values yielded by the interaction equations for each of the methods outlined in this example:

Table 6.20 Comparison of Interaction Equation Values

Story	Column Designation	Interaction Eq. Value using K_{K_n}	Interaction Eq. Value using K_{R_L}	Interaction Eq. Value using Notional Load
Bottom	A1	0.809	0.902	0.773
	A2	0.613	0.689	0.594
	A3	0.553	0.625	0.546
	A4	0.902	0.946	0.888
	B1	0.807	0.839	0.839
	B2	0.729	0.729	0.756
	B3	0.729	0.729	0.756
	B4	0.809	0.842	0.838
	C1	0.805	0.838	0.817
	C2	0.729	0.729	0.756
	C3	0.729	0.729	0.757
	C4	0.808	0.843	0.818
	D1	0.746	0.834	0.688
	D2	0.597	0.660	0.593
	D3	0.539	0.599	0.547
	D4	0.886	0.974	0.854
Middle	A1	0.348	0.350	0.330
	A2	0.347	0.347	0.385
	A3	0.233	0.233	0.272
	A4	0.491	0.493	0.526
	B1	0.348	0.350	0.381
	B2	0.540	0.540	0.567
	B3	0.540	0.540	0.567
	B4	0.348	0.351	0.381
	C1	0.358	0.361	0.390
	C2	0.540	0.540	0.567
	C3	0.540	0.540	0.567
	C4	0.358	0.361	0.390
	D1	0.279	0.282	0.256
	D2	0.345	0.345	0.384
	D3	0.234	0.234	0.273
	D4	0.436	0.439	0.466

6.4 - Stability of Four-Story, Four-Bay Composite PR Frame

To study the different approaches to stability calculations in PR frames that are controlled by drift, an example from a trial design by Leon [AISC, 1996] will be discussed (Fig. 6.6). This frame is a four-story, four bay composite frame with story heights of 13'-6", 32' bays and 30' frame spacing.

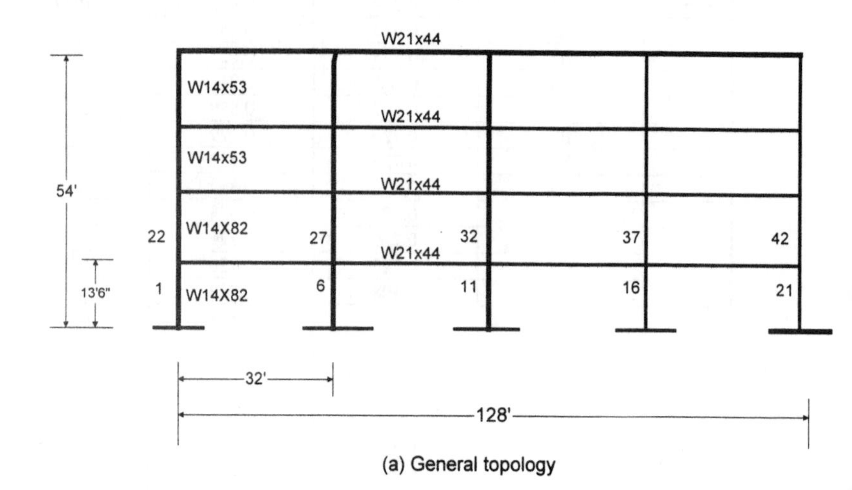

Figure 6.6 Composite partially-restrained frame for Example 6.4.

All girders are fully composite W2lx44 A572 with an effective I = 2370 in.^{4} This effective I is a combination of the positive and negative moments of inertia of the composite beam [Leon, 1994; AISC, 1996]. Each girder carries a tributary width of 30 ft., with three intermediate beams framing into the girders at $L/4$, $L/2$, and $3L/4$. Columns in the first two stories are W14x82; columns in the upper two stories are W14x53. All beams and columns are A572 (F_y = 50 ksi).

The loading for this frame is given in Fig. 6.7. The dead loads were taken as 100 psf, while the live loads were assumed as 50 psf and appropriately reduced. In this type of design the gravity loads (1.0D) are applied to the bare steel frame assuming that all the connections are pinned. This is because the PR characteristics of these connections arise from the composite action of the floor slab which is not present until the concrete hardens. The computer program used to carry out the analysis automatically incorporates this effect. Once the concrete has hardened and composite PR action is activated, the reduced live loads with their appropriate load factor plus any additional dead load above the 1.0D load factor are applied based on tributary areas to both the columns and the locations of the framing beams.

The complete loading for the beams and columns at this stage is shown in Fig. 6.7(a), while the resulting moments are shown in Table 6.21. It is important to note that because of the sequence of construction and the load case being used, only the loads corresponding to $0.2D + 0.5L$ are actually applied to the connections. As shown in Table 6.21, this results in relatively small moments at the connections.

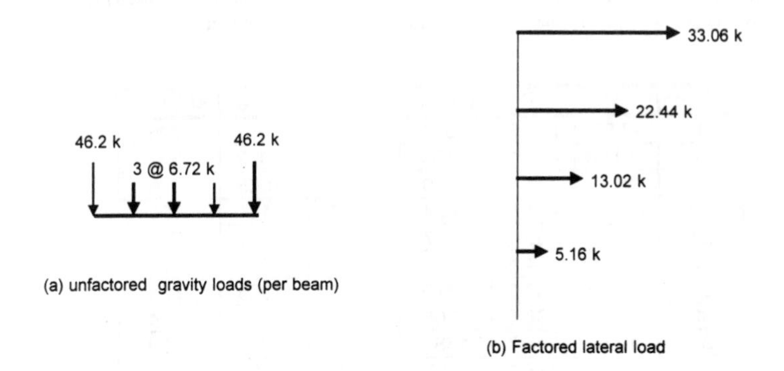

Figure 6.7 Loads acting on the structure.

Column	V	P	$M\,C$ bot	$M\,C$ top	$M\,B$ left	$M\,B$ right
1	1.13	222.3	-63	-120	0	282
6	-0.17	455.2	8	19	-656	605
11	0.00	451.3	0	0	-588	588
16	0.17	455.2	-8	-120	-605	656
21	-1.13	222.3	63	120	-282	0
22	1.95	166.7	-162	-154	0	304
27	-0.38	341.3	32	27	-637	598
32	0.00	338.5	0	0	-591	591
37	1.13	341.3	-32	-27	-598	637
42	-1.95	116.7	162	154	-304	0

Notes:
V = shear in the column, kips
P = axial load, kips
$M\,C$ = moment in the column (top and bottom), kip-in
$M\,B$ = moment in the beams (left and right of the column), kip-in

Table 6.21 Analysis results for $0.5L$ case

The lateral loads, shown in Fig. 6.7(b), are then applied and the results superimposed. The results are shown in Table 6.22. Table 6.23 shows the change in moments (*dM*) as the lateral loads were applied. These will be used in the calculations of dM_f / dM_n. Note that for purposes of this example the dM_f / dM_n terms will be taken as the total change in moments due to the application of the loads to the composite frame rather than as the change in moments during the last load step. This assumption is correct only if the behavior of the frame is linear and is used here for illustrative purposes only. In general, this assumption will be unconservative.

Column	*V*	*P*	*M C* bot	*M C* top	*M B* left	*M B* right
1	-7.85	203.8	1073	305	0	-725
6	-19.08	450.1	2103	1220	-2002	-897
11	-19.28	451.4	2110	1245	-2065	-886
16	-18.33	461.8	2054	1152	-2104	-629
21	-9.14	239.1	1137	467	-1106	0
22	-4.65	154.4	420	452	0	-672
27	-19.15	337.9	1679	1679	-1965	-894
32	-19.53	338.6	1706	1711	-2059	-874
37	-17.99	345.7	1581	1592	-2086	-629
42	-7.21	178.1	640	661	-1097	0

Table 6.22 Analysis results for the 1.0*E* load case

Column	*V*	*P*	M C bot	M C top	*M B* left	*M B* right
1	-8.98	-18.55	1136	425	0	-1007
6	-18.91	-5.06	2095	1201	-1346	-1502
11	-19.28	0.12	2110	1245	-1477	-1475
16	-18.50	6.59	2062	1272	-1499	-1285
21	-8.01	16.76	1074	347	-824	0
22	-6.60	-12.33	582	606	0	-976
27	-18.77	-3.41	1647	1652	-1328	-1492
32	-19.53	0.14	1706	1711	-1468	-1465
37	-19.12	4.37	1613	1619	-1488	-1266
42	-5.26	61.36	478	507	-793	0

Table 6.23 - Changes in moments

For this analysis two types of PR connections will be used: a composite PR connection for the interior connections and a steel (top and seat angle) connection for the exterior columns. The moment-rotation curves for these PR connections are shown in Fig 6.8. The behavior of the connections will be idealized by straight segments connecting the squares shown also in Fig. 6.8.

For all calculations the initial stiffness ($R_{k,i}$) will be taken as equal to the secant stiffness at 2.5 mRad for the composite PR connection ($R_{k,i}$ = 988,000 kip-in/Rad) and the steel connection ($R_{k,i}$ = 320,000 kip-in/Rad). It should be noted that the composite PR connection used here is substantially stiffer than a typical PR steel connection. This is due primarily to the presence of slab steel and the torquing of all bolts used in the connection. The design of this type of frame [AISC, 1996] is predicated on obtaining a high initial stiffness so that frame behavior in the service range is linear. This, of course, is not a prerequisite for PR frame design but only an artifice to simplify the analysis.

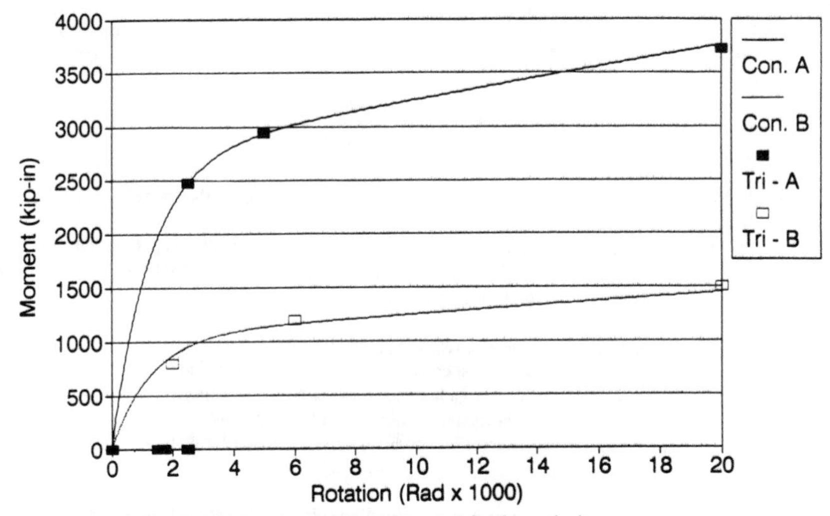

Figure 6.8 Moment-rotation curves used for the analysis.

The complete load-deflection response for the frame under a non-proportional loading corresponding to ($1.0D + 0.5L + XE$), where X is the earthquake load factor, is shown in Fig. 6.9. It can be seen that the frame behavior at service loads is essentially linear and that only slight non-linear behavior can be noticed at the factored level (load factor $X = 1.0$). The analysis was conducted using a program that accounted for both the tri-linear connection characteristics and the second order effects. The equivalent lateral loads due to the earthquake loading case will be used in this example and corresponds to the moments shown in Tables 6.21 through 6.23. The earthquake case was chosen since it results in higher drifts and axial loads, and thus potentially in a more critical stability condition.

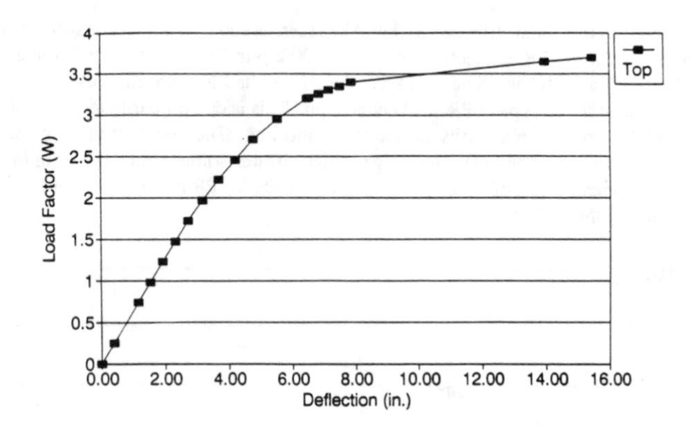

Figure 6.9 Lateral load deflection response for the frame.

Tables 6.24 through 6.27 show the summary of the stability calculations applying Eqs. (3.17) and (2.25) for several variations of the three procedures described in Section 3.4.1. The seven cases shown in Tables 6.25 and 6.26 correspond to:

Case 1 -Unloading stiffness equal to $R_{k,i} = R_{kser}$ and loading stiffness equal to zero. This is the simplest, but probably most conservative approach. The value of R_{kser} can be obtained directly from a database of a moment-rotation curve by inspection. Because for the spring constant is zero for the beams framing at the top of columns 1 and 22, these are considered leaner columns and no K factor is computed for them in Tables 6.25 through 6.27.

Case 2 - Stiffnesses are taken as the corresponding stiffnesses from the tri-linear curves used in the analysis (Fig. 6.8). Thus there are three possible discrete values of stiffness for the connections (R_{kser}, R_{kf}, and R_{ku}) as shown in Fig. 3.2(a). R_{kser} was used for the unloading side and R_{kf} was used for the loading side. For this case dM_f / dM_n was taken as zero. This is slightly more complicated than Case 1 in that two stiffnesses are needed. However, the stiffnesses are obtained directly from the moment-rotation curves and not from the frame analysis, so little additional computational overhead is involved.

Case 3 - The tangent stiffness to the curves shown in Fig. 6.9 for the given moment was used for the loading side. In this case dM_f / dM_n was taken as zero. R_{kser} was used for the unloading side. This is the most accurate approach but requires that a step-by-step analysis be carried out in order to determine the instantaneous stiffness.

Col	*Ic,b*	*Ic,t*	*Ib,l*	*Ib,r*	*Lc*	*Lg*	*Kc,l*	*Kc,r*	*Rk,l*	*Rk,r*	*I'b,l*	*I'b,r*	*GA*	*GB*	*Kn*	*Pu*	Sum *Pu*	*I/Kn^2*	*KKn*
1	Fixed	882	0	2370	162	384	0	320	0.00	1.79	0	442	1.00	7.62	1.83	203.75	1806.2	263.4	2.30
6	Fixed	882	2370	2370	162	384	68.6	988	0.38	5.52	134	768	1.00	3.74	1.62	450.14	1806.2	336.1	1.55
11	Fixed	882	2370	2370	162	384	68.6	988	0.38	5.52	134	768	1.00	3.74	1.62	451.42	1806.2	336.1	1.55
16	Fixed	882	2370	2370	162	384	68.6	988	0.38	5.52	134	768	1.00	3.74	1.62	461.79	1806.2	336.1	1.53
21	Fixed	882	2370	2370	162	384	75	0	0.42	0.00	145	0	1.00	23.23	2.1	239.06	1806.2	200.0	2.13
22	882	541	2370	2370	162	384	0	320	0.00	1.79	0	442	7.62	5.80	2.52	154.37	1354.6	85.2	2.90
27	882	541	2370	2370	162	384	68.6	988	0.38	5.52	134	768	3.74	2.84	1.9	337.89	1354.6	149.9	1.96
32	882	541	2370	2370	162	384	68.6	988	0.38	5.52	134	768	3.74	2.84	1.9	338.64	1354.6	149.9	1.96
37	882	541	2370	2370	162	384	68.6	988	0.38	5.52	134	768	3.74	2.84	1.9	345.67	1354.6	149.9	1.94
42	882	541	2370	0	162	384	75	0	0.42	0.00	145	0	23.23	17.66	4.17	178.06	1354.6	31.1	2.70

Ic,b = moment of inertia of bottom column (in^4)
Ic,t = moment of inertia of top column (in^4)
Ib,l = moment of inertia for beam framing from left (in^4)
Ib,r = moment of inertia for beam framing from right (in^4)
Lc = length of column (in)
Lg = length of beam (in)
Kc,l = stiffness of left connection (kip-in/rad x 10-3)
Kc,r = stiffness of left connection (kip-in/rad x 10-3)
I'b,l = modified left beam moment of inertia
I'b,r = modified right beam moment of inertia
GA, GB = stiffness factors
Kn = nomograph K
Pu = column axial load
Sum *Pu* = total story axial load
I/Kn^2 = product of I/Kn for each column for use in Eq. (2.25)
KKn = effective length based on story concept
Note susbcripts "*b*" and "*t*" refer to top and bottom, while "*l*" and "*r*" refer to left and right

Table 6.24 - Computations for Case 2.

Col	Case 1	Case 2	Case 3	Case 4	Case 5	Case 6	Case 7
1	---	2.30	2.06	---	2.31	1.94	2.08
6	1.70	1.55	1.39	1.59	1.55	1.31	1.40
11	1.70	1.55	1.38	1.59	1.55	1.31	1.40
16	1.68	1.53	1.37	1.57	1.53	1.29	1.38
21	2.33	2.13	1.90	2.19	2.13	1.79	1.92
22	---	2.90	2.29	---	2.61	2.49	2.02
27	2.10	1.96	1.55	1.84	1.77	1.68	1.36
32	2.10	1.96	1.55	1.84	1.77	1.68	1.36
37	2.08	1.94	1.53	1.82	1.75	1.66	1.35
42	2.89	2.70	2.13	2.53	2.43	2.32	1.88

Table 6.25 - K_{Kn} values.

Col	Case 1	Case 2	Case 3	Case 4	Case 5	Case 6	Case 7
1	---	0.81	1.02	---	0.81	1.15	1.00
6	0.68	0.81	1.02	0.77	0.81	1.15	1.00
11	0.68	0.81	1.02	0.77	0.81	1.15	1.00
16	0.68	0.81	1.02	0.77	0.81	1.15	1.00
21	0.68	0.81	1.02	0.77	0.81	1.15	1.00
22	---	0.48	0.77	---	0.59	0.65	1.00
27	0.42	0.48	0.77	0.55	0.59	0.65	1.00
32	0.42	0.48	0.77	0.55	0.59	0.65	1.00
37	0.42	0.48	0.77	0.55	0.59	0.65	1.00
42	0.42	0.48	0.77	0.55	0.59	0.65	1.00

Table 6.26 - Comparison for K_{Kn} to the FR case.

Case 4 - Similar to Case 1 but with approximate dM_f / dM_n correction. The correction is labeled approximate because the necessary moments were taken from an analysis that assumed a linear rather than a continuously changing stiffness. The errors due to this simplification are small. This corresponds to Procedure 3 in Section 3.4.1.

Case 5 - Similar to Case 2 but with approximate dM_f / dM_n correction. This corresponds to Procedure 2 in Section 3.4. 1.

Case 6 - Similar to Case 3 but with approximate dM_f / dM_n correction. This corresponds to Procedure 1 in Section 3.4. 1.

Case 7 - FR connections with approximate dM_f / dM_n correction.

Table 6.24 shows the complete set of calculations for Case 2 where a simplified tri-linear model was used. Table 6.25 shows the K_{Kn} values after application of Eq. (2.25) to the different of the cases described above. For comparison purposes the case of a FR frame with similar members is also shown.

Table 6.25 shows that there is a significant spread on the computed K-factors depending on what stiffnesses are assumed. A comparison of the Cases 1, 5, and 6 shows that for the second story columns Case 1 (the simplified approach) gives a substantially larger K-factor (K = 2. 10) than the tri-linear method (Case 5 = 1.77). The first story columns do not give a good idea of the spreads in K since the fixed base provides significant additional restraints.

Table 6.26 shows the effect of the connection flexibility on P_{cr}. The table shows the ratio of $1/K_{Kn}^2$ normalized to the FR connection case (Case 7). Large losses of capacity can be noted for all cases, but particularly for Procedure 3.

Col	Case 1	Case 2	Case 3	Case 4	Case 5	Case 6	Case 7
1	---	0.71	0.89	---	0.71	1.00	0.87
6	0.59	0.71	0.89	0.67	0.71	1.00	0.87
11	0.59	0.71	0.89	0.67	0.71	1.00	0.87
16	0.59	0.71	0.89	0.67	0.71	1.00	0.87
21	0.59	0.71	0.89	0.67	0.71	1.00	0.87
22	---	0.74	1.18	---	0.91	1.00	1.53
27	0.64	0.74	1.18	0.84	0.91	1.00	1.53
32	0.64	0.74	1.18	0.84	0.91	1.00	1.53
37	0.64	0.74	1.18	0.84	0.91	1.00	1.53
42	0.64	0.74	1.18	0.84	0.91	1.00	1.53

Table 6.27 - Comparison for K_{Kn} to the FR case.

Table 6.27 shows a comparison similar to that in Table 6.26, but normalized to Case 6 which can be considered the more accurate solution to the problem.

The seven load cases used in this example could be considered as representing the range of possible options in design, from the very simple one of Case 1 to the most accurate one of Case 6. The results of the example show that the simplest assumptions (Case 1) can lead to over-conservative designs for cases where the real structure has not undergone "yielding" at the service load levels. In this example, where the frame design was geared to obtaining essentially linear behavior in that range, one would expect that Case 2 would give better results. For cases of less stiff PR frames with steel-only connections, it is expected that Case 1 would give better results. This underscores the fact that the reliability of the moment-connection used has a large influence on the results, since for connections where softening takes place near the service load levels the results will be very sensitive to the local changes in stiffness.

The results for Case 4 show the importance of the dM_f / dM_n correction. The results for Cases 5 and 6 are better than the others but they require advanced analysis capabilities and great care in the post-processing of the data. Therefore the results of this example indicate that there does not seem to be a simple procedure that will provide consistent, safe, and economic results for checking the stability of columns in PR frames.

The computational overhead in the more accurate procedures is large and leads to two possible implementation strategies. One is to develop post-processing routines capable of handling the *K*-factor calculations. The second, and preferable option, is to avoid all the *K*-factor calculations and conduct a second-order, non-linear analysis. Since the latter would be needed in any case to develop the data needed to implement a *K*-factor calculation correctly, it seems that the use of advanced analysis is the only rational solution for the design of PR frames.

References

AISC [1993], "Load and Resistance Factor Design Specification for Structural Steel Buildings," Second Edition, December 1, 1993, AISC, Chicago, Illinois.

Chen, W. F. and Lui, E. M. [1991], *Stability and Design of Steel Frames*, CRC Press, Boca Raton, Florida.

LeMessurier, W. J. [1977], "A Practical Method for Second-Order Analysis. Part 2: Rigid Frames," *Engineering Journal*, AISC, Second Quarter, 49-67.

Leon, R. T. [1994], "Composite Semi-Rigid Construction," *Engineering Journal*, AISC, 31(2), 57-67.

Leon, R. T., Hoffman, J. J., and Staeger, T. [1996], "Partially Restrained Composite Connections: A Design Guide," Steel Design Guide Series, AISC, Chicago, Illinois.

White, D. W., and Hajjar, J. F. [1991], "Application of Second-Order Elastic Analysis in LRFD: Research to Practice." *Engineering Journal*, AISC, 28(4), 133-148.

Ziemian, R. D. [1990], "Advanced Methods of Inelastic Analysis in the Limit States Design of Steel Structures," Ph.D. Dissertation, School of Civil and Environmental Engineering, Cornell University, Ithaca, New York, August.

Ziemian, R. D., McGuire, W., and Deierlein, G. G. [1992], "Inelastic Limit States Design: Part I - Planar Frame Studies," *Journal of Structural Engineering*, ASCE, 118(9), 2532-2549.

Chapter 7

Conclusions

This report has discussed three procedures for assessing the stability of beam-columns when designing unbraced or braced steel frame structures. The derivations, step-by-step design procedures, and accuracy of two effective length approaches, both outlined in Commentary Chapter C of the Second Edition of the AISC LRFD Specification [AISC 1993], have been detailed. In addition, a notional load (or equivalent imperfection) approach has been presented within the context of the LRFD Specification. Chapter 5 also provided criteria for determining when stability need not be considered directly during design. This chapter provides conclusions on these approaches for stability design.

The two effective length procedures, labeled in Chapter 2 as K_{K_n} and K_{R_L}, both provide an efficient means of accounting for the stability of a steel column, and they are compatible with the requirement within the AISC Specifications that an effective length factor must be used when computing column axial strength, P_n. As shown in Chapters 2, 3, and 6, each approach has its strengths and weaknesses. Summary comments and conclusions on these two effective length approaches to stability design follow:

1. Both techniques for computing effective length are based on an assessment of the buckling capacity of each story in the frame. For this assessment, it is assumed that, as the load on a story is increased, the stronger columns in a story restrain the weaker columns from buckling (in the plane of the story), until the story as a whole buckles in a sidesway mode. Compared to using the AISC nomographs directly, story-based approaches generally provide a more accurate assessment of both the elastic and inelastic stability characteristics of columns in unbraced frame structures. Story-based approaches are especially beneficial when assessing the stability of structures in which a substantial percentage of the gravity load is resisted by "leaning" columns, or in frames in which the stiffness parameter $L\sqrt{P/EI}$ varies significantly among the beam-columns.

2. For both approaches, it is neither necessary, nor in fact desirable, to compute the effective length factors themselves for incorporation into the expression for λ_c -- i.e., in the final term of Eq. (2.2): $\lambda_c = (KL/r\pi)\sqrt{F_y/E}$, where the K factor may be either elastic or inelastic. Rather, it is most efficient to compute the elastic or inelastic critical loads directly for incorporation into the middle term of Eq. (2.2): $\lambda_c = \sqrt{\hat{\tau}P_y/P_{e\hat{\tau}}}$, using for the denominator either $P_{e\hat{\tau}(K_n)}$ from Eq. (2.25c) or $P_{e\hat{\tau}(R_L)}$ from Eq. (2.28c).

3. As discussed in Sections 2.1 and 2.2, column inelasticity at incipient buckling may be incorporated into either effective length procedure through the use of an inelastic stiffness reduction factor, τ, computed from Eq. 2.7. This computation is iterative, as shown in Eq.

2.7, or it may be approximated in a non-iterative computation, giving a value refereed to in this report as $\hat{\tau}$. A common approximate value for $\hat{\tau}$ is obtained if P_u/ϕ_c is used in place of P_n in Eq. (2.7a), or in place of $P_{e\tau}$ in Eq. (2.7b). While clearly beneficial, computing an inelastic stiffness reduction factor for each column in a frame is cumbersome -- often engineers neglect this computation in present-day practice. The inelastic stiffness reduction may be taken as 1.0 if elastic critical loads are to be used.

4. To insure the integrity of these two effective length approaches, the limits of Eqs. (2.73) or (2.74) for $P_{e\hat{\tau}(K_n)}$ or K_{K_n}, respectively, and Eqs. (2.67) and (2.68) for $P_{e\hat{\tau}(R_L)}$ or K_{R_L}, respectively, must be enforced. For most practical frames, it is rare that well proportioned columns will breach this limit. However, for a frame for which these limits must be checked throughout the design calculations, the efficiency of the story-based approaches degrades, particularly for the more efficient K_{R_L} procedure.

5. In Section 2.3.2, it is also shown that, using either of these effective length approaches, it is conservative to compute the first term of the interaction equation ($P_u/\phi_c P_n$) for *the non-leaning column with the largest value of* P_u/P_y in a story, and to use this value as the first term of the interaction equation for *all* the remaining non-leaning columns in the same story (for a given direction of story sidesway). This approach to stability design has the potential for providing substantial savings in design computations compared with computing an effective length factor (either story-based, system-based, or directly from the AISC nomographs) individually for every column in the story. Of course, the minimum limiting values of effective length given in Eqs. (2.67) and (2.73) must be enforced whenever one uses these effective length procedures, although these limits would rarely be breached in common structural framing systems.

6. In Sections 2.3.2, it is shown that the two story buckling procedures reviewed differ only in the manner in which they compute $\sum_{non-leaner} P_{cr(story)}$. Accurate assessment of this story buckling resistance is central to the accuracy, or inaccuracy, of story-based effective length procedures. In addition, Section 2.4.3 indicates that, using either of these two story-based procedures, the error in the values of effective length computed for every column in a story is identical (when compared to the values of effective length computed from a system buckling analysis). Other columns in a story do not, for example, "compensate" for one column's unconservative effective length factor by having a conservative effective length factor. This reinforces the need to insure that $\sum_{non-leaner} P_{cr(story)}$ is computed accurately, or conservatively.

7. Chapter 3 outlines a comprehensive technique for assessing stability of framing systems that use partially-restrained connections. The technique proposed is amenable to hand computations, thus retaining the spirit of present-day effective length computations. As mentioned in Chapter 3, more elaborate techniques based upon the use of the computer and an

appropriate nonlinear connection model have potential for providing a more thorough assessment of stability for these types of structures.

The notional load approach described in Chapter 4 provides an alternative to effective length methods of stability assessment. In addition to presenting a basic description of the notional horizontal load technique, Chapter 4 and its appendices outline in detail a careful calibration of the notional load magnitude. This calibration and the step-by-step procedures outlined in that chapter insure either accurate or conservative results when assessing in-plane buckling, out-of-plane buckling, and the behavior of spatial beam-columns (i.e., those bent biaxially) for both single-story and multi-story frame construction, all within the context of the AISC LRFD Specification. While not as mature as the effective length procedures, the notional load approach may be considered to be as comprehensive as the former procedures in terms of its applicability to the design of a wide variety of steel frame structures, especially those composed of wide-flange members. As has been noted in Chapter 4, the notional load approach or variants of it are used in many of the modern limit states steel design standards worldwide. Summary comments and conclusions regarding the notional load approach include the following:

1. In the notional load approach, the applied notional loads generate additional bending moments in the columns which approximate the effects of initial imperfections and residual stresses on the column stability. Calibration of the magnitude of these notional loads is required to insure accurate assessment of these imperfections. In the case of pure gravity loading, these additional bending moments are still required, essentially to simulate the moments caused by actual frame out-of-plumbness imperfections when gravity load is applied to the structure. Effective length approaches, on the other hand, are rooted in bifurcation analysis, and these additional column bending moments are not generated as part of these approaches. Some argue that stability assessment should be tied directly to member or frame bifurcation, especially for assessing stability of members subjected, in theory, solely to axial force. Others argue that all non-leaning columns in a frame are subjected to some combination of axial force and bending moment due to the inevitable out-of-plumbness imperfections arising from the erection process, and thus the notional load approach provides a more appropriate representation of frame behavior. In this vein, it should be noted that the notional loads actually amplify not only the column bending moments, but all of the forces and bending moments in the lateral resistance system, a circumstance which some feel reasonably represents frame behavior. In the end, proponents of the notional load technique feel that the small amplification of the frame's forces and bending moments is a small price to pay for simplicity.

2. In fact, the chief advantage of the notional load approach is its simplicity and the brevity of calculations. Chapter 4 provides a "tiered" calibration for the value of the notional load, with "simple", "modified" and "refined" calibrations being described. The simple and modified calibrations have an important advantage as compared to nomograph-based effective length techniques: no G factors need to be computed -- in fact, no member-level stability computations of any sort need to be done, short of computing P_n for a column based upon its actual unbraced length. In addition, inelastic stiffness reduction is accounted for implicitly

within the calibration of the notional load value. The refined notional load calibration was undertaken to reduce the conservatism associated with the simple and modified approaches for heavily restrained beam-columns bent in double-curvature bending. Compared to the simple and modified approaches, additional computations are required for the refined notional load approach. However, the total level of computational effort required for the refined approach is no more than that associated with the calculation of story buckling loads.

3. When using the notional load approach, no additional checks need to be carried out to verify the applicability of the method, as it is not required to distinguish between a sway and a non-sway buckling mode. In other words, no limiting values such as those imposed in Eqs. (2.67) and (2.73) need to be checked on a member-by-member level. Thus, the notional load technique accounts implicitly for the behavior of leaning columns, and of non-leaning columns that are largely braced by the other column in the story. This is a key advantage of this approach.

These three approaches may be compared further. The computation of K_{K_n}, or, equivalently, $P_{e\hat{\tau}(K_n)}$, is based on the use of the AISC nomographs, and its associated G factors, to determine the relative stiffnesses of the columns in a story. Consequently, of the three techniques discussed in this report, this approach is based on theory that is the most familiar to practicing engineers. In addition, the use of G factors is advantageous for accounting for more complex relations between beam and column stiffness. For example, as shown in Chapter 3, K_{K_n} forms the basis for computing a story-based effective length factor for frames having partially-restrained connections. The committee is not prepared to present approaches for assessing stability of PR frames within the context of using either K_{R_L} or a notional load approach, since the details of how one should model a PR frame for the required sidesway analyses remains a subject of research and debate.

The chief advantage of using K_{R_L} versus K_{K_n} is its simplicity. Consider Eqs. (2.28c) and (2.2) to compute λ_c: $P_{e\hat{\tau}(R_L)} = \lambda_{R_L} P_u \;\Rightarrow\; \lambda_c = \sqrt{P_y / P_{e\hat{\tau}(R_L)}} = \sqrt{P_y / \left(\lambda_{R_L} P_u\right)}$. The story buckling parameter λ_{R_L} is computed only once per story for each potential global direction of story buckling. The computations then required *per column* are simply to compute the ratio P_u/P_y, and, as indicated earlier, the first term of the interaction needs to be computed only for the column with the largest value of P_u/P_y. The computation of the elastic story buckling parameter λ_{R_L}, in turn, requires summing the column axial forces across the story to obtain $\sum_{all} P_u$ (a computation required in some form for all approaches discussed in this report), computing R_L, the ratio of axial force in the leaning columns to the total axial force in the story, and computing $\sum_{non-leaner} HL/\Delta_{oh}$, a relatively simple calculation based upon a first-order sidesway analysis. The computation of K_{K_n} still requires calculating appropriate G factors, including any necessary corrections (e.g., see Eq. (2.11)). Computation of G factors is always a cumbersome column-by-column computation, and is error prone for complex structures.

In addition, it has been shown [Hajjar and White 1994] that K_{R_L} is consistently more accurate than K_{K_n} if uncorrected G factors (e.g., not using Eq. (2.11)) are used for computing K_{K_n}. As outlined in Section 2.4.2, the P-δ effect is approximated in the computation of K_{K_n} as per the assumptions of the AISC nomograph, while K_{R_L} assumes a P-δ effect that is conservative in all but extreme cases. Using the results from a first-order analysis to compute Δ_{oh} for the calculation of K_{R_L} is not a foolproof or "exact" means of assessing the P-δ behavior of a frame, but it does capture the nuances of the story's sidesway behavior, just as do the more "exact" system buckling (eigenvalue) analysis approaches. In addition, shear flexibility of the beam and column elements may be incorporated into the analyses used in the K_{R_L} and notional load approaches, while the computation of K_{K_n} inherently assumes Euler-Bernoulli beam theory.

In short, the K_{R_L} approach provides a highly accurate assessment of member and story stability, yet it requires few computations per beam-column -- in the opinion of the committee, remarkably fewer than the use of the AISC nomographs. This is also a primary advantage of using the notional horizontal load approach.

Finally, Chapter 5 has isolated conditions in which stability need not be assessed rigorously for any of the columns in a given story. In particular, if the B_2 factor for a story is less than 1.11, and the story is sufficiently stiff (i.e., the story parameter S_L from Eq. (5.27) is less than 2.25, a value which is rarely breached in practice), then an effective length factor of 1.0 may be used for assessing in-plane stability of all columns in this story of the unbraced frame, without the use of a notional horizontal load. The maximum error obtained by using $K = 1$ in these cases is approximately 6%, and simply limiting the permissible value of the interaction equation to values slightly less than 1.0 insures a safe design. With many unbraced frames governed by stiffness rather than strength, the rigorous derivation of Ch. 5 provides a powerful approach for stability assessment.

The direct use of the AISC nomographs has served the structural engineering profession well for nearly half a century, and their use remains the mainstay of stability design in the United States. Yet numerous new approaches to stability assessment based on first- or second-order elastic analysis/design methodologies have received much attention in the research community and have begun to be adopted by specification committees worldwide. This committee feels that the three techniques described in this document, coupled with solid techniques for determining when frame structures are not stability-critical, do provide strong alternatives for stability design. The approaches are more comprehensive than the direct use of the AISC nomographs, in that they are applicable to a wider range of modern-day frame structures. At the same time, the K_{R_L} and notional load approaches discussed in this document are, quite often, less cumbersome to use than the nomographs.

As discussed at the beginning of this report, in Section 1.1, there remain scores of issues relating to column stability which have not as yet been addressed satisfactorily in the literature (e.g., stability of columns subjected to seismic excitation, and accounting properly for the presence of concrete floor slabs when assessing column stability). It is the hope of the committee that this report provides useful guidelines for the implementation of these contemporary techniques into design, and that it provides the next stepping stone for the continued enhancement of stability design procedures for frame structures.

APPENDIX A

OVERVIEW OF PROCEDURES FOR SYSTEM BUCKLING ANALYSIS

For most ordinary rectangular "shear-type" building frames, the simpler story-based procedures tend to produce answers for design that are at least as good as that obtained from a system buckling analysis. However, for general and "non-regular" framing, a system buckling analysis may be necessary for accurate determination of the overall frame stability behavior. Some of the most common approaches for elastic system buckling analysis are outlined below. At the end of this discussion, the extension of these elastic procedures to inelastic system buckling analysis is described.

Linear Buckling Analysis

The system buckling analysis utilized to obtain $\lambda_{system} P_u$ in Eq. (2.22) (with $\hat{\tau} = 1$) is generally termed a linear buckling analysis. This type of analysis seeks the lowest value of the load parameter λ_{system} for which the determinant of the global structure stiffness matrix vanishes,

$$\det[K] = 0 \tag{A.1}$$

In this procedure, the global stiffness matrix [K] is based on the original, undeformed geometry. In other words, changes in the geometry as the loads are increased on the structure are ignored, and the buckling load is associated with a bifurcation from this ideal state. The global stiffness matrix is obtained by assembly of element stiffness matrices, which may be developed analytically based on the conventional member stability functions or approximately based on finite element interpolation of the displacements.

Element Stiffness Matrices for the Stability Function Approach

For the typical six degree-of-freedom frame element used in two-dimensional frame analysis and shown in Fig. A.1, the element matrices for the stability function approach may be expressed as:

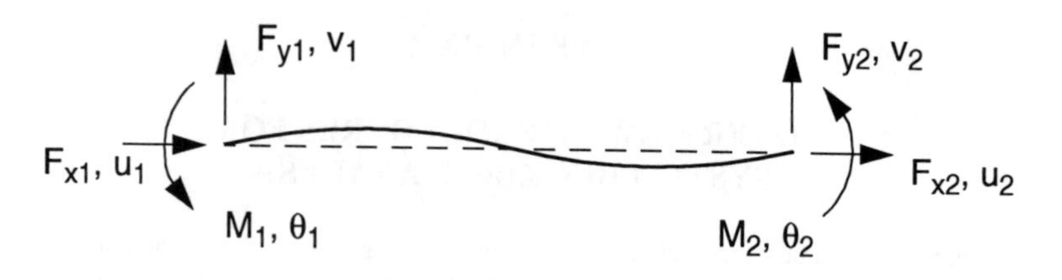

Figure A.1. Two-dimensional frame element.

$$[\mathrm{k}]=\begin{bmatrix} \frac{EA}{L} & 0 & 0 & -\frac{EA}{L} & 0 & 0 \\ & \frac{2(S_{ii}+S_{ij})}{L^2}-\frac{P}{L} & \frac{(S_{ii}+S_{ij})}{L} & 0 & \frac{-2(S_{ii}+S_{ij})}{L^2}+\frac{P}{L} & \frac{(S_{ii}+S_{ij})}{L} \\ & & S_{ii} & 0 & \frac{-(S_{ii}+S_{ij})}{L} & S_{ij} \\ & sym. & & \frac{EA}{L} & 0 & 0 \\ & & & & \frac{2(S_{ii}+S_{ij})}{L^2}-\frac{P}{L} & \frac{-(S_{ii}+S_{ij})}{L} \\ & & & & & S_{ii} \end{bmatrix} \tag{A.2}$$

where S_{ii} and S_{ij} are the member flexural stiffness coefficients obtained from the elastic beam-column stability functions [Livesley and Chandler, 1956], and P is positive in compression. For compressive P, these coefficients may be expressed as

$$S_{ii} = \frac{EI}{L}\left[\frac{\kappa L sin(\kappa L) - (\kappa L)^2 cos(\kappa L)}{2 - 2cos(\kappa L) - \kappa L sin(\kappa L)}\right] \tag{A.3a}$$

$$S_{ij} = \frac{EI}{L}\left[\frac{(\kappa L)^2 - \kappa L sin(\kappa L)}{2 - 2cos(\kappa L) - \kappa L sin(\kappa L)}\right] \tag{A.3b}$$

The parameter κL is defined as

$$\kappa L = \pi\sqrt{|P|/P_e} = \pi\sqrt{|P|L^2/\pi^2 EI} = L\sqrt{|P|/EI} \tag{A.4}$$

Equation (2.26) may be written in a more compact form as

$$[\mathrm{k}]=\frac{EI}{L}\begin{bmatrix} A/I & 0 & 0 & -A/I & 0 & 0 \\ & 12\Phi_1/L^2 & 6\Phi_2/L & 0 & -12\Phi_1/L^2 & 6\Phi_2/L \\ & & 4\Phi_3 & 0 & -6\Phi_2/L & 2\Phi_4 \\ & & & A/I & 0 & 0 \\ & & & & 12\Phi_1/L^2 & -6\Phi_2/L \\ & & & & & 4\Phi_3 \end{bmatrix} \quad \text{(A.5)}$$

The terms Φ_1 through Φ_4 are related to the elastic beam-column stability functions and are provided by Chen and Lui [1987]. All these terms are functions of κL (Eq. (A.4)). Different functions must be used for Φ_1 through Φ_4 in Eq. (A.5) (or for S_{ii} and S_{ij} in Eqs. (A.3)) for axial tension and axial compression. Also, these functions are indeterminate when P is exactly equal to zero, in which case, the values for Φ_1 through Φ_4 should be taken as 1.0. That is, the stiffness matrix commonly employed for first-order elastic analysis is recovered if the axial force is zero.

The Continuous Eigenvalue Problem

Since the stability functions are continuous nonlinear functions of the member axial force P for axial tension or axial compression, the solution of Eq. (A.1) may be regarded as a *continuous* eigenvalue problem and requires special procedures [Hancock, 1991]. However, the resulting solutions are exact and do not require subdivision of the members into multiple elements to achieve sufficient accuracy. The reader is referred to [Wittrick and Williams, 1971] for reliable and accurate eigenvalue routines for *continuous* eigenvalue problems. These routines are based on the Sturm sequence property [Bathe, 1982], and ensure that the lowest eigenvalue is always found. They do not require the calculation of the determinant of the structure stiffness matrix. However, they are computationally intensive since they require reformulation of the global stiffness at every iteration within the analysis.

Element Stiffness Matrices for the Finite Element Approach

If the finite element approach is employed, the element stiffness matrices may be expressed as the sum of the ordinary first-order elastic stiffness matrix, $[k_e]$, and the *geometric stiffness matrix*, $[k_g]$. The $[k_e]$ matrix is given by Eq. (A.5) if all the terms Φ_1 through Φ_4 are set equal to 1.0. Furthermore, the element geometric stiffness matrix may be written as

$$
\left[k_g\right] = P \begin{bmatrix} 0 & 0 & 0 & 0 & 0 & 0 \\ & 6/5L & 1/10 & 0 & -6/5L & 1/10 \\ & & 2L/15 & 0 & -1/10 & -L/30 \\ & & & 0 & 0 & 0 \\ & sym. & & & 6/5L & -1/10 \\ & & & & & 2L/15 \end{bmatrix} \qquad (A.6)
$$

where P is taken as negative for compression. It is interesting to note that ($[k_e]$ + $[k_g]$) can be obtained directly from Eq. (A.5) by using a Taylor series expansion for Φ_1 through Φ_4 and retaining only the first two terms [Chen and Lui, 1987].

The Linear Eigenvalue Problem

When the finite element approach is used, the resulting eigenvalue problem is *linear* and takes the form

$$
\det\left(\left[K_e\right] - \lambda_{system}\left[K_g\right]\right) = 0 \qquad (A.7)
$$

where $[K_e]$ is the structure's global linear elastic stiffness (based on the original undeformed geometry) and $[K_g]$ is the structure's global geometric stiffness matrix based on the axial forces P_u in each element. Detailed discussions of linear eigenvalue routines for buckling analysis are given in [Bathe, 1982] and [Hancock, 1984]. It is important to note that the approximations in the finite element approach that result in the linearization of the stiffness equations in terms of P also generally result in the need to subdivide the members into multiple elements to achieve an accurate buckling solution. This is particularly true if more than one point of inflection occurs within a member's length in its buckling mode. However, it can be shown that the use of three elements per member is sufficient to achieve less than one percent error in the element stiffness matrices for the most severe case of an axially loaded strut with full rotational and lateral restraint at its ends (in which case the elastic critical load of the member is $P_{cr} = 4\pi^2 EI/L^2$) [Allen and Bulson, 1980; White and Hajjar, 1991].

Inelastic System Buckling Analysis

All of the above discussions have assumed that λ_{system} is calculated as the load parameter associated with elastic buckling of the structural system from its undeflected geometry. An inelastic system buckling analysis may be conducted using the same equations, procedures, and concepts discussed above if all the terms EI and EA are replaced by $\hat{\tau}EI$ and $\hat{\tau}EA$. If $\hat{\tau}$ is computed as the exact τ value associated with the axial forces $\lambda_{system} P_u$ (i.e., substituting $\lambda_{system} P_u / \phi_c$ into Eqs. (2.7a) for P_n or into Eqs. (2.7b) for $0.877P_{e\tau}$), the global equations are such that the eigenvalue problem is no longer linear, both for the finite element approach as well as for the stability function approach. Procedures for this type of analysis are discussed in [Ziemian, 1990] and [Hancock, 1991].

References

Allen, H.G., and Bulson, P.S. [1980], *Background to Buckling,* McGraw-Hill, New York, NY, 582 pp.

Bathe, K.J. [1982], *Finite Element Procedures for Engineering Analysis*, Prentice Hall, Englewood Cliffs, NJ, 735 pp.

Chen, W.F. and Lui, E.M. [1987], *Structural Stability: Theory and Implementation*, Elsevier, New York, 490 pp.

Hancock, G.J. [1984], "Structural Buckling and Vibration Analyses on Microcomputers," *Civil Engineering Transactions*, Institution of Engineers Australia, Vol. CE26, No. 4, 327-331.

Hancock, G.J. [1991], "Second-Order Elastic Analysis Solution Techniqus and Verification," Structural Analysis to AS4100, A Two Day Post-Graduate Course, Nov. 6 & 7, pp. 3-1 to 3-29.

Livesley, R.K. and Chandler, D.B. [1956], *Stability Functions for Structural Frameworks*, Manchester University Press.

White, D.W., and Hajjar, J.F. [1991], "Application of Second-Order Elastic Analysis in LRFD: Research to Practice," *Engineering Journal*, AISC, 28(4),133-148.

Wittrick, W.H. and Williams, F.W. [1971], "A General Algorithm for Computing Natural Frequencies of Elastic Structures," *Quarterly Journal of Mechanics and Applied Mathematics*, Vol. XXIV, Part 3, 264-285.

Ziemian, R.D. [1990], "Advanced Methods of Inelastic Analysis in the Limit States Design of Steel Structures," Ph.D. Dissertation, School of Civil and Environmental Engineering, Cornell University, Ithaca, NY, August.

APPENDIX B

DERIVATION OF STORY BUCKLING LOAD FOR A STORY HAVING COLUMNS OF UNEQUAL LENGTH

Sections 2.3.2 and 2.4 of this report derived two story-based effective length procedures which are included in the Second Edition of the AISC LRFD Commentary Chapter C [AISC, 1993]. The form of the equations presented in the Commentary presumes all columns in the story have equal length. This appendix rederives these equations for a story having columns of unequal length. In addition, it derives in detail the manner in which the P-δ effect is addressed through the use of C_L. A value for C_L that is ordinarily conservative is assumed in the derivation of K_{R_L} (see Section 2.4.2.4). All the derivations assume elastic behavior. Column inelasticity may be accounted for by replacing EI in all calculations by τEI, where τEI may be considered as the "effective elastic" section rigidity (see Section 2.2).

B.1 P-Δ Effect on Story Buckling Load

Consider the structure shown in Fig. B.1, in which the leaning column j on the right is loaded with a gravity load P_{uj}, and the fixed-base cantilever column i, on the left, is subjected to zero gravity load. The two columns are considered here to be of different lengths.

As both Yura and LeMessurier showed (and as discussed in Section 2.4.2), column i, which is fully fixed at its base, must resist the shear forces created by the leaner column as the structure sways. When the structure is subjected to sidesway due to a lateral load H_i, it displaces an amount Δ_{ph} to its final equilibrated position, where Δ_{ph} is the elastic *second-order* displacement of the structure. This displacement equals the first-order sidesway displacement, Δ_{oh}, plus an additional deflection due to geometric nonlinearity in the system. Specifically in this case, the geometric nonlinearity is due to the P-Δ effect caused by the axial force in the leaner column acting through its lateral displacement. Therefore, as is seen in Fig. B.2, the axial force P_{uj} displaced an amount Δ_{ph} creates a shear $P_{uj}\Delta_{ph}/L_j$ in the leaner column to counterbalance the second-order couple, $P_{uj}\Delta_{ph}$.

This shear is transferred into the girder as an axial force, to maintain equilibrium at joint C, and this force is then transferred into the cantilever column as a shear to maintain equilibrium at joint B (note that it is assumed here that the lateral displacement of both ends of the girder are identical). Therefore, the shear in the cantilever column is equal to $H_i + P_{uj}\Delta_{ph}/L_j$, which equals the first order shear plus the shear due to the P-Δ effect. The corresponding moment at the bottom of column i equals the first order moment plus the frame's P-Δ moment $\left(H_i + P_{uj}\Delta_{ph}/L_j\right)L_i = H_iL_i + \left(P_{uj}\Delta_{ph}L_i/L_j\right)$. Fig. B.3b shows the first-order bending moment

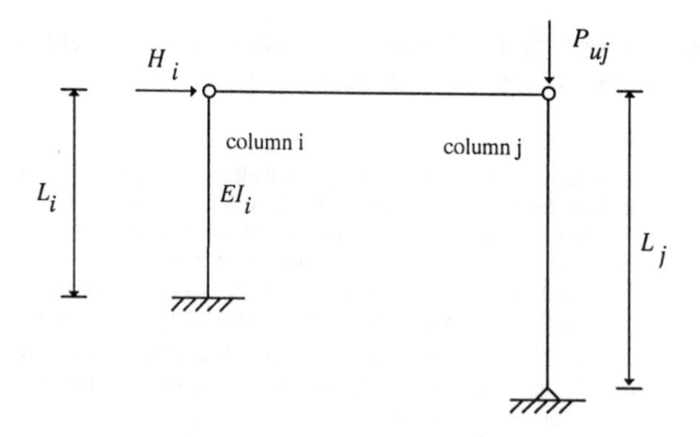

Figure B.1 One-Bay, One-Story Frame with Axial Load Applied to Leaner Column

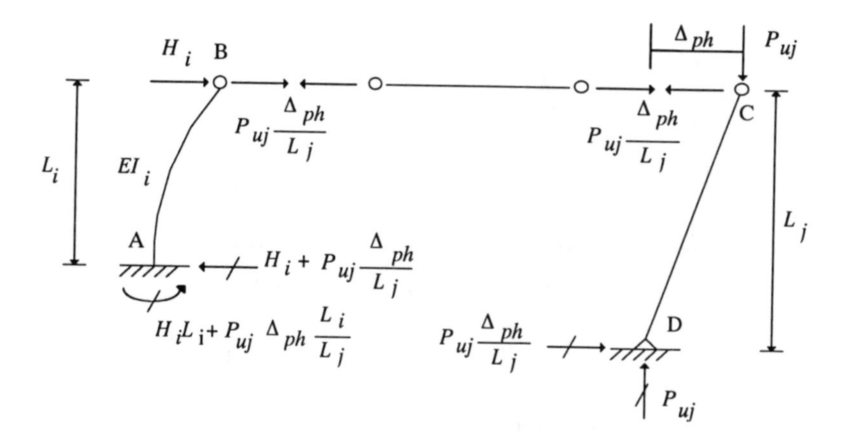

Figure B.2 Transfer of P-Δ Effect from Leaner Column to Rigidly-Connected Column

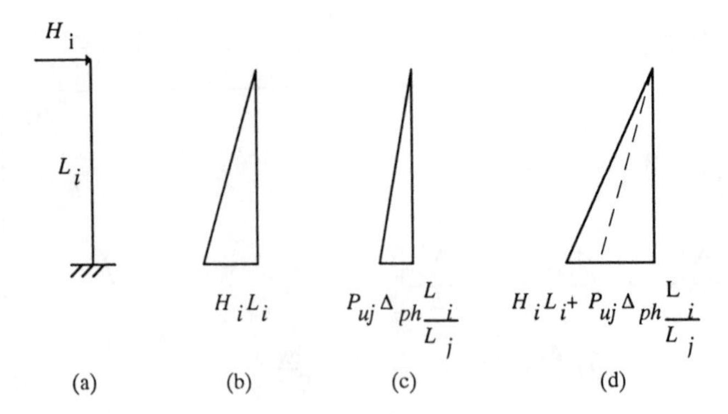

Figure B.3 Flexural Behavior of Rigidly-Connected Column Due to P-Δ Effect from Leaner Column diagram of the cantilever column, and Fig. B.3c the additional second-order bending moment. Fig. B.3d shows the total bending moment in the rigidly-connected column.

As long as neither column buckles, and assuming elastic behavior, the final, equilibrated lateral deflection of the structure may be found from moment-area principles. For this simple case, this second-order lateral deflection also equals the shear in column i divided by its first-order sidesway stiffness. Since this column is not subjected to any axial force, its final elastic sidesway resistance equals its elastic first-order resistance from Eq. (2.33), $3EI_i/L_i^3$. Therefore:

$$\Delta_{ph} = \left(H_i + \frac{P_{uj}}{L_j}\Delta_{ph}\right)\frac{L_i^3}{3EI_i} = \left(H_i + \frac{P_{uj}}{L_j}\Delta_{ph}\right)\frac{1}{P_{Li}/L_i} = \frac{H_i}{P_{Li}/L_i} + \frac{(P_{uj}/L_j)\Delta_{ph}}{P_{Li}/L_i} \tag{B.1}$$

Solving for Δ_{ph} in Eq. (B.1), we obtain:

$$\Delta_{ph} = \frac{H_i}{P_{Li}/L_i - P_{uj}/L_j} \tag{B.2}$$

Eq. (B.2) is an expression for the *second-order* elastic displacement of this simple structure, including its P-Δ effect. If the standard assumptions embedded in this analysis (i.e., cross sections which are initially plane and normal to the centroidal axis of the member remain so during loading, etc.) are satisfied, Eq. (B.2) is exact for this structure. Again, this is because the cantilever column is not subjected to any direct axial force (i.e., its stiffness is unaffected by the loading).

From Eq. (B.2) it may be seen that Δ_{ph} approaches infinity when P_{uj} is increased by a factor λ_{story} until:

$$\frac{\lambda_{story}P_{uj}}{L_j} = \frac{P_{Li}}{L_i} \tag{B.3}$$

The buckling load of this frame may then be expressed as:

$$P_{cr_i} = \lambda_{story}P_{ui} = \frac{P_{Li}/L_i}{P_{uj}/L_j}(P_{ui}) \tag{B.4}$$

B.2 Combined P-Δ and P-δ Effects on Story Buckling Load

Consider next the second-order behavior of the same structure, but with column i loaded axially instead of the leaner column (Fig. B.4). The leaner column now contributes nothing to the behavior of the structure. Consequently, the second-order behavior of the cantilever in Fig. 2.6, Section 2.4.2, may be considered individually. Fig. B.5a is a free-body diagram of the deformed

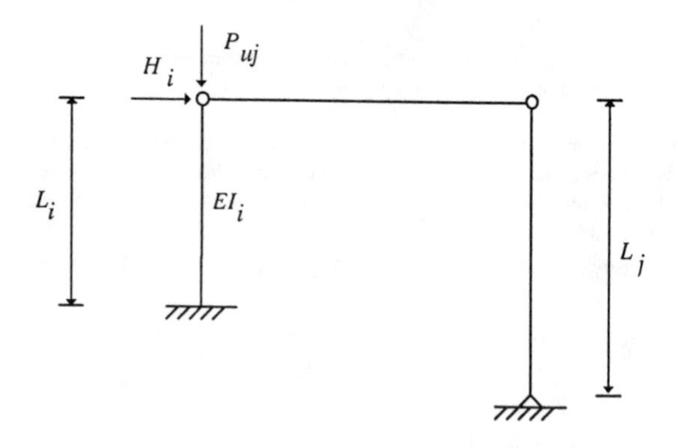

Figure B.4 One-Bay, One-Story Frame with Axial Load Applied to Rigidly-Connected Column

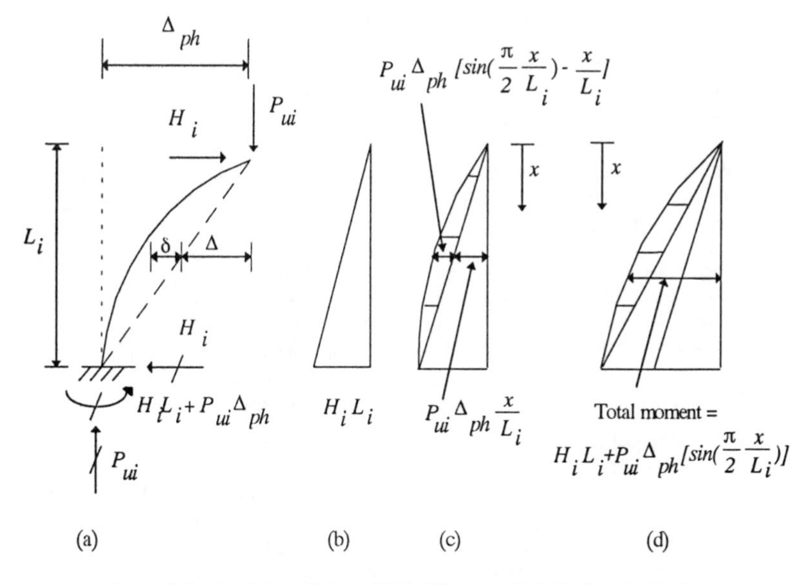

Figure B.5 Combined P-Δ and P-δ Effects in Axially Loaded Column

column. Fig. B.5b shows the first-order bending moment diagram, while Fig. B.5c shows the bending moment diagram due to the additional second-order moment.

Note the difference between the second-order moment diagrams in Figs. B.3c and B.5c. The axially loaded cantilever column has a second-order moment not only due to the axial force acting through the rotation of its chord (i.e., $P_{ui}\Delta_{ph}$), but also due to the axial force acting through the curvature of the member. This moment corresponds to the cross-hatched region in Figs. B.5c through B.5d. The curvature causes the displacement δ relative to the chord, shown in Fig. B.5a, and the extra moment P-δ, shown in Fig. B.5c, corresponds to the P-δ effect. The total moment on the member is shown in Fig. B.5d.

As described in Section 2.4.2, in keeping with LeMessurier's basic approach, it is necessary to determine the first-mode sidesway buckled shape of this structure, after which its buckling load may be calculated. In particular, this buckled shape may be approximated by considering the second-order effects on the structure's deflected shape. As P_{ui} approaches the buckling load of the isolated cantilever column, the lateral deflection approaches the shape of a sine curve, as shown by Euler [1759]. In other words, as P_{ui} increases, the deflected shape of the cantilever would approximately equal $\Delta_{ph} \sin \dfrac{\pi x}{2L_i}$ (using an origin at the top of the column for convenience, so that $\Delta = 0$ when $x = 0$ and $\Delta = \Delta_{ph}$ when $x = L_i$, as shown in Fig. B.5a). Thus, for this case, the deflected shape is assumed to equal the buckled shape of a cantilever loaded purely by concentric axial force, rather than a shape represented by a cubic polynomial (i.e., the first-order deflected configuration of a cantilever subjected to lateral tip load H.

Based upon this deflected shape, the first- and second-order bending moment diagrams may be constructed for this cantilever (Figs. B.5b through B.5d). The second-order tip deflection may then be determined using moment-area principles, which yields:

$$\Delta_{ph} = \frac{1}{EI_i}\int_0^{L_i}\left(H_i L_i \frac{x}{L_i}\right)x dx + \frac{1}{EI_i}\int_0^{L_i}\left(P_{ui}\Delta_{ph}\frac{x}{L_i}\right)x dx + \frac{P_{ui}\Delta_{ph}}{EI_i}\int_0^{L_i}\left[\sin\left(\frac{\pi}{2}\frac{x}{L_i}\right) - \frac{x}{L_i}\right]x dx$$

$$= \frac{H_i L_i^2}{L_i}\frac{L_i^2}{3EI_i} + \frac{P_{ui}\Delta_{ph}L_i}{L_i}\frac{L_i^2}{3EI_i} + \frac{P_{ui}\Delta_{ph}}{P_{Li}}\frac{3}{L_i^2}\int_0^{L_i}\left[\sin\left(\frac{\pi}{2}\frac{x}{L_i}\right) - \frac{x}{L_i}\right]x dx$$

$$\Delta_{ph} = \frac{H_i}{P_{Li}/L_i} + \frac{P_{ui}\Delta_{ph}/L_i}{P_{Li}/L_i} + \frac{P_{ui}\Delta_{ph}/L_i}{P_{Li}/L_i}\left[\frac{3}{(\pi/2)^2} - 1\right] \tag{B.5}$$

and therefore

$$\Delta_{ph} = \frac{H_i}{P_{Li}/L_i} + \frac{P_{ui}\Delta_{ph}/L_i}{P_{Li}/L_i} + \frac{P_{ui}\Delta_{ph}/L_i}{P_{Li}/L_i} C_{Li} \tag{B.6}$$

where C_{Li} is LeMessurier's *Clarification Factor* [LeMessurier, 1977] for column i. One may observe that the third term, which accounts for the P-δ effect, is directly proportional to the second term, which accounts for the P-Δ effect. This third term did not occur in Eq. (B.2), since the rigidly-connected column of Fig. B.1 experienced no P-δ effect.

For the case of Fig. B.4, as may be seen from Eq. (B.5) :

$$C_{Li} = \frac{3}{(\pi/2)^2} - 1 = 0.216 \tag{B.7}$$

Multiplying the numerator and the denominator of the radical by $\frac{EI_i/L_i^2}{EI_i/L_i^2}$ gives:

$$C_{Li} = \frac{3EI_i/L_i^2}{(\pi/2)^2 \; EI_i/L_i^2} - 1 = \frac{P_{Li}}{P_{e\hat{t}(nomo)i}} - 1 \tag{B.8}$$

where $P_{e\hat{t}(nomo)i}$ is the Euler buckling load of this cantilever based upon the use of the sidesway uninhibited nomograph effective length factor of 2.0. It is an approximation to use a nomograph effective length factor, unless of course all of the assumptions of the nomograph are met (as outlined in Section 2.2.1). This assumption is made to simplify the procedure for calculating C_{Li}, as will be seen below.

In addition, when invoking the moment-area theorem in Eq. (B.5), it was assumed that the deflected shape of the column subjected to large P_{ui} is a sine curve, thus resulting in the second-order bending moment diagram of Fig B.5d. However, as mentioned earlier, if the axial force in column i approaches zero, the deflected shape would approach the form of a cubic polynomial, not a sine curve. For this case, the third term of Eq. (B.5) becomes, from the moment-area theorem:

$$\frac{P_{ui}\Delta_{ph}}{EI_i} \int_0^L \left\{ \left[\left(\frac{x}{L_i}\right) - \left(\frac{x}{L_i}\right)^3 \right] x \right\} dx = \frac{P_{ui}\Delta_{ph}}{P_{Li}} \frac{3}{L_i^2} \int_0^L \left\{ \left[\left(\frac{x}{L_i}\right) - \left(\frac{x}{L_i}\right)^3 \right] \frac{x}{2} \right\} dx \tag{B.9}$$

and therefore:

$$C_{Li} = \frac{3}{L_i^2}\int_0^L \left\{\left[\left(\frac{x}{L_i}\right)^2 - \left(\frac{x}{L_i}\right)^4\right]\frac{x}{2}\right\}dx = \frac{1}{5} = 0.2 \tag{B.10}$$

Thus, one may see that C_{Li} actually varies for the cantilever column depending upon the magnitude of the axial force in the column: it equals 0.2 for small P_{ui}, and it equals 0.216 for large P_{ui}.

Of course, for a general frame, it is not practical to determine the exact displaced shape of each column. Also, the above variation in C_{Li} is small. Therefore, LeMessurier assumed C_{Li} to be independent of the axial force in the column, with this value based upon the sidesway displacement of the column taking the form of a sine curve. This procedure is simply more convenient and practical than accounting properly for the dependence of C_{Li} on the magnitude of the axial force (i.e., accounting for the variation in a column's P-δ effect with the level of axial force), which would require an iterative nonlinear analysis that captures the true buckled shape of the structure. With LeMessurier's approach, one instead approximates the column's second-order displaced shape, if a second-order analysis is being performed, or its buckled shape, if a buckling analysis is being performed, by a sine curve to incorporate the P-δ effect. LeMessurier showed that it is quite accurate and generally conservative to use the sine curve rather than the cubic displacement curve when representing the P-δ effect since it corresponds to the more critical case of the axial force being relatively high. He also showed that C_{Li} generally ranges from 0 to 0.216 for a beam-column having arbitrary end restraints [LeMessurier, 1977]. However, C_{Li} may range even larger than 0.216 in extreme circumstances (in particular, when a column has a negative G factor at one or both ends, or when a column's axial force exceeds $\pi^2 EI/L^2$; see Section 2.2.1 for a discussion of these phenomena) [LeMessurier, 1993].

Now, consider Eq. (B.6) again. Solving for Δ_{ph} gives:

$$\Delta_{ph} = \frac{H_i}{P_{Li}/L_i - P_{ui}/L_i - C_{Li}P_{ui}/L_i} \tag{B.11}$$

As with Eq. (B.2), the buckling load of this column equals the load at which the denominator of Eq. (B.11) equals zero (i.e., at which the lateral displacement equals infinity). For the case of Eq. (B.11), this load occurs when P_{ui} is increased by a factor λ_i until:

$$\lambda_i \frac{P_{ui}}{L_i} = \frac{P_{Li}}{L_i} - \lambda_i \frac{C_{Li}P_{ui}}{L_i} \tag{B.12}$$

and therefore, after solving for λ_i in Eq. (B.12):

$$P_{cr_i} = \lambda_i P_{ui} = \frac{\frac{P_{Li}}{L_i}}{\frac{P_{ui}}{L_i} + \frac{C_{Li} P_{ui}}{L_i}} (P_{ui}) = \frac{P_{Li}}{(1 + C_{Li})} \tag{B.13}$$

Comparing Eqs. (B.4) and (B.13), one may observe that the destabilizing effect of the P-δ moments on the buckling load of this structure take the form of $1 + C_{Li}$ in the denominator.

LeMessurier next generalized the definitions for P_L and C_L to a column with any end conditions by introducing the stiffness factor β, which accounts for the effects of the column's elastic restraints:

$$P_{Li} = \beta_i \frac{EI_i}{L_i^2} \tag{B.14}$$

Then, using a suitably general form of $P_{e\hat{\tau}(nomo)i}$:

$$P_{e\hat{\tau}(nomo)i} = \frac{\pi^2 EI_i}{(K_{ni} L_i)^2} \tag{B.15}$$

C_{Li} may be obtained from Eqs. (B.8), (B.14), and (B.15):

$$C_{Li} = \frac{P_{Li}}{P_{e\hat{\tau}(nomo)i}} - 1 = \frac{\beta_i K_{ni}^2}{\pi^2} - 1 \tag{B.16}$$

where, as described in Section 2.4.2.3, β_i is given by:

$$\beta_i = \frac{6(G_A + G_B) + 36}{2(G_A + G_B) + G_A G_B + 3} \tag{B.17}$$

When the gravity load is applied entirely to the leaner column, as in Eqs. (B.1) through (B.4), C_{Li} equals 0.2 for the cantilever column. This is an extreme example: since the rigidly-connected column is subjected to no axial force, it actually experiences no P-δ effect (recall the moment diagrams of Fig. B.3). The column is acting as a translational spring in this case, and its shape is described as a cubic polynomial (see Eq. (B.10)). However, if for consistency one used Eq. (B.16) to calculate C_{Li} for that column, thus assuming a sinusoidal displaced shape, a conservative value of 0.216 is obtained for the cantilever column. This is regardless of the level of axial force that is applied to the column. Note also that C_{Li} is always zero for a leaning column since the curvature is zero in a leaning column.

As described in Section 2.4.2.4, it is evident from this derivation that, because of second-order effects, the buckling load of any story in an unbraced frame varies depending upon how the gravity forces are distributed among the columns in that story (see the discussion of Section 2.4.2.3).

B.3 Derivation of Story-Based Effective Length Including Columns of Unequal Length

For a frame having several multiple rigidly-connected and/or leaner columns which may be of different lengths, Eqs. (B.4) and (B.13), which express the buckling load of a story having only one rigidly-connected column, generalize to:

$$\sum_{all}\lambda\frac{P_u}{L}=\sum_{non-leaner}\frac{P_L}{L}-\sum_{non-leaner}\lambda\frac{C_L P_u}{L} \tag{B.18}$$

or, by rearranging terms:

$$\sum_{all}\lambda\frac{P_u}{L}+\sum_{non-leaner}\lambda\frac{C_L P_u}{L}=\sum_{non-leaner}\frac{P_L}{L} \tag{B.19}$$

Based upon the assumption outlined in Sections 2.3.2 and 2.4 that $\lambda=\lambda_{story}$ for all columns in the story, Eq. (B.19) becomes:

$$\sum_{all}\lambda\frac{P_u}{L}+\sum_{non-leaner}\lambda\frac{C_L P_u}{L}=\sum_{non-leaner}\frac{P_L}{L}\;\Rightarrow\;\lambda_{story}=\frac{\displaystyle\sum_{non-leaner}\frac{P_L}{L}}{\displaystyle\sum_{all}\frac{P_u}{L}+\sum_{non-leaner}\frac{C_L P_u}{L}} \tag{B.20}$$

Therefore, analogous to the derivation of Section 2.3.2:

$$\left[P_{e\hat{t}(story)}=\frac{\pi^2 EI}{\left(K_{story}L\right)^2}\right]=\left[\lambda_{story}P_u=\frac{\displaystyle\sum_{non-leaner}\frac{P_L}{L}}{\displaystyle\sum_{all}\frac{P_u}{L}+\sum_{non-leaner}\frac{C_L P_u}{L}}(P_u)\right] \tag{B.21}$$

One may recognize Eq. (B.21) as an alternative form of Eq. (2.16), where in Eq. (B.21) the buckling load of the story, $\sum_{non-leaner}P_{cr}$ is represented by the right hand side of Eq. (B.18).
Following the derivation presented in Section B.2, one can then see that a more general expression for the story buckling parameter, analogous to that used in Eq. (2.16), is as follows:

$$\lambda_{story} = \frac{\displaystyle\sum_{non-leaner} \frac{P_{cr}}{L}}{\displaystyle\sum_{all} \frac{P_u}{L}} \tag{B.22}$$

Equation (B.21) then generalizes to:

$$\left[P_{e\hat{\tau}(story)} = \frac{\pi^2 EI}{\left(K_{story} L\right)^2} \right] = \left[\lambda_{story} P_u = \frac{\displaystyle\sum_{non-leaner} \frac{P_{cr(story)}}{L}}{\displaystyle\sum_{all} \frac{P_u}{L}} (P_u) \right] \tag{B.23}$$

and solving for K, one obtains a general expression for story-based effective length factors, accounting for stories having columns of unequal length:

$$K_{story} = \sqrt{\frac{1}{P_u} \frac{\pi^2 EI}{L^2} \frac{\displaystyle\sum_{all} \frac{P_u}{L}}{\displaystyle\sum_{non-leaner} \frac{P_{cr}}{L}}} \tag{B.24}$$

The effective length factor K_{K_n} and its corresponding buckling load, $P_{e\hat{\tau}(K_n)}$, follow directly from this formulation:

$$\lambda_{K_n} \sum_{all} \frac{P_u}{L} = \sum_{non-leaner} \frac{P_{e\hat{\tau}(nomo)}}{L} = \sum_{non-leaner} \frac{\pi^2 EI}{\left(K_n L\right)^2} \frac{1}{L} \Rightarrow \lambda_{K_n} = \frac{\displaystyle\sum_{non-leaner} \frac{\pi^2 EI}{\left(K_n L\right)^2} \frac{1}{L}}{\displaystyle\sum_{all} \frac{P_u}{L}} \tag{B.25}$$

$$P_{e\hat{\tau}(K_n)} = \frac{\pi^2 EI}{\left(K_{K_n} L\right)^2} = \lambda_{K_n} P_u = \frac{\displaystyle\sum_{non-leaner} \frac{\pi^2 EI}{\left(K_n L\right)^2} \frac{1}{L}}{\displaystyle\sum_{all} \frac{P_u}{L}} (P_u) \tag{B.26}$$

$$K_{K_n} = \sqrt{\frac{1}{P_u}\frac{I}{L^2}\frac{\sum_{all}\frac{P_u}{L}}{\sum_{non-leaner}\frac{I}{\left(K_n L^2\right)^2}\frac{1}{L}}} \tag{B.27}$$

Similarly, the effective length factor K_{R_L} and its expression for buckling load, $P_{e\hat{\tau}(R_L)}$ follow from Eqs. (B.23) and (B.24). As explained in Section 2.3.2.4, based upon the customary range of values exhibited by C_L, the amount of axial force which a story may withstand usually falls within the following range:

$$(0.82)\sum_{non-leaner}\frac{P_L}{L} \leq \sum_{non-leaner}\frac{P_{cr(story)}}{L} < \sum_{non-leaner}\frac{P_L}{L} \tag{B.28}$$

and expression for $P_{e\hat{\tau}(R_L)}$ and K_{R_L} follow:

$$\left[P_{e\hat{\tau}(R_L)} = \frac{\pi^2 EI}{\left(K_{R_L} L\right)^2}\right] = \left[\lambda_{R_L} P_u = \frac{\sum_{non-leaner}\frac{P_L}{L}\left(0.85 + 0.15R_L\right)}{\sum_{all}\frac{P_u}{L}}\left(P_u\right)\right] \tag{B.29}$$

$$K_{R_L} = \sqrt{\frac{1}{P_u}\frac{\pi^2 EI}{L^2}\frac{\sum_{all}\frac{P_u}{L}}{\frac{\sum_{non-leaner} H}{\Delta_{oh}}\left(0.85 + 0.15R_L\right)}} \tag{B.30}$$

where

$$R_L = \frac{\sum_{leaner} P_u / L}{\sum_{all} P_u / L} \tag{B.31}$$

and

$$\sum_{non-leaner}\frac{P_L}{L} = \frac{\sum_{non-leaner} H}{\Delta_{oh}} \tag{B.32}$$

and $\sum_{leaner} P_u$ represents the summation of axial force in the leaner columns of the story.

Appendix B -- References

AISC [1993], *Load and Resistance Factor Design Specification for Structural Steel Buildings*, Second Edition, AISC, Chicago, Illinois.

AISC [1994], *Manual of Steel Construction - Load and Resistance Factor Design*, Second Edition, AISC, Chicago, Illinois.

LeMessurier, W. J. [1977], "A Practical Method for Second-Order Analysis. Part 2: Rigid Frames," *Engineering Journal*, AISC, Second Quarter, 49-67.

LeMessurier, W. J. [1993], "Discussion of the Proposed LRFD Commentary to Chapter C of the Second Edition of the AISC Specification," Presentation made to the ASCE Technical Committee on Load and Resistance Factor Design, April 18, 1993, Irvine, California.

APPENDIX C

DERIVATION OF STORY BUCKLING CAPACITY FOR A STORY HAVING WEAK COLUMNS

Section 2.4.4 derived limits on the allowable magnitudes of story-based effective length factors or, alternately, axial buckling capacities. These limits are imposed to insure conservative assessment of story stability: The limits are summarized below:

1. For $P_{e\hat{\tau}(K_n)}$:

$$P_{e\hat{\tau}(K_n)} \leq 1.6 P_{e\hat{\tau}(nomo)} \tag{C.1}$$

where $P_{e\hat{\tau}(nomo)} = \dfrac{\pi^2 EI}{(K_n L)^2}$

2. For K_{K_n}:

$$K_{K_n} \geq \sqrt{\frac{5}{8}} K_n \tag{C.2}$$

where K_n is the effective length factor obtained from the AISC LRFD sidesway uninhibited nomograph [AISC, 1993].

3. For $P_{e\hat{\tau}(R_L)}$:

$$P_{e\hat{\tau}(R_L)} \leq 1.7 P_L = 1.7 \frac{HL}{\Delta_{oh}} \tag{C.3}$$

Using K_{K_n} and K_{R_L} as outlined in Section 2.4, Figs. 2.16 and 2.19 through 2.22 plot the total capacity of their respective frames for various distributions of axial force across the story. As described in Section 2.4.4, for some distributions of axial force in these frames, the weaker column is loaded such that it breaches the applicable limit (Eqs. (C.1) through (C.3)). This appendix derives an expression for the buckling capacity of a frame once the weaker column is loaded such that it has breached its allowable limit of capacity. It is this expression that is used to generate the plots in Sections 2.4.4 and 2.4.6 once one of the columns has breached its limit of capacity.

To begin this derivation, consider the unsymmetric frame of Fig. C.1. For this discussion, assume $P_{e\hat{\tau}(R_L)}$ is the approach used to compute story stability (the discussion based upon $P_{e\hat{\tau}(K_n)}$ is similar). The first order sidesway stiffness of this frame may be written as:

$$\sum_{non-leaner} \frac{P_L}{L} = \frac{c_k P_{Lk}}{L_k} \tag{C.4}$$

where the right column is labeled as k (and the left as j) in both the figure and the above equation, and where c_k is some constant greater than 2 (since the right column is weaker than the left). This expression may be used to compute $P_{e\hat{\tau}(R_L)}$ to assess story stability by substituting it expression into Eq. (B.29) from Appendix B, and recognizing that R_L equals zero for the frame in Fig. C.1:

$$P_{e\hat{\tau}(R_L)} = \frac{\pi^2 EI}{\left(K_{R_L} L\right)^2} = \lambda_{R_L} P_u = \frac{\sum\limits_{non-leaner} \frac{P_L}{L}\left(0.85 + 0.15 R_L\right)}{\sum\limits_{all} \frac{P_u}{L}} \left(P_u\right) \tag{B.29}$$

$$P_{e\hat{\tau}(R_L)k} = \frac{P_{uk}}{\sum\limits_{all} \frac{P_u}{L}} \sum_{non-leaner} \frac{P_L}{L}(0.85) = \frac{P_{uk}}{\sum\limits_{all} \frac{P_u}{L}} \frac{c_k P_{Lk}}{L_k}(0.85) \leq 1.7 P_{Lk} \tag{C.5}$$

These expressions account for columns of unequal length. The load ratio on the right column that causes the limit of Eq. (C.3) to be breached is obtained from Eq. (C.5):

$$\frac{P_{uk_{(limit)}}}{\sum\limits_{all} \frac{P_u}{L}} = \frac{1.7 P_{Lk}}{c_k \frac{P_{Lk}}{L_k}(0.85)} \quad \Rightarrow \quad \frac{\frac{P_{uk_{(limit)}}}{L_k}}{\sum\limits_{all} \frac{P_u}{L}} = \frac{2}{c_k} < 1 \tag{C.6}$$

If more load is shifted to the right column (while holding the total axial force in the story constant), thus loading the right column in excess of $P_{uk_{(limit)}}$, the right column's capacity should be considered to be constant at a value of $1.7P_{Lk}$. This is because beyond $P_{uk_{(limit)}}$, the right column is assumed to be buckling in a braced mode, and thus it is presumed to have reached its maximum capacity. Any additional load applied to the right column does not affect its capacity. Of course, once the right column breaches this limit, it should almost always be redesigned [Squarzini and Hajjar, 1993].

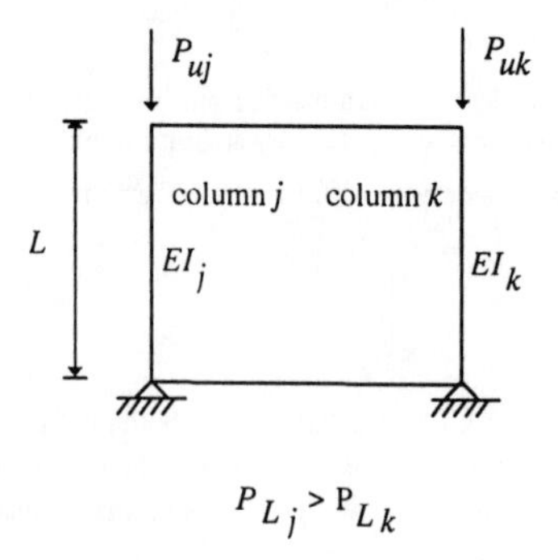

Figure C.1 One-Bay, One-Story Frame with Axial Load Distributed Between Two Rigidly-Connected Columns

However, to investigate here the effectiveness and behavior of P_{eR_L} once the right column has breached its limit, the member is not redesigned in this example. Instead, once P_{uk} exceeds $P_{uk_{(limit)}}$, the frame's total buckling capacity is computed based upon the right column having a capacity equal to $1.7P_{Lk}$. Recall that the frame's original total buckling capacity is computed as:

$$P_{e\hat{\tau}(R_L)j} + P_{e\hat{\tau}(R_L)k} = \sum_{non-leaner} P_{e\hat{\tau}(R_L)} = \frac{P_{uj}}{\sum_{all} \frac{P_u}{L}} \left[\sum_{non-leaner} \frac{P_L}{L}(0.85) \right] + \frac{P_{uk}}{\sum_{all} \frac{P_u}{L}} \left[\sum_{non-leaner} \frac{P_L}{L}(0.85) \right] \tag{C.7}$$

If the right column were to reach its imposed limiting capacity (i.e., $P_{eR_Lk} = 1.7P_{Lk}$), then Eq. (C.7) is no longer a valid estimation of the story's total capacity. A new total capacity must be calculated, and Eq. (C.7) becomes:

$$\frac{P_{uj}}{\sum_{all} \frac{P_u}{L}} \left[\sum_{sidesway\_buckling} \frac{P^{new}_{cr(story)}}{L} \right] + 1.7P_{Lk} = \sum_{sidesway\_buckling} P^{new}_{cr(story)} \tag{C.8}$$

where $\sum_{sidesway\_buckling}$ equals the summation of the axial force of the rigidly-connected columns which have *not* breached their respective limits (Eq. (C.3)). For the simple frame of Fig. C.1, the load on the left column may be expressed as a function of x, the distance of the "moving load" from the left column (see Fig. 2.14): $P_{uj} = (1-x)\sum_{all} P_u$. Therefore,

$$\frac{(1-x)\sum_{all} P_u}{\sum_{all} \frac{P_u}{L}} \left[\sum_{sidesway\_buckling} \frac{P^{new}_{cr(story)}}{L} \right] + 1.7P_{Lk} = \sum_{sidesway\_buckling} P^{new}_{cr(story)} \tag{C.9}$$

If the columns are of equal length:

$$\sum_{sidesway\_buckling} P^{new}_{cr(story)} = \frac{1.7P_{Lk}}{x} \tag{C.10}$$

This expression for total frame buckling capacity only varies with the gravity load distribution (i.e., x) and not with the total story stiffness, although it is also directly related to the limited capacity of the right column. Eq. (C.10) is plotted in Figs. 2.16 and 2.19 through 2.22. For the case shown in the figure, it controls when x is larger than approximately 0.5. Note that the envelope created by the constant total capacity predicted by Eq. (C.7) (before the weak right column breaches its limit) and the total capacity predicted by Eq. (C.10) (after the limit is breached) remains conservative for all values of x in all of these figures. If the limit were not imposed, Eq. (C.7) would overpredict the capacity of this frame as x approaches 1 for these frames.

For the approaches outlined in Sections 2.4.1 and 2.4.2, the total story buckling capacity is assumed to be independent of the distribution of axial force in the rigidly-connected columns in the story if no buckling capacity limit is breached for any of the columns (e.g., if all columns in the story are of equal length, the total capacity is taken as $\sum_{non-leaner} P_{e\hat{\tau}(nomo)}$ for the approach described in Section 2.4.1, and it equals $\sum_{non-leaner} P_L(0.85+0.15R_L)$ for the approach described in Section 2.4.2). This capacity is then assumed to be "distributed" to the different columns in the story *based upon the amount of axial force* in each of the story's rigidly-connected columns (as indicated in Eqs. (2.25) and (2.28)). However, once one or more of the columns in the story buckle in a braced mode, the actual buckling capacity of the frame drops off noticeably, as shown in Fig. 2.16 of Chapter 2. Therefore, once one or more of the columns in the story breach their capacity limits, LeMessurier insured through the imposition of these limits that the total buckling capacity of the story correspondingly decreases. Eq. (C.9) may be generalized to a form appropriate for an arbitrary multi-column story in a structure:

$$\frac{R_{RC}\sum_{all} P_u}{\sum_{all} \frac{P_u}{L}} \left[\sum_{sidesway\_buckling} P_{cr(story)}^{new} \right] + \sum_{braced\_buckling} 1.7 P_L = \sum_{sidesway\_buckling} P_{cr(story)}^{new} \tag{C.11}$$

or, if all columns in the story are of equal length:

$$\sum_{sidesway\_buckling} P_{cr(story)}^{new} = \frac{\sum_{braced\_buckling} 1.7 P_L}{1 - R_{RC}} \tag{C.12}$$

where:

$$R_{RC} = \frac{\sum_{all} P_u - \sum_{braced\_buckling} P_u}{\sum_{all} P_u} = 1 - \frac{\sum_{braced\_buckling} P_u}{\sum_{all} P_u} = \frac{\sum_{sidesway\_buckling} P_u}{\sum_{all} P_u} \tag{C.13}$$

The ratio R_{RC} equals the proportion (between 0 and 1) of the axial force which is taken by the rigidly-connected columns that are *not* buckling in a braced mode. The term $\sum_{braced\_buckling} P_u$ equals the summation of the axial force of the remaining rigidly-connected columns (namely, all rigidly-connected columns in the story which have breached the limit of Eq. (C.3), and which thus are *possibly* buckling in a braced mode). Of course, if a column breaches its limit, it should in fact be redesigned, after which a new total buckling capacity of the frame should be recomputed anyway [Squarzini and Hajjar, 1993].

References

AISC [1993], *Load and Resistance Factor Design Specification for Structural Steel Buildings*, Second Edition, AISC, Chicago, Illinois.

Squarzini, M. J. and Hajjar, J. F. [1993], "An Evaluation Of Proposed Techniques For Predicting Column Capacity," Structural Engineering Report No. ST-93-3, Department of Civil and Mineral Engineering, University of Minnesota, Minneapolis, Minnesota.

APPENDIX D

BUCKLING AND NONLINEAR ELASTIC BEHAVIOR OF A SWAY COLUMN

A sway column of length L and with zero translational restraint and elastic rotational end-restraints (Fig. D.1) is used as the vehicle for the theoretical investigation of the notional load approach to the assessment of column stability in unbraced frames. In the present analysis, the column is assumed to be perfectly plumb and is subjected to an axial load P and a notional lateral load N as in Fig. D.1. The rotational stiffnesses of the elastic restraints at ends A and B of the column are denoted α_A and α_B respectively. These elastic restraints may be expressed conveniently in non-dimensional form as

$$\begin{aligned} \overline{\alpha}_A &= \alpha_A L/EI \\ \overline{\alpha}_B &= \alpha_B L/EI \end{aligned} \tag{D.1}$$

in which EI is the flexural rigidity of the column section.

D.1 Elastic Buckling

In determining the elastic buckling load of a member, only the axial force in the member is considered. Therefore, ignoring the notional load N in Fig. D.1, the differential equation expressing equilibrium of the column in the buckled position is

$$EI\frac{d^2v}{dz^2} + Pv - \alpha_A\theta_A = 0 \tag{D.2}$$

in which θ_A is the rotation of the column at end A. By solving this differential equation in conjunction with appropriate boundary conditions, it can be shown that the condition for elastic buckling can be expressed as

$$\overline{\alpha}_A\overline{\alpha}_B + (\overline{\alpha}_A + \overline{\alpha}_B)\left(\frac{\pi}{K}\right)\cot\left(\frac{\pi}{K}\right) - \left(\frac{\pi}{K}\right)^2 = 0 \tag{D.3}$$

in which K is the *elastic* effective length factor for the sway column.

A modified column slenderness parameter $\lambda_{c(L)}$ based on the *actual* member length can be defined as

$$\lambda_{c(L)} = \sqrt{\frac{P_y}{P_E}} = \frac{1}{\pi}\left(\frac{L}{r}\right)\sqrt{\frac{F_y}{E}} \tag{D.4}$$

in which P_E is the Euler buckling load

$$P_E = \frac{\pi^2 EI}{L^2} \tag{D.5}$$

P_y is the yield load, F_y is the yield stress, E is the elastic modulus and r is the radius of gyration of the cross-section for the relevant axis of buckling. A modified effective column slenderness parameter λ_c, based on the *effective* length, can therefore be defined as

$$\lambda_c = \frac{1}{\pi}\left(\frac{KL}{r}\right)\sqrt{\frac{F_y}{E}} = K\lambda_{c(L)} \tag{D.6}$$

D.2 Inelastic Buckling

It was shown in Chapter 2 that the column strength curve of the AISC LRFD Specification can be expressed in the following unified form

$$P_n = 0.877\frac{\pi^2 EI\tau}{(KL)^2} = \frac{0.877\tau}{\left(K\lambda_{c(L)}\right)^2}P_y = \frac{0.877\tau}{\lambda_c^{\,2}}P_y \tag{D.7}$$

in which K is the *inelastic* effective length factor and τ is termed the *inelastic stiffness reduction factor*. Calculation of K and τ entails the iterative simultaneous solution of the *inelastic* buckling equation for the elastically restrained sway column (analogous to the elastic buckling equation of Eq. (D.3))

$$\frac{\overline{\alpha}_A}{\tau}\frac{\overline{\alpha}_B}{\tau} + \left(\frac{\overline{\alpha}_A}{\tau} + \frac{\overline{\alpha}_B}{\tau}\right)\left(\frac{\pi}{K}\right)\cot\left(\frac{\pi}{K}\right) - \left(\frac{\pi}{K}\right)^2 = 0 \tag{D.8}$$

the AISC LRFD column strength equation

$$\frac{P_n}{P_y} = \begin{cases} 0.658^{\lambda_c^{\,2}} & \text{for } \lambda_c < 1.5 \\ 0.877/\lambda_c^{\,2} & \text{for } \lambda_c \geq 1.5 \end{cases} \tag{D.9}$$

and the inelastic stiffness reduction factor equation

$$\tau = \begin{cases} 1.0 & \text{for } P_n/P_y \leq 0.390 \\ \dfrac{1}{0.877\ln 0.658}\left(\dfrac{P_n}{P_y}\right)\ln\left(\dfrac{P_n}{P_y}\right) & \text{for } P_n/P_y > 0.390 \end{cases} \tag{D.10}$$

for the three quantities K, τ and P_n/P_y.

D.3 Elastic Nonlinear Behaviour

In the AISC LRFD and other specifications, elastic *second-order* bending moments are required for use in the interaction equation for beam-column strength. These second-order bending moments can be determined directly from an elastic second-order analysis of the structure or alternately from an elastic *first-order* analysis of the structure in conjunction with "moment amplification factors". In the notional load approach, computation of the strength of an axially loaded compression member requires the elastic second-order moments arising from the notional lateral loads to be determined. For the sway beam-column considered in this appendix, the exact second-order bending moments can be evaluated as set out below.

The *first-order* bending moment distribution $M_f = M_f(z)$ along the member for the beam-column problem depicted in Fig. D.1 can be expressed in closed form as

$$M_f(z) = NL\Phi \tag{D.11}$$

in which

$$\Phi = \frac{1}{2}\left[\frac{2(z/L)(\overline{\alpha}_A + \overline{\alpha}_A\overline{\alpha}_B + \overline{\alpha}_B) - \overline{\alpha}_A(2+\overline{\alpha}_B)}{\overline{\alpha}_A + \overline{\alpha}_A\overline{\alpha}_B + \overline{\alpha}_B}\right] \tag{D.12}$$

The first-order end-moments are therefore given by

$$\begin{aligned} M_{fA} &= NL\Phi_A \\ M_{fB} &= NL\Phi_B \end{aligned} \tag{D.13}$$

in which Φ_A and Φ_B are defined in terms of the end restraints by

$$\begin{aligned} \Phi_A &= \frac{1}{2}\left[\frac{\overline{\alpha}_A(2+\overline{\alpha}_B)}{\overline{\alpha}_A + \overline{\alpha}_A\overline{\alpha}_B + \overline{\alpha}_B}\right] \\ \Phi_B &= \frac{1}{2}\left[\frac{\overline{\alpha}_B(2+\overline{\alpha}_A)}{\overline{\alpha}_A + \overline{\alpha}_A\overline{\alpha}_B + \overline{\alpha}_B}\right] \end{aligned} \tag{D.14}$$

The first-order relative lateral deflection Δ_f of the ends of the member is given by

$$\Delta_f = \frac{HL^3}{12EI}\left[\frac{4\overline{\alpha}_A + \overline{\alpha}_A\overline{\alpha}_B + 4\overline{\alpha}_B + 12}{\overline{\alpha}_A + \overline{\alpha}_A\overline{\alpha}_B + \overline{\alpha}_B}\right] \tag{D.15}$$

Determination of the elastic *second-order* bending moment distribution $M = M(z,P)$ entails the solution of the differential equation of equilibrium

$$EI\frac{d^2v}{dz^2} + Pv - \alpha_A\theta_A + Nz = 0 \tag{D.16}$$

By solving Eq. (D.16) and incorporating appropriate boundary conditions, it can be shown that the distribution of elastic second-order bending moment in the beam-column of Fig. D.1 is given by

$$M(z,P) = NL\Omega \tag{D.17}$$

in which

$$\Omega = \frac{(\overline{\alpha}_A\overline{\alpha}_B\cos\mu L - \overline{\alpha}_A\overline{\alpha}_B - \overline{\alpha}_A\mu L\sin\mu L)\cos\mu z + (\overline{\alpha}_A\overline{\alpha}_B\sin\mu L + \overline{\alpha}_B\mu L + \overline{\alpha}_A\mu L\cos\mu L)\sin\mu z}{\mu L\left((\overline{\alpha}_A + \overline{\alpha}_B)\mu L\cos\mu L + \left(\overline{\alpha}_A\overline{\alpha}_B - (\mu L)^2\right)\sin\mu L\right)} \tag{D.18}$$

and

$$\mu L = \pi\sqrt{\frac{P}{P_E}} \tag{D.19}$$

The elastic second-order end-moments M_A and M_B are therefore are given by

$$\begin{aligned} M_A &= NL\Omega_A \\ M_B &= NL\Omega_B \end{aligned} \tag{D.20}$$

in which Ω_A and Ω_B are defined in terms of the end restraints $\overline{\alpha}_A$ and $\overline{\alpha}_B$ and axial force parameter μL as

$$\begin{aligned} \Omega_A &= \frac{\overline{\alpha}_A(\mu L\sin\mu L - \overline{\alpha}_B\cos\mu L + \overline{\alpha}_B)}{\mu L\left[(\overline{\alpha}_A + \overline{\alpha}_B)\mu L\cos\mu L + \left(\overline{\alpha}_A\overline{\alpha}_B - (\mu L)^2\right)\sin\mu L\right]} \\ \Omega_B &= \frac{\overline{\alpha}_B(\mu L\sin\mu L - \overline{\alpha}_A\cos\mu L + \overline{\alpha}_A)}{\mu L\left[(\overline{\alpha}_A + \overline{\alpha}_B)\mu L\cos\mu L + \left(\overline{\alpha}_A\overline{\alpha}_B - (\mu L)^2\right)\sin\mu L\right]} \end{aligned} \tag{D.21}$$

It is now possible to define general "amplification factors" δ_A, δ_B, for the end bending moments in an elastically restrained sway beam-column

$$\delta_A = \frac{\Omega_A}{\Phi_A}$$
$$\delta_B = \frac{\Omega_B}{\Phi_B} \tag{D.22}$$

These amplification factors account for the second-order effects of the loads acting on the member in its displaced and deformed configuration, thus enabling the elastic second-order moments to be expressed in terms of first-order moments simply as

$$M_A = M_{fA}\delta_A$$
$$M_B = M_{fB}\delta_B \tag{D.23}$$

The second-order relative lateral deflection Δ of the ends of the member is given by

$$\Delta = \frac{HL^3}{EI}\left[\frac{\Omega_A + \Omega_B - 1}{(\mu L)^2}\right] \tag{D.24}$$

in which Ω_A and Ω_B are defined by Eq. (D.21).

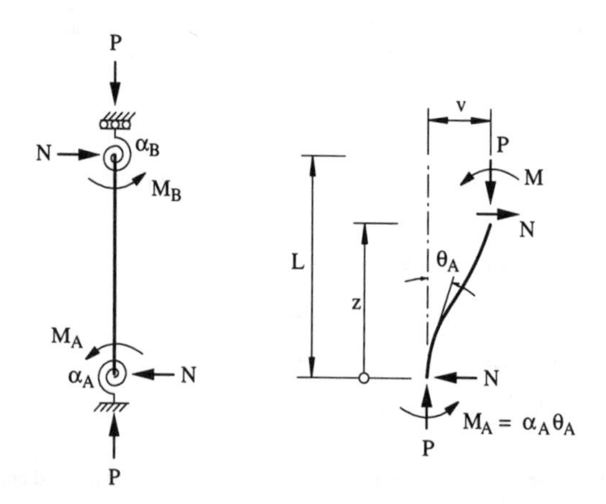

Fig. D.1 Rotationally restrained sway column with zero translational restraint

APPENDIX E

THEORETICAL ANALYSIS OF LEANING COLUMN FRAME

The leaning column frame illustrated in Fig. E.1 and subjected to gravity and lateral loads is an extreme case for the story buckling load concept and is therefore investigated here using the notional load approach. For simplicity it is assumed that both columns are of equal length; for generality, the cross-sectional properties, yield stresses and applied gravity loads may differ. In the following, all quantities relating to the left-hand (restrained) column, henceforth termed column 1, are indicated by the subscript "1". Similarly, all quantities relating to the right-hand (leaning) column, hereafter termed column 2, are indicated by the subscript "2". All lateral restraint for the frame is thus provided by column 1, which is assumed to be restrained rotationally by girders such that the rotational stiffnesses provided to the base and top of the column are given by α_A and α_B. As in Appendix D, these rotational stiffnesses α_A and α_B can be conveniently non-dimensionalized with respect to the length L and flexural rigidity EI_1 of column 1, giving

$$\overline{\alpha}_A = \frac{\alpha_A L}{EI_1} \qquad \overline{\alpha}_B = \frac{\alpha_B L}{EI_1} \tag{E.1}$$

The aim of this appendix is to explain the procedures which can be used to deduce the "strength" of the frame under gravity loads using the notional load and effective length approaches. The frame can therefore be visualized as being loaded *proportionally* by the gravity loads P_1 and P_2, and the lateral load H, until the ultimate strength ("failure") of the "critical member" or system occurs. For convenience, the applied gravity loads may be written in terms of a single load parameter P as

$$P_1 = \rho_1 P \qquad P_2 = \rho_2 P \tag{E.2}$$

with $\rho_1+\rho_2 = 1$ and therefore $P_1+P_2 = P$. For a given lateral load H, an axial strength parameter for the frame can therefore be defined by the ratio

$$\gamma = \frac{P_{u1} + P_{u2}}{P_{y1} + P_{y2}} = \frac{\sum P_{ui}}{\sum P_{yi}} \tag{E.3}$$

in which P_i is the applied gravity load acting on column i at the ultimate strength of the frame, and P_{yi} is the squash load of column i, with the summation extending over both columns.

E.1 Elastic First-Order Analysis

Assuming axial inextensibility, and that the girders are much longer than the columns, the elastic first-order lateral deflection of the eaves may be deduced from Eq. (D.15) of Appendix D as

$$\Delta_f = \frac{HL^3}{12EI_1}\left[\frac{4\overline{\alpha}_A + \overline{\alpha}_A\overline{\alpha}_B + 4\overline{\alpha}_B + 12}{\overline{\alpha}_A + \overline{\alpha}_A\overline{\alpha}_B + \overline{\alpha}_B}\right] \tag{E.4}$$

The elastic first-order end-moments may similarly be deduced from Eq. (D.13) of Appendix D as

$$\begin{aligned} M_{fA} &= HL\Phi_A \\ M_{fB} &= HL\Phi_B \end{aligned} \tag{E.5}$$

in which Φ_A and Φ_B are defined by Eq. (D.14).

E.2 Elastic Second-Order Analysis

Assuming axial inextensibility of all members, small displacements, and that the beam is much longer than the columns, the exact elastic second-order distribution of axial forces, end-shears and end-moments are as given in Fig. E.2. It can be seen that the restrained column 1 in Fig. E.2 is precisely the sway beam-column depicted in Fig. D.1 with an axial load (termed P in Fig. D.1) of $P_1 = \rho_1 P$, and a lateral load (termed N in Fig. D.1) of $H + P_2\Delta/L = H + \rho_2 P\Delta/L$. Using Eq. (D.20), the elastic second-order end-moments M_A and M_B in column 1 may be thus be expressed in the form

$$\begin{aligned} M_A &= (H + P_2\,\Delta/L)L\Omega_A = (H + \rho_2 P\,\Delta/L)L\Omega_A \\ M_B &= (H + P_2\,\Delta/L)L\Omega_B = (H + \rho_2 P\,\Delta/L)L\Omega_B \end{aligned} \tag{E.6}$$

in which Δ is the elastic second-order lateral deflection of the eaves as indicated in Fig. E.2, and Ω_A and Ω_B are defined by Eq. (D.21) as a function of the end-restraints $\overline{\alpha}_A$ and $\overline{\alpha}_B$ and the axial force parameter

$$\mu_1 L = L\sqrt{\frac{P_1}{EI_1}} = L\sqrt{\frac{\rho_1 P}{EI_1}} \tag{E.7}$$

To compute Δ, Eq. (D.24) of Appendix D furnishes

$$\Delta = \frac{(H + \rho_2 P\,\Delta/L)L^3}{EI_1}\left[\frac{\Omega_A + \Omega_B - 1}{(\mu_1 L)^2}\right] \tag{E.8}$$

which can be solved for Δ to yield

$$\Delta = \frac{HL}{P}\left[\frac{\Omega_A + \Omega_B - 1}{1 - \rho_2(\Omega_A + \Omega_B)}\right] \tag{E.9}$$

Substituting Eq. (E.9) into Eqs. (E.6) then yields the following expressions for the elastic second-order moments at the base and top of column 1:

$$M_A = \frac{HL\rho_1\Omega_A}{1 - \rho_2(\Omega_A + \Omega_B)}$$
$$M_B = \frac{HL\rho_1\Omega_B}{1 - \rho_2(\Omega_A + \Omega_B)} \tag{E.10}$$

E.3 Elastic Buckling Analysis for the Sway Mode

The elastic buckling load of the system for the sway mode of failure can be interpreted as that load at which the elastic second-order lateral deflection Δ increases without bound. The elastic buckling condition is therefore defined by setting the denominator of Eq. (E.9) to zero, that is

$$1 - \rho_2(\Omega_A + \Omega_B) = 0 \tag{E.11}$$

in which Ω_A and Ω_B are defined in terms of the effective length factor K_1 pertaining to column 1 as

$$\Omega_A = \frac{\bar{\alpha}_A\left((\pi/K_1)\sin(\pi/K_1) - \bar{\alpha}_B\cos(\pi/K_1) + \bar{\alpha}_B\right)}{(\pi/K_1)\left[(\bar{\alpha}_A + \bar{\alpha}_B)(\pi/K_1)\cos(\pi/K_1) + \left(\bar{\alpha}_A\bar{\alpha}_B - (\pi/K_1)^2\right)\sin(\pi/K_1)\right]}$$
$$\Omega_B = \frac{\bar{\alpha}_B\left((\pi/K_1)\sin(\pi/K_1) - \bar{\alpha}_A\cos(\pi/K_1) + \bar{\alpha}_A\right)}{(\pi/K_1)\left[(\bar{\alpha}_A + \bar{\alpha}_B)(\pi/K_1)\cos(\pi/K_1) + \left(\bar{\alpha}_A\bar{\alpha}_B - (\pi/K_1)^2\right)\sin(\pi/K_1)\right]} \tag{E.12}$$

The frame elastic critical load P_e, defined here as the sum of the applied axial loads P_1 and P_2 at elastic critical sidesway buckling, is then given by

$$P_e = \frac{1}{\rho_1}\frac{\pi^2 EI_1}{(K_1 L)^2} \tag{E.13}$$

If the axial load in the leaning column (column 2) at sidesway buckling is interpreted as a "critical load", it is possible to define a "pseudo" effective length factor $\overline{K}_2$ pertaining to this column as

$$\overline{K}_2 = K_1\sqrt{\frac{\rho_1}{\rho_2}}\sqrt{\frac{EI_2}{EI_1}} \tag{E.14}$$

However, this is not a "real" K-factor in the conventional sense because it is associated with failure of the *lateral resisting system* of the frame (i.e., column 1), rather than flexural buckling of the leaning column itself. Leaning columns are designed appropriately using an effective length factor of unity. Accordingly, $K_2 = 1$ has been assumed for the purposes of assessing the strength of the leaning column frames in Chapter 4.

E.4 Inelastic Buckling Analysis for the Sway Mode

If inelastic effective length factors are to be computed, the relevant buckling condition is

$$1-\rho_2(\Omega_A+\Omega_B)=0 \tag{E.15}$$

in which Ω_A and Ω_B are defined in terms of the effective length factor K_1 and the inelastic stiffness reduction factor τ_1, both pertaining to column 1, as

$$\Omega_A = \frac{(\overline{\alpha}_A/\tau_1)\left((\pi/K_1)\sin(\pi/K_1)-(\overline{\alpha}_B/\tau_1)\cos(\pi/K_1)+(\overline{\alpha}_B/\tau_1)\right)}{(\pi/K_1)\left[\left((\overline{\alpha}_A/\tau_1)+(\overline{\alpha}_B/\tau_1)\right)(\pi/K_1)\cos(\pi/K_1)+\left((\overline{\alpha}_A/\tau_1)(\overline{\alpha}_B/\tau_1)-(\pi/K_1)^2\right)\sin(\pi/K_1)\right]}$$

$$\Omega_B = \frac{(\overline{\alpha}_B/\tau_1)\left((\pi/K_1)\sin(\pi/K_1)-(\overline{\alpha}_A/\tau_1)\cos(\pi/K_1)+(\overline{\alpha}_A/\tau_1)\right)}{(\pi/K_1)\left[\left((\overline{\alpha}_A/\tau_1)+(\overline{\alpha}_B/\tau_1)\right)(\pi/K_1)\cos(\pi/K_1)+\left((\overline{\alpha}_A/\tau_1)(\overline{\alpha}_B/\tau_1)-(\pi/K_1)^2\right)\sin(\pi/K_1)\right]}$$

(E.16)

Calculation of K_1, τ_1, and the associated column strength P_{n1}, is an iterative procedure and follows the guidelines outlined in Section D.2 with the exception that the member buckling equation (Eq. (D.8)) is replaced by the frame buckling equation (Eq. (E.15)) applicable in the present context.

The frame inelastic critical load $P_{e\tau}$, defined here as the sum of the applied axial loads P_1 and P_2 at inelastic sidesway buckling, is then given by

$$P_{e\tau} = \frac{1}{\rho_1}\frac{\pi^2\tau_1 EI_1}{(K_1 L)^2} \tag{E.17}$$

Analogously to Eq. (E.14), the "pseudo" effective length factor $\overline{K}_2$ for the leaning column is deduced as

$$\overline{K}_2 = K_1 \sqrt{\frac{\rho_1}{\rho_2}} \sqrt{\frac{\overline{\tau}_2 EI_2}{\tau_1 EI_1}} \tag{E.18}$$

in which $\overline{\tau}_2$ is the "pseudo" inelastic stiffness reduction factor for the leaning column, given by

$$\overline{\tau}_2 = \begin{cases} 1.0 & \text{for } \overline{P}_{n2}/P_{y2} \le 0.390 \\ \dfrac{1}{0.877 \ln 0.658} \left(\dfrac{\overline{P}_{n2}}{P_{y2}}\right) \ln\left(\dfrac{\overline{P}_{n2}}{P_{y2}}\right) & \text{for } \overline{P}_{n2}/P_{y2} > 0.390 \end{cases} \tag{E.19}$$

in which $\overline{P}_{n2} = 0.877 \rho_2 P_{e\tau}$.

As for elastic buckling, $\overline{K}_2$ is not a "real" K-factor in the conventional sense because it is associated with inelastic sidesway buckling of the lateral resisting system of the frame, rather than inelastic flexural buckling of the leaning column itself. The assumption of $K_2 = 1$ is appropriate for assessing the strength of the frame.

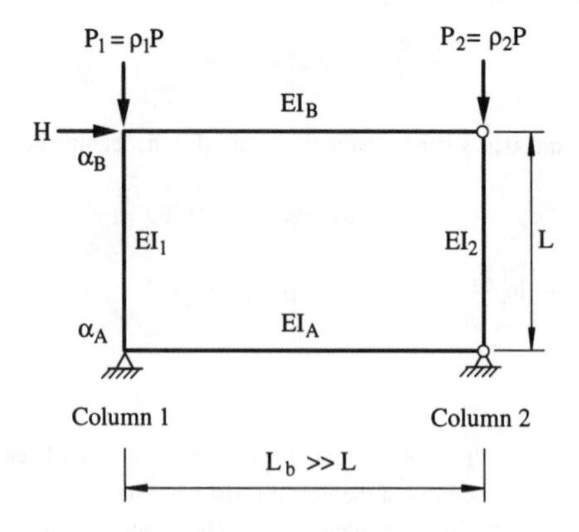

Fig. E.1 Leaning column frame

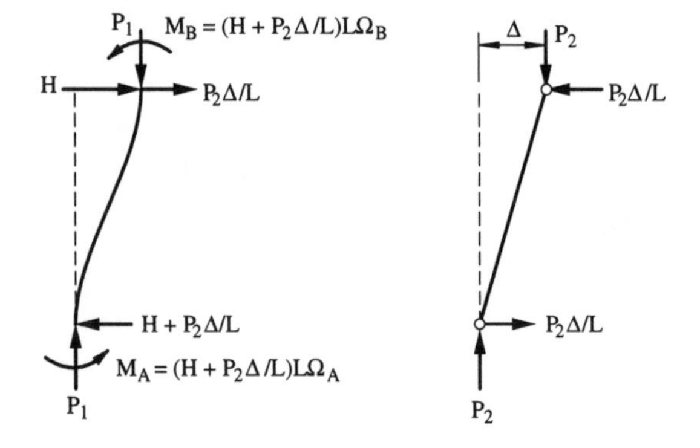

Fig. E.2 Second-order statics of leaning column frame

APPENDIX F

REFINED NOTIONAL LOAD CALIBRATION

F.1 Introduction

In the "simple" notional load approach, the notional lateral load N assumed to act at a particular story level was universally taken as $\xi = 0.005$ times the total gravity load Q acting at that story level. In the vast majority of cases, this calibration has been shown to be either accurate or conservative, with the greatest degree of conservatism being in frames comprising columns of low to intermediate slenderness which are bent in reverse-curvature and have significant rotational end-restraints. It is likely that such columns will be prevalent in unbraced multistory frames, especially in the lower stories. This appendix presents a refined calibration of the notional load parameter ξ which will significantly reduce or eliminate the conservatism of the simple approach; such a refinement will be particularly advantageous for the application of the notional load approach to multistory frames.

Considering for the minute the calibration of the notional load for a single unbraced beam-column as undertaken in Section 4.6, the following observations provide motivation and guidance for the invention of a refined expression for ξ:

- As the yield stress of the member increases, so too does the notional load required for precise calibration against the AISC LRFD column curve.
- Irrespective of the length of the column, as the lateral stiffness of the sway member approaches that of one restrained rigidly against rotation at both ends (i.e., as the effective length factor tends to unity), the required notional load tends to zero.
- Irrespective of the magnitude of the end restraints (and hence the effective length factor), as the actual length of the member approaches zero, so too does the required notional load.

As well as the above facts, it may be surmised from a probabilistic argument that the notional load parameter ξ applicable to multistory frames should reduce as the number of bays in a story increases. The justification for this supposition, which has parallels with the question of how the number of bays in a story should influence the story out-of-plumbness imperfection included in advanced analysis (see Section 4.11.2), is outlined in Section F.4.

Proceeding directly from the above observations, it is assumed here that the refined notional load parameter $\xi = \xi_R$ which is generally applicable to a complete story can be expressed in the form

$$\xi = \xi_R = \xi_0 k_y k_S k_\lambda k_c \qquad \text{(F.1)}$$

in which ξ_0 is the "basic" or "simple" notional load parameter of 0.005 employed in the simple notional load approach, and k_y, k_S, k_λ and k_c are adjustment or refinement factors accounting for the yield stress of the columns in the story, the lateral stiffness of the story, the (effective) slenderness of the story, and the number of columns in the story, respectively.

Initially in the context of a single member, Section F.2 calibrates each of the adjustment factors independently. The effectiveness of the various calibrations and the validity of the interaction between the factors implied by Eq. (F.1) is then investigated in Section F.3 for a wide range of parameters. Section F.4 extends the definitions of the adjustment factors from a single sidesway uninhibited member to a complete story.

F.2 Calibration for a Single Member

Analogously to the simple notional calibration undertaken in Section 4.6.2, the refined calibration considered in this appendix is initially developed for a single unbraced beam-column member with zero translational restraint and elastic rotational end-restraints (see also Appendix D). Accordingly, only the first three adjustment factors given in Eq. (F.1)—the yield stress adjustment factor k_y, the lateral stiffness adjustment factor k_S and the slenderness adjustment factor k_λ—are considered in the calibrations and numerical verifications expounded in this section. Extensions of the definitions of the various adjustment factors to a complete story of a frame are presented in Section F.4.

The first-order lateral stiffness of a single member with elastic end-restraints defined by G-factors G_A and G_B can be quantified conveniently in at least three ways. From Eq. (D.15) of Appendix D, the first-order lateral deflection Δ_f of the ends of the member may be expressed in terms of G-factors by

$$\Delta_f = \frac{1}{12\gamma}\frac{HL^3}{EI} \tag{F.2}$$

in which

$$\gamma = \frac{G_A + G_B + 6}{4G_A + 2G_A G_B + 4G_B + 6} \tag{F.3}$$

For the limiting case of rigid rotational end-restraints ($G_A = G_B = 0$), $\gamma = 1$ and hence the corresponding first-order lateral deflection is

$$\Delta_{f00} = \frac{1}{12}\frac{HL^3}{EI} \tag{F.4}$$

From the perspective of practical design, if one knew the G-factors pertaining to a particular sway column, then the quantity γ as defined in Eq. (F.3) would be an appropriate definition of lateral stiffness. On the other hand, if one wished to avoid all G-factor calculations, as is indeed is highly desirable in the notional load approach and a prime reason for its development, a sidesway analysis could be performed to yield the first-order "inter-story" sidesway deflection Δ_f. Another useful definition of lateral stiffness is then the ratio of the reference deflection Δ_{f00} for rigid restraints to the actual deflection Δ_f,

$$\overline{\Delta} = \frac{\Delta_{f00}}{\Delta_f} \tag{F.5}$$

As yet a third alternative, one can quantify the lateral stiffness of a member from the bending moment distribution and lateral deflection Δ_f determined from a first-order sidesway analysis. For the limiting case of rigid end-restraints at both ends, the bending moments at the ends of the member attain their maximum value of

$$M_{f00} = \frac{6EI\Delta_f}{L^2} \tag{F.6}$$

A definition of the lateral stiffness of the member is

$$\overline{M} = \frac{M_{fA} + M_{fB}}{2M_{f00}} \tag{F.7}$$

in which M_{fA} and M_{fB} are the first-order moments at ends A and B, respectively, of the member. It can be shown easily that γ given by Eq. (F.3), $\overline{\Delta}$ given by Eq. (F.5), and $\overline{M}$ given by Eq. (F.7) are all precisely equivalent definitions of lateral stiffness S, that is $S = \gamma = \overline{\Delta} = \overline{M}$.

Finally, it is also interesting to note the following approximate relation for a sway member

$$K_e \approx \frac{1}{\sqrt{S}} \tag{F.8}$$

in which K_e is the elastic effective length factor determined from the member buckling equation (Eq. (D.3)).

For single columns so chosen that the adjustment factors k_λ and k_y are unity (the definitions of these factors are given later), the effect of lateral stiffness S on the required magnitude of the notional load to match the AISC LRFD column curve (where the effective slenderness is based on the *inelastic* effective length) is shown in Fig. F.1. The results in this figure correspond to the two bounding cases for sway columns with zero translational restraint, namely: zero moment at one end, maximum at the other (single-curvature bending, $\beta = 0$); and, equal and opposite moments at the ends (reverse-curvature bending, $\beta = 1$). All sway columns in practical frames are likely to have their end-moment ratio somewhere between these two extremes. For practical application in the notional load approach, the curves shown in Fig. F.1 have been bounded above with the approximate bi-linear expression

$$k_S = \begin{cases} 1.0 & \text{for } 0.0 \le \sqrt{S} \le 0.5 \\ 2\left(1-\sqrt{S}\right) & \text{for } 0.5 \le \sqrt{S} \le 1.0 \end{cases} \tag{F.9}$$

To calibrate the length adjustment factor k_λ, the approximation of Eq. (F.8) is first applied to define an approximate effective slenderness parameter λ_c as

$$\lambda_c \approx \frac{\lambda_{c(L)}}{\sqrt{S}} \tag{F.10}$$

with the calibration then undertaken assuming the yield stress refinement factor $k_y = 1.0$.

The variation of the theoretically required notional load with the approximate elastic effective slenderness parameter of Eq. (F.10) for various end-restraint conditions is shown in Fig. F.2(a) for single-curvature ($\beta = 0$) cases, and in Fig. F.2(b) for double-curvature ($\beta = 1$) cases. Although it can be seen in these figures that the theoretical curves vary somewhat haphazardly, it is evident that the general trend is for the required notional load to increase along with the elastic effective slenderness. Such a result is also intimated by Eqs. (4.19) and (4.20) of Chapter 4, at least for very slender columns. Since it is impractical (and may indeed not be rational) to devise a simple expression for k_λ to approximate all the curves shown in Fig. F.2(a) and (b), it is instead proposed to support the general trend evident in the figures by adopting the following simple formula for the length adjustment factor k_λ

$$k_\lambda = \frac{\lambda_{c(L)}}{\sqrt{S}} \tag{F.11}$$

Aside from its simplicity, this expression for k_λ is mostly on the conservative side, especially in the low and intermediate slenderness range.

Considering now the yield stress adjustment parameter k_y, it is hinted by Eqs. (4.19) and (4.20) that such a parameter may be formulated rationally in terms of the material parameter $\sqrt{F_y/E}$. Combining this notion with the results given in Table 4.2(b) for a cantilever column with a rigid base (for which $k_\lambda = 1.0$ and $k_S = 1.0$) gives

$$k_y = 22\sqrt{\frac{F_y}{E}} \tag{F.12}$$

as a simple and effective formula which modifies the notional load appropriately for the effects of different yield stresses.

To summarize, the notional load parameter ξ for a single member is expressed in refined format as

$$\xi = \xi_R = \xi_0 k_S k_\lambda k_y \tag{F.13}$$

where

$$\xi_0 = 0.005$$

$$k_S = \begin{cases} 1.0 & \text{for } 0.0 \le \sqrt{S} \le 0.5 \\ 2\left(1-\sqrt{S}\right) & \text{for } 0.5 \le \sqrt{S} \le 1.0 \end{cases}$$

$$k_\lambda = \frac{\lambda_{c(L)}}{\sqrt{S}} \tag{F.14}$$

$$k_y = 22\sqrt{\frac{F_y}{E}}$$

If the above proposal is judged to be unduly complex, then the following guidelines and simplifications can be adopted at the cost of some conservatism (in the majority of cases), or an acceptable degree of unconservatism (for some atypically slender columns):

1. $\xi = 0.005$ can be adopted universally, provided the steel yield stress is not greater than $F_y = 65$ ksi (450 MPa).

2. The extreme simplicity of the yield stress adjustment parameter k_y is such that it is recommended that it be applied at all times. If the yield stress adjustment alone is applied, then it is sufficiently accurate to universally adopt notional load parameters of $\xi = 0.004$, 0.0045 and 0.005 for yield stresses F_y of 36 ksi (250 MPa), 50 ksi (350 MPa) and 65 ksi (450 MPa), respectively. Application of the yield stress correction factor alone furnishes the "modified" notional load parameter, as described in Section 4.6.3.

3. The lateral stiffness parameter k_S can be conservatively taken as unity.

4. The slenderness adjustment parameter k_λ can be conservatively taken as unity for the majority of practical columns. For very slender columns, adoption of $k_\lambda = 1.0$ may evince a small (probably acceptable) degree of unconservatism compared to the corresponding strengths determined using inelastic effective length factors in conjunction with the AISC LRFD column curve.

F.3 Investigation of Calibration Accuracy

As detailed in the previous section, the refined expression for the notional load parameter ξ is defined uniquely by the length of the column, the end-restraints and the yield stress. The refined notional load parameter is not a function of the degree of elastic second-order amplification. For a particular member, a single value of ξ can therefore be used for any combination of axial force and moment.

Using the refined notional load parameter ξ, the AISC LRFD beam-column interaction equation can be solved (iteratively) to deduce the nominal strength P_n of a sway column, where the axial resistance term $P_{n(L)}$ is based on the actual column length. Through comparison with the AISC LRFD column curve, which was shown previously to be an accurate indicator of advanced analysis column strengths when the effective slenderness parameter λ_c is determined using the correct *inelastic* effective length, a practical assessment of the accuracy of the refined notional load calibration can be effected.

For columns free at one end ($G_B = \infty$, end-moment ratio $\beta = 0$), the column strength curves derived from the refined notional load approach for various values of G_A (0.0, 6.0, 20.0) are shown in Fig. F.3(a). The corresponding percentage deviations of the refined notional load column curves from the AISC LRFD column curve are depicted in Fig. F.3(b). It should be pointed out that the results shown in Fig. F.3, and also in Fig. F.4 discussed later, are independent of the yield stress by virtue of the yield stress adjustment parameter k_y which was expressed in the form $k_y = \alpha\sqrt{F_y/E}$, α being a constant set equal to 22 by calibration. It can be seen in Fig. F.3 that the refined notional load approach infers column strengths which are very close to the AISC LRFD column curve for all end-restraint conditions over the full range of slenderness. The deviation of the notional load results from the AISC LRFD column curve is generally within 5 % and often much less. On the conservative side, the maximum discrepancy appears to be around 8.6 % for a fairly slender cantilever column with a rigidly restrained base.

For columns with equal elastic end-restraints ($G_A = G_B$, $\beta = 1$) the column strength curves implied by the refined notional load approach for a range of restraint stiffnesses are shown in Fig. F.4(a), with the corresponding percentage deviations from the AISC LRFD curve indicated in Fig. F.4(b). For the case of rigid rotational restraints at both ends, for which the actual and effective column lengths are equal, the theoretical requirement of zero notional load is correctly inferred using the refined notional load approach on account of the lateral stiffness refinement parameter k_S evaluating to zero. For heavily restrained sway columns with equal end-restraints ($G_A = G_B = 0.6$), the refined notional load results are slightly conservative over the whole slenderness range, but by no more than 5 %. For sway columns with medium and light end-restraints ($G_A = G_B = 6.0, 20.0$), the deduced notional load strengths are slightly (within 5 %) conservative in the low and intermediate slenderness ranges, with a small degree of unconservatism apparent at higher slendernesses. This unconservatism at high slendernesses compared to the AISC LRFD column curve is not a significant concern since the corresponding column strength curves generated from advanced analysis of physically imperfect members are also above the AISC LRFD curve in this slenderness range (see Figs. 4.12 and 4.13). The column strength curves deduced from the refined notional load approach thus appear to be more consistent with the advanced analysis results than with the AISC LRFD curve.

The results shown in Figs. F.3 and F.4 and the accompanying discussion above thus indicate that the proposed refinement procedure for the notional load approach infers column strengths which are within 5 % of those prescribed by the AISC LRFD Specification (based on the correct inelastic column effective length) for the overwhelming majority of columns likely to be encountered in practice. This represents a substantial improvement in accuracy compared to the simple and modified notional load approaches, particularly for columns bent in double-curvature which is a common occurrence in multistory frames.

F.4 Extension of the Refined Notional Load Approach to a Complete Story

The preceding exposition on the calibration and verification of the refined notional load approach was limited to a single sway column. Three notional load refinement parameters were introduced which account in some approximate way for the yield stress of the column (k_y), the lateral stiffness of the column (k_S) and the (effective) slenderness of the column (k_λ). Of greater and more practical interest, however, is the extension of this philosophy to a complete story of an unbraced frame. Upon rational determination of an appropriate value of ξ, the notional lateral load N acting on a story is given by $N = \xi \Sigma P$, in which ΣP represents the sum of all gravity loads acting on that particular story. As outlined in Section F.1 and Eq. (F.1), when a complete story is being considered it is relevant to consider a fourth refinement parameter k_c, which accounts for the number of columns per plane. The rationale for and definition of this fourth parameter k_c is described in Section F.4.1 following, while the extension of the other three parameters (k_y, k_S and k_λ) to a complete story is discussed in Section F.4.2.

F.4.1 Influence of the Number of Bays

Although it may be reasonable for the advanced analysis of a *one-bay* frame to assume that both columns and hence the whole story is out-of-plumb by an amount corresponding to the erection tolerance specified in design standards, this assumption of maximum out-of-plumbness for all columns becomes increasingly statistically unjustifiable as the number of bays, or columns per plane, increases. It would thus seem plausible from a probabilistic viewpoint for the level of story out-of-plumbness modeled in plastic zone advanced analysis to decline as the number of bays increases. Similar comments also apply to the magnitude of the notional load parameter ξ used in the notional load approach.

For consistency with the advanced analysis recommendations described in Section 4.11.2, the expression for the notional load adjustment parameter k_c is given by

$$k_c = \sqrt{0.5 + 1/c} \quad (\text{but } k_c \leq 1.0) \tag{F.15}$$

in which c is the number of columns per plane.

F.4.2 Story-Based Definitions of Refinement Parameters for Yield Stress, Lateral Stiffness and Slenderness

The yield stress adjustment factor k_y defined by Eq. (F.12) is straightforwardly applicable to a complete story as invariably each of the constituent columns has the same yield stress. If this is not the case, some appropriate "average" yield stress over all the restrained (non-leaning) columns in the story can be used to evaluate k_y.

As far as the lateral stiffness adjustment factor k_S is concerned, it is most convenient from a practical viewpoint if the lateral stiffness indicator S is determined from the results of a first-order lateral load analysis, in which the lateral loads acting at each story level are equal to some (fixed) proportion of the gravity loads acting on the corresponding story. A first-order analysis of this type furnishes the inter-story drift Δ_f for each story, together with the bending moments M_{fAj} and M_{fBj} at the bottom and top, of each column j in the story. Through an extension of Eq. (F.7), the story lateral stiffness parameter S is defined by

$$S = \frac{\sum_{\text{all}} \left(M_{fAj} + M_{fBj} \right)}{2 \sum_{\text{all}} M_{f00j}} \tag{F.16}$$

in which the summation j is taken over all the columns in the story (including leaning columns), and M_{f00j} is defined as

$$M_{f00j} = \frac{6EI_j \Delta_f}{L_j^2} \tag{F.17}$$

with L_j and EI_j being the length and flexural rigidity, respectively, of column j. The M_{fAj} and M_{fBj} in Eq. (F.16) are the first-order moments at the ends of the column j, both taken positive when the member is bent in reverse-curvature. As defined by Eq. (F.16), values of first-order lateral story stiffness are in the range $0 \leq S \leq 1$. A value of $S = 0$ corresponds to zero lateral stiffness, and $S = 1$ corresponds to the stiffness achieved when the top and bottom of all columns in the story are connected to flexurally rigid, axially inextensible beams. Once S has been determined, the story lateral stiffness parameter k_S remains defined by Eq. (F.9). It is pertinent to recall at this time that a first-order lateral load analysis of the type described above is also used in the effective length approach to facilitate the calculation of modified G-factors based on the far-to-near end-moment ratio.

The effective slenderness adjustment parameter k_λ is assumed to be defined for a complete story by

$$k_\lambda = \frac{1}{c_r \sqrt{S}} \sum_{\text{restrained}} \lambda_{c(L)j} \tag{F.18}$$

in which the summation j is taken over the c_r restrained (non-leaning) columns in the story. The quantity $\lambda_{c(L)j}$ is a modified length parameter for column j defined by

$$\lambda_{c(L)j} = \frac{1}{\pi}\left(\frac{L_j}{r_j}\right)\sqrt{\frac{F_y}{E}} \tag{F.19}$$

where r_j is the radius of gyration of column j for the relevant buckling axis.

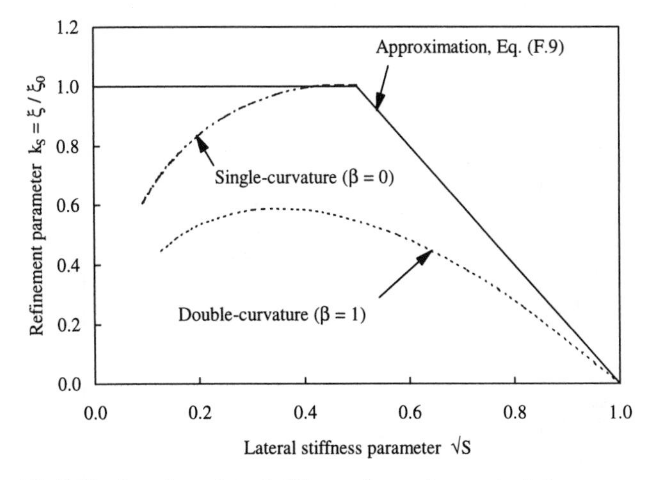

Fig. F.1 Calibration of story lateral stiffness refinement parameter k_S for a sway column under the conditions of $k_y = k_\lambda = 1$

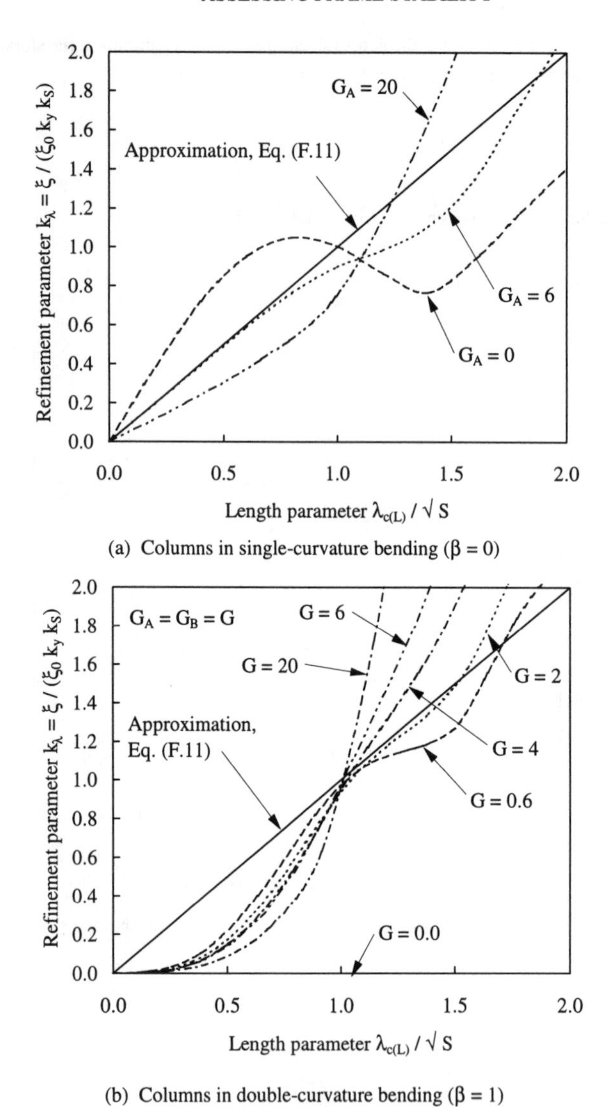

(a) Columns in single-curvature bending ($\beta = 0$)

(b) Columns in double-curvature bending ($\beta = 1$)

Fig. F.2 Variation of theoretically required notional load with slenderness and end-restraint

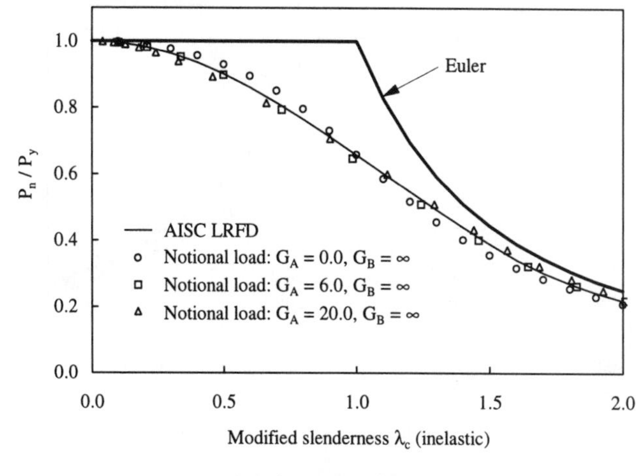

(a) Column strength curves

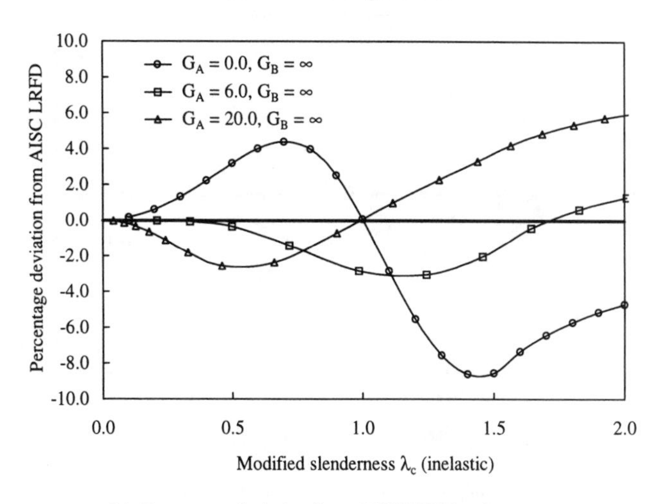

(b) Percentage deviation from AISC LRFD column curve

Fig. F.3 Column strength curves and corresponding deviations from AISC LRFD derived using refined notional load approach for sway columns in single-curvature bending ($\beta = 0$)

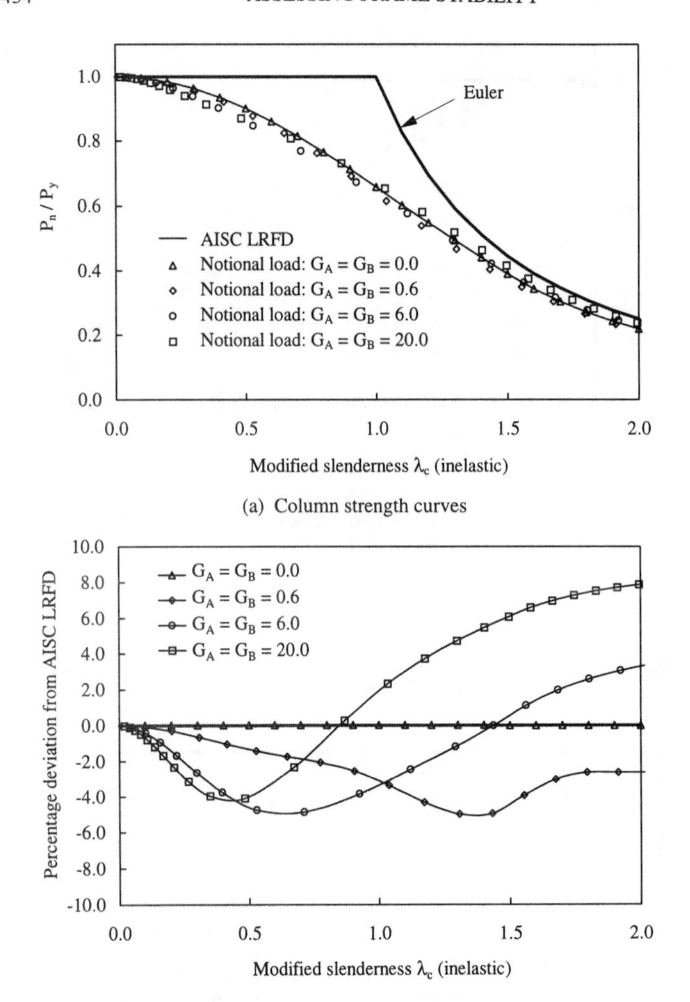

(a) Column strength curves

(b) Percentage deviation from AISC LRFD column curve

Fig. F.4 Column strength curves and corresponding deviations from AISC LRFD derived using refined notional load approach for sway columns in double-curvature bending ($\beta = 1$)

INDEX